Dynamics of Settlement Patterns in the Shekhawati Region of Rajasthan

Prehistoric to early historic periods with special reference to ancient mining and metal processing activities

Kishore Raghubans

BAR International Series 2671
2014

Published in 2016 by
BAR Publishing, Oxford

BAR International Series 2671

Dynamics of Settlement Patterns in the Shekhawati Region of Rajasthan

ISBN 978 1 4073 1313 9

BAR Publishing is the trading name of British Archaeological Reports (Oxford) Ltd. British Archaeological Reports was first incorporated in 1974 to publish the BAR Series, International and British. In 1992 Hadrian Books Ltd became part of the BAR group. This volume was originally published by Archaeopress in conjunction with British Archaeological Reports (Oxford) Ltd / Hadrian Books Ltd, the Series principal publisher, in 2014. This present volume is published by BAR Publishing, 2016.

Printed in England

BAR
PUBLISHING

BAR titles are available from:

BAR Publishing
122 Banbury Rd, Oxford, OX2 7BP, UK
EMAIL info@barpublishing.com
PHONE +44 (0)1865 310431
FAX +44 (0)1865 316916
www.barpublishing.com

In the sweet memory of My Parents

Late. Shri. Keshavlal Raghubans

Late. Smt. Vijay-Lakshmi Raghubans

CONTENTS

Acknowledgements

This research has brought me close to many people in my life and generated thoughts of the real sense of being. I am grateful to all of them.

I would like to thank specifically Dr S.V.P. Halakatti of the Archaeological Survey of India, who inspires my work. Among my many friends I would like to acknowledge Sunil, Vaishali, Paromita, Arun, Tamal, Lajwanti, Kajal and Rajesh, and, at Dharwad, Ravi Yeligar, Arlali Math, Grish, Jayraj, Tarachand, Jagannath, Rishi and Abhishek.

I am also grateful to the many kind people who provided help and support during my long fieldwork trip, including especially Madanlal Meena at Nim-kaThana. Basavraj Badiger and Basavaraj Mayachari from Karnataka provided inspiration for many of the illustrations used in this book.

Dr Raj Kumar Hans has been particularly important in terms of the development of my academic life. My late father Shri Keshavlal gave me the freedom to pursue my lengthy academic work and was silently by my side in all my endeavours. My late mother Smt. Vijay-Lakshmi Raghubans, would, I hope, be proud to see me reach this stage in my career. Among my immediate family and those close to me I wish to acknowledge Shanta Didi, Shaku Di, Asha Di, Usha Di and Sean. My brothers Shyam Dai and Mohan Dai provided constant support. My nieces and nephews Akash, Arman, Sonal, Sneha, Sean, Deepak, Gayatri, Simran, Viraj, Adarsh have helped me relax in times of stress.

My life partner, Kriti, has lovingly endured all those long hours of me sitting away at my desk; her computer skills have been indispensible. And our two little angels, Tisa and Siddhant Sai, are stimulating and constant joys.

Finally, I would like to dedicate this work to the ancient miners of the Shekhawati region – they shaped history and civilization from the substance of the earth.

Dr Kishore Raghubans, March 2014

LIST OF ILLUSTRATIONS

CHAPTER: 1

CHAPTER 2

CHAPTER 3

CHAPTER 4

CHAPTER 5

APPENDIX 1

APPENDIX 2

CHAPTER 1 - INTRODUCTION

Present is a reflection of past. When we say, we have seen in present, in reality it is the reflection of the past. Nothing is present it is mere perception of the past, what we have understood or perceived. There is no one history but there are many. When we say history, it is our perception. Archaeology is our interpretation to the past. Perception of the past, changes with present. When we say about civilization – culture – society, it is the present perception of the past, which is not the reality because the past changes with the present. Therefore, the change has to be accepted. What we say of the present it may change in future, which will change the past. The change has to be documented and archived time-to-time, which may give the clear picture of the past in terms of History or Archaeology, which is need of this time. Past has to be re-thinking in the present, which has also become past.

Thought on Present and Past

Introduction

It gives brief outline to the present trend of research on settlement patterns; research carried out in the study area and its implications for the settlement pattern studies. The previous work done, which relies on the published and unpublished data relating to archaeological, geological and metallurgical studies by the Central and State government bodies' viz. Archaeological Survey of India, Geological Survey of India and State Archaeology Department of Rajasthan and the work of Indian and foreign scholars from different universities. And finally discuss about the aims of the research, which focuses on the study of settlement-pattern studies and research methodology carried out during exploration and classification of sites.

1.1 Research Area

The Rajasthan, earlier known as Rajputana came into existence on 30th March 1949. The territory of Rajasthan is confined between north latitude 23°31' to 30°12'N and east longitude 69°30'E to 78°15' E, within the total area of about 3,42,239 Sq. Kms. The extreme outer district boundary lines of Ganganagar (north), Jaisalmer (west), Banswara (south), and Dholipur (east) join it. The state boundary is bounded on the west and northwest by Pakistan; on the north and northeast by Punjab and Uttar Pradesh; and on the southeast and southwest by Madhya Pradesh and Gujarat (Imperial Gazetteer of India 1909) (Figure 1.1).

The research area is confined between north latitude 27°15′ to 28°15′ and east longitude 75°30′ to 76°00′, within the total area of about 5,500 Sq. Kms in parts of Sikar and Jhunjhunun districts of Rajasthan in Shekhawati region. The Shekhawati is a semi-arid region located in the northeastern part of Rajasthan. It encompasses the administrative districts of Sikar and Jhunjhunun. From the administrative and geographical point of view, Shekhawati is limited to Sikar and Jhunjhunun districts only. Its area is 13,784 Sq. Kms. The extreme outer district boundary lines of Churu, Jhunjhunun, and Mahendragarh bound the Sikar district on north; Churu and Nagaur on west; Jaipur and Nagaur on south and Jaipur on east. Similarly, Churu touches Jhunjhunun district on north; Bhiwani and Mahendraagrh on east and Sikar on west and south (Sikar District Gazetteers Agarwal 1978; Jhunjhunun District Gazetteer Gupta 1984). (Figure 1.2 and 1.3).

1.2 Present Trend of Research

The research design is structured in such a way that could allow to develop inferences concerning settlement location, function, distribution and trend in settlement density at a regional scale in Rajasthan the State of India, with a view to understand the ecological adaptation and cultural changes through prehistoric to early historic period.

In order to understand the part played by the local communities, inhabiting resource rich areas on the cultures of Greater Indus and Ganga-Yamuna Doab region through various periods of human history a study of Rajasthan in Shekhawati region was undertaken. It is worthy to emphasis that this research is therefore a regional study, and not one directed towards an understanding of only the area directly surveyed. During the fieldwork, constant efforts have been made to gather data within the framework, which would allow inference at regional scale.

From the study of the previous data available, it appears that it is the product of an unstructured or haphazard archaeological work where little attention was given to sampling procedures and to the craft indicators and variables needed to understand settlement pattern and function of such settlements. It is therefore not possible to draw conclusions from such data. The resolution of such questions requires fresh explorations. After an extensive examination of archaeological data, geographical factors and location of copper rich resource areas it was decided to undertake detail fieldwork between North latitude 27°15′ to 28°15′; East longitude 75°30′ to 76°00′; which fall within the limits of Sikar and Jhunjhunun district of Rajasthan in Shekhawati region.

1.3 Previous Work

Advantages to work in Rajasthan emerge at various levels. One of the primary advantages is that there is already substantial body of published and unpublished archaeological data on which the present research is founded. Nearly 223 settlements of various periods have already been located in the present research area, though the details of various variables necessary to carry out any meaningful settlement studies were lacking. Nevertheless, this archaeological data has proved useful as a starting point for the research. Besides, nearly 23 sites of different chronological periods have been excavated in Rajasthan, which provides basis for chronological framework for the periodization of the settlements.

The territory of Rajasthan is supposed to be a "melting–pot" of human history. The physical distribution of mountain ranges and arid zones in Rajasthan have influenced hominid colonization and dispersal patterns through time. The earliest inhabitants of South Asia belonged to the hominid species *Homo erectus*, the ancestors of modern human *Homo sapiens*. The stone tools using hominids lived in the Mewar region of Rajasthan over two million years ago where the oldest stone tools have been discovered in the Chambal, Gambhiri and Berach basins. Temporary camps of Middle Palaeolithic, Upper Palaeolithic and Mesolithic hunters-gatherers have been found in Mewar and Marwar regions of Rajasthan.

There is a profuse distribution of Chalcolithic settlements all over Rajasthan, consists of Harappan, Ahar and Ganeshwar-Jodhpura culture sites. The Aravalli Mountain ranges of Rajasthan has numerous concentrations of copper, zinc, lead and silver ores. Near these ore concentrations, many old smelting activity areas have been located.

The present research area also has many Early Historic sites i.e. Painted Grey Ware sites represent early Iron Age culture sites and Rang Mahal Ware related sites represent the early centuries of Common Era. Though these sites fall outside the core area of these cultures it is interesting to know that what kind of variables played part in the settlement location and the specific changes and adjustments, which took place after the introduction of iron technology.

Another advantage to work in Rajasthan emerges in the availability of paleo-environmental data. At present considerable data has accumulated concerning this region. Many scholars like Allchin, (Allchin *et.al*: 1978), Singh (Singh *et.al*: 1972) and Misra (Misra: 1989) have provided sufficient information.

Besides, with the recent improved studies in metallurgy (Cradock: 1995), ethno-archaeology and experimental-archaeology (Bhan *et.al:* 1999) and pyro-technology (Miller: 1994) has indicated the type of major craft indicators and patterns need to be recognized at a metal processing settlement.

1.4 Aims of the Research

None of the previous studies in the research area has given sufficient information on the geomorphic features and their relationships with the archaeological evidence. Thus, the present study offers the first attempt in this direction to:

i. understand and reconstruct the dynamics of settlement patterns from Prehistoric to Early-historic periods in a selected region of Rajasthan in the context of ecological and cultural factors;

ii. use the methods of regional analysis to develop models for explaining economic and functional relations between settlements. Economic development is understood by analyzing variations in style and technologies used for certain artifacts like ceramics, lithics and metals. Functional differences in terms of raw material resources, smelting sites, processing sites and possible interaction between these are looked into. It also aims to understand geomorphic adaptation of archaeological sites and their possible interaction with mining areas and metal-processing-activity areas.

1.5 Technical Methodologies

It deals with technical methods; the discussion of which, to my understanding, becomes very important for a critical appreciation of the theme and the chosen area.

The technical method adopted during exploration was site location through scrutinizing satellite imagery and topographic maps juxtaposed with the relevant literature. The site and its area came to be marked by using GPS and by tracing the approach route from the center point i.e. Nim-ka-Thana (Khetri Maudh). The sampling procedure of the relevant artifacts collected randomly within the density area was followed by the documentation of geomorphic features associated with the site. Periodization of sites is based on comparative study of lithic assemblages, ceramics and minor antiquities and the association of these sites with the ancient mining and metal processing activity areas.

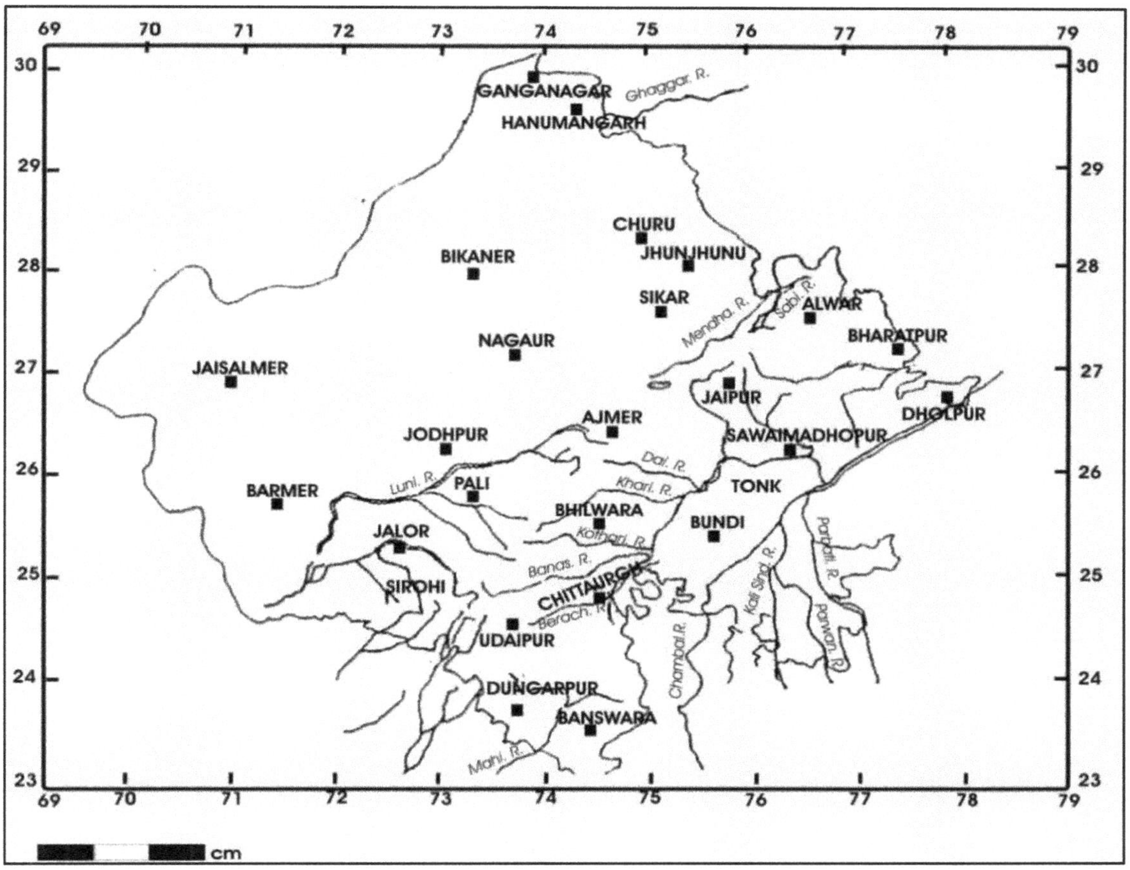

Figure 1.1: Map of Rajasthan

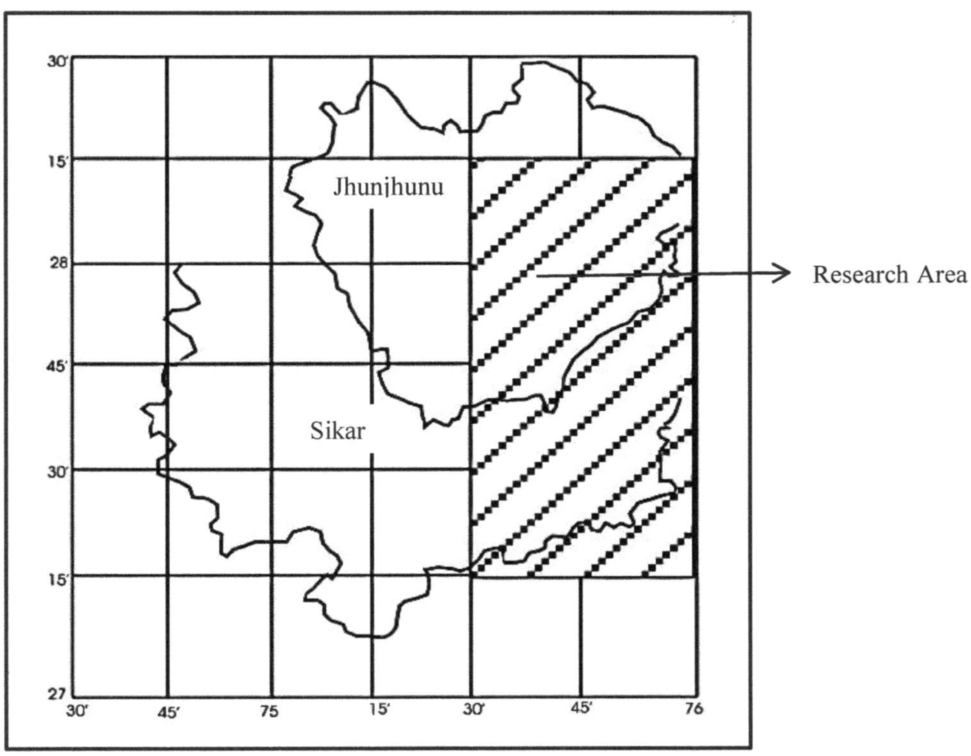

Figure 1.2: Map of Research Area

Figure 1.3: Satellite Imagery of Research Area

CHAPTER 2 - ENVIRONMENT

Environment study highlights brief account of the environmental setting viz. geology, geomorphology, drainage system, general parameter of major river (Kantli River), minor rivers / streams, characteristic of Kantli River, generalized chronological sequence of geomorphic units in Kantli block with geology and soil type is analyze by duly incorporating the strategy of adaptations followed during different cultural periods.

2.1 Geology of Rajasthan

The predominantly rocky and arid state of Rajasthan is unique from the point of view of its geology. Geologically, this state represent the northwestern part of Peninsular India comprising one of the oldest mountain chains 'The Aravalli Mountain Range', which possibly mark the line dividing older groups of rocks to the east and younger group of rocks to the west. Apart from this, it is perhaps the only state in India possessing a stratigraphy representing all the eras of the Geological Time Scale (Roy *et.al.* 1998). The region of Rajasthan is characterized by a wide range of litho-stratigraphic units, belonging to Proterozoic, Palaeozoic, Mesozoic and Cenozoic (Tertiary and Quaternary) eras.

Proterozoic rocks are predominantly confine to the mountain ranges and peneplains of eastern Rajasthan, categorized as Bhilwara, Aravalli, Delhi and Vindhyan Supergroup of rocks and associated intrusive. The four-fold classification assigned by A.M. Heron (1917, 1932, 1936, and 1953) has provided the basic framework for the Proterozoic rocks of Rajasthan. This classification envisages an overall evolution of rocks through three orogenic cycles, which still forms the basis for the stratigraphic classification. Later on the work carried out on similar lines by Raja Rao (1967), Raja Rao *et.al.* (1971), Crawford (1970), Roy (1988), Gupta *et.al.* (1980 and 1992) etc. has given sufficient information on Proterozoic rocks.

Palaeozoic rocks forming Aravalli and Delhi supergroup of rocks form the eastern most boundary and the Malani Igneous suite as the southern boundary. The arid terrain of western Rajasthan formed an extensive basinal setup from Cambrian to Tertiary. These major depositional basins viz. (i) Marwar Basin - Cambrian. (ii) Bap Badhaura Remnant - Permian. (iii) Lathi Basin - Jurassic. (iv) Jaisalmer Basin - Mesozoic and Tertiary. (v) Palana Ganganagar shelf - Tertiary, and (vi) Barmer Basin - Cretaceous. The Tertiary sediments accumulates an enormous thickness of sedimentary sequence of marine and deltaic environment (Gangadhar, K. 1996).

Mesozoic rocks in Rajasthan are represented by the sedimentary sequences belonging to the Lathi, Jaisalmer, Barmer basins, and the Palana-Ganaganagar shelf. These Mesozoic sediments are seen overlying the rocks of the Malani Igneous Suite, the Marwar Supergroup and the Bap Badhaura remnants of Permian age. They cover a vast stretch extending from Barmer in the south to Mahendragarh in the north and Jaisalmer in the west to Lathi in the east (Dasgupta, 1975).

The **Jurassic rocks** forming north-northeast to south-southwest freshwater sedimentary sequence in western Rajasthan has been designated as the Lathi formation. These sedimentary with an aggregate thickness of more than 400m comprise coarse grained and current bedded sandstone, conglomerates, lithic arentites, siltstone and shale with ferrugenised bands and haematitic nodules. They are also characterized by an abundance of dicot fossil woods (La Touche, 1902).

The **Cretaceous rocks** represent marine sedimentary sequence of Jaisalmer and Barmer basins. In the Jaisalmer Basin, these Cretaceous sediments are comprised of Jaisalmer limestone, Baisakhi shale, Bedesar sandstone, Parihar sandstone and Abu limestone. The aggregate thickness of sedimentary exceeds 700m and is characterized by numerous faunal fossil assemblages. In the Barmer Basin, the Cretaceous rock is represented by solitary outcrops the 'Fatehgarh sandstone' overlying the Lathi sandstone. These rocks comprise ferruginous sandstone, conglomerate, phosphatic sandstone and mudstone, with an aggregate thickness of 50m (Dasgupta 1975).

Cenozoic rocks, the sedimentary sequences representing the Paleocene and Eocene periods of Cenozoic era in western Rajasthan depict the deltic conditions followed by marine transgression. In the Jaisalmer Basin, the territories comprise the sedimentary sequence of – Sanu sandstone, Khuiala limestone and Bandah limestone. In the Barmer basin the Tertiary sediments having an aggregate thickness of 400m are represented by Akli Bentonite, Mandal sandstone, and Kapurdi Fukller's earth. The Palana-Ganganagar shelf comprises the Marh sandstone and Joira Fuller's earth, with an average thickness of 350m. The litho-stratigraphic succession of the Tertiary sequences for all the three depositional basins in western Rajasthan. The quaternary sediments of fluvial, aeolian, lacustrine, and residual origin overlie tertiary sedimentary sequences as well as the older lithostratigraphic units.

2.2 Geology of Research Area

The Shekhawati region comprised of lithostratigraphic sequences of Proterozoic and Cenozoic eras. Proterozoic rocks comprise of Delhi supergroup of rocks and Post-Delhi Igneous Intrusive. Delhi supergroup of rocks have been dated around 2000–800 mya. Lying unconfromable over the older sediments of Bhilwara and Aravalli supergroup, the Delhi supergroup rocks form a major constituent of the Aravalli mountain range. These marine clastics with subordiantes chemnogenic sediments are characterized by a number of concordant or discordant intrusive as well as extrusive phases. The Delhi Supergroup of rocks were affected by multiple episodes of folding viz., DF1 (Delhi supergroup folding) (AF2) (Aravalli supergroup folding), DF2 (AF3), and DF3 (AF4) (Gangadhar, K. 1996) and polyphase

metamorphism. The sedimentary and volcanic rocks of the Delhi supergroup are found to have recrysallied first under regional metamorphism (upto amphiboliate fancies), which was later superimposed by the thermal metamorphism (hornblende fancies). The important granite bodies within the Delhi Supergroup include Erinpuira granites, Sender granites, Ambaji granites and Kishangarh granites. The Delhi Supergroup of rocks is further classified into:
(i) Older Alwar Formation with main rock types are quartzities, schist, and conglomerate.
(ii) Younger Ajabgarh Formation with main rock types are schist, gneisses, calc-gneisses, dolomitic marble, amphibolite, and phyllite.

The Post Delhi Igneous Intrusive activities followed by the uplift of Delhi super group of rocks during the last stage of DF2 was the major one, affecting large areas of Aravalli proto-continent. The leuco-granites of Bairath, Sikar, Sendra, Erinpura, Ambaji, and Godhra covering the entire mountain range are testimony to this massive event, which took place around 900 ± 50mya and these are found through the Alwar and Ajabgarh groups (Gangadhar, K. 1996). The Cenozoic era is comprise of Quaternary (Recent to Sub recent) consisting of sand, alluvium, etc. (Figure 2.1).

2.3 Quaternary Geology of Rajasthan

The arid and semi-arid terrain characterized by rich assemblages of quaternary sediments. These sediments are of residual fluvial, lacustrine, and aeolian nature and have been deposited by a combination of surficial processes, under the prevailing climatic conditions during the past. The quaternary sediments exhibit an aggregate thickness of more than 300m, deposited over the pre-quaternary basement rocks (Roy *et.al.* 1998).

With the dawn of the twentieth century, Thar Desert has attracted a fair number of scientific researchers. A few noteworthy researches have been carried out by La Touche (1902), Holland and Christie (1909), Auden (1952), Ghosh (1952), Krishna (1952), etc. However, the arid terrain of Rajasthan has become a center of attraction from the point of view of understanding the early Pleistocene epoch. Resultantly Holocene climatic vicissitude, voluminous data have been generated on the numerous aspects of Quaternary period. Few important aspects, on which valuable contribution has been made, include archaeology, palynology / palaeoclimate, geomorphology and neotectonism.

The accumulation of thick pile of varied Quaternary sediments in Thar Desert can be attributed to numerous factors related to the pre-quaternary topographic and structural configuration. Neogene-quaternary tectonism and the past climate change. These factors in turn were responsible for controlling the intensity of various surficial processes, governing an overall modification in the terrain. Studies carried on Quaternary sediments established that the climatic changes in Thar region were fluctuating between humid and arid (Misra *et.al.*1988, Singh *et.al.*1974).

2.4 Quaternary Geology of Research Area

The Shekhawati region exhibits vast accumulation of quaternary sediments of fluvial and aeolian origin. Distribution wise, the aeolian sediments cover northern and central part of the region. The fluviatile sediments are well exposed within the river valleys of the present day drainage systems. The continuity of these fluviatile in the form of flood plains is concealed under a thick cover of windblown sands of recent and sub-recent period.

2.5 Geomorphology of Rajasthan

The most striking geomorphic feature of Rajasthan is the **Aravalli hills**, which intersect the region into two halves from northeastern extremity near Khetri to southwestern extremity near Mount Abu, though the hills are traceable as far as Delhi. The $3/5^{th}$ of the region lies to the northwestern limit and $2/5^{th}$ of the region lies to the southeastern limit. The Aravalli hills form the main and the most important hill range having an irregular width. It does not form a continuous hill range and it is dissected at many places. It runs continuously from southwest of Sirohi to the northwest of Khetri for a distance of about 550 kms. Further north, it extends up to Delhi as isolated hillocks. From Delhi to the plains of Gujarat, it has a length of 650 kms, and has summits exceeding 1225m in height (Roy *et.al.* 1998).

The Aravalli hills forms four main hill ranges:

i. The northeastern hill range continues upto northwestern part of Jaipur district and parts of Alwar district. The Alwar hills next to the Kumbhalgarh summit in the Mewar region attain an average height of 560-670m. The intermittent valleys being generally wide and along Babai (780m), Kho (920m) and Raghunathgarh (1055m) are some of the summits located on this hill range.
ii. The central Aravalli range subdivided into Shekhawati hill range and the Marwar hill range. The Shekhawati hills attain an average height of 400m, which extend from Sambhar Lake in south to Singhana in Jhunjhunun district in north. The hills of the Marwar region separate the Marwar plains from the Mewar plateau and cluster near Ajmer. The highest summit of this hill range is Taragarh (873m) having steep western slopes.
iii. The hills of Mewar occupy the entire area of Dungarpur and Udaipur districts and the eastern parts of Sirohi district. This northeast-southwest trending hill range forms a major water divide as the Great Indian Water Divide, just south of Ranakpur in Pali district. West of this divide, the drainage flows to join the Narmada River system whereas the drainage in the east forms tributaries of the Banas-Chambal river system. Except for the Mount Abu hill, the highest landforms of the Aravalli range forming a plateau, locally known as Bohrat, forms a part of the Mewar hill range. This plateau occurs between Kumbhalgarh

and Gogunda and constituted of quartzite. The relief of the Mewar hills is dependent on the lithological character of the constituting rocks. Its south-central part is a monotonously rolling country dissected by ramifying shallow valleys with smoothly rounded knolls.

iv. Mount Abu hill range covers an area of 150 Sq. kms and is around 1200m high above the msl. The highest summit in Rajasthan, i.e. Gurushikar (1727m) is located on the Abu batholith of the Erinpura granite. Other summits of this hill range are Sher (1597m) and Achalgarh (1380m).

The erosional and depositional processes in Rajasthan forms several distinct terrains, which mainly evolved during late tertiary and quaternary periods as degradational and aggradational units as follows:

Western plain is a great sandy tract region, cover the area about 175000 Sq. kms lying to the west of the Aravalli's and extending up to the Indus basin. This sandy tract region further sub-divided into two major units, namely:
(i) The western arid zone divided into desert plain or Marusthali and dune free plains.
(ii) The eastern semi-arid zone includes the Ghaggar basin in the north, the Shekhawati plains in the northeast, the Luni basin in the southeast, the Nagaur uplands covering major parts of Jodhpur and Barmer, and the western parts of Jalore, Sirohi, and Pali districts.

Eastern plain is bounded to the northeast, east, and southeast of the Aravalli hill range. The Aravalli range upto north of Udaipur demarcates its western limit. In the southeastern boundary lies the Vindhyan plateau. It is mainly, formed by the pediments and flood plains of the major rivers: Chambal, Banas, and Mahi. The northern extension of this plain in Bharatpur and Morena districts joins three major basins, namely, Chambal, Banas, and Mahi.

Southeastern plateau is generally known as the Hadauti plateau. It is confined to the southern and southeastern parts of Rajasthan, extending into greater parts of Jhalawar and Kota districts. This plateau has an average height of 500m above msl and is dotted with isolated hillocks. This plateau in Rajasthan occurs in the upper catchment of the Chambal River to the southeast of the Mewar plains. It merges with the Malwa plateau in the east. Geomorphologically, it can be divided into Vindhyan scarp land and Deccan Lava plateaus.

2.6 Geomorphology of Research Area

The Shekhawati region display mosaic of land features, which forms five distinct units: rocky highlands, pediment zone, rocky peneplains, alluvial plains, and aeolian plains (Figure 2.2).

Rocky Highlands represent discontinuous hills and ridges belonging to the Aravalli hill range. The hills are more than 500m in height; Raghunathgarh hill (1050m msl), Danta Ramgarh (692m msl), Ganwari hill (846m msl), Dokan hill (798m msl), Khetri hill (790m msl), Kakariya hill (978m msl), etc. The rocky highland area characterized by narrow to wide V-shaped valleys, numerous small streams, and rivulets.

Pediment Zone is restricted to a narrow belt running along the hill trends, which is characterized by moderate to steep slopes and is comprised of colluvial as well as dunal sands. Intensive tectonism and erosional activity by water action has developed innumerable deep gullies. These gullies are more pronounce in the areas where aeolian materials predominate. The more prone example areas are around Nim ka Thana, Madhgarh, Jarli, Sunari, Hanumansagar, Makari, Amarsar, Bagar, Guhala, etc.

Rocky Peneplain shows flat to undulatory peneplanation surface with a thin veneer of soils. The area around Nawalgarh, Udaipurwati, Danta Ramgarh, etc., falls within the rocky peneplains.

Alluvial Plains characterized by a rich assemblage of recent to sub-recent alluvial materials comprising admixtures of silt, sandy gravel and pebbles. These alluvial deposits is distributed over a vast stretch of area all along the present day active fluvial systems (Newer alluvium) and the buried flood plains of the defunct and buried palaeochannels (Older alluvium). Alluvium varies from highly angular bouldery gravels near the hills to coarse sand, which is mostly rework under aeolian action. The alluvium of fluviatile origin consists of boulder, gravel, pebble, coarse sand, and clay. The older alluvium are found at depths and also in the riverbeds as strongly cemented basal conglomerate, but the younger alluvium is found near the land surface along the present river channels and is devoid of any cementing matrix. The absence of kankar formed insitu in the younger alluvium also distinguishes it from the older alluvium clay forms only a minor ingredient of alluvium in the present day stream channels. The quaternary fluvial sediments of the Kantli river basin into three litho-units:

i. The younger fluvial unit, confined to either bank of the present day Kantli channel and comprising the present day river sands.
ii. The older fluvial unit, comprising pedoclastic pedogenised, coarse to fine grained sand and silty sand with gravels and pebbles, and
iii. The oldest fluvial unit, comprising calcareous conglomerate with sand and silty sand.

Aeolian Plains are sand generally overlies the alluvium exhibiting high variation in its thickness and seems to have the control of pre-existing topographic configuration, which exhibits various dune patterns. These wind-blown sands are characterized by their non-stratified, very fine to medium-grained, sub-angular to sub-rounded nature. Compositionally, these aeolian are dominated by quartz with subordinate proportions of ferruginous, calcareous and micaceous impurities. The different types of dunes observed in the research area are parabolic, barchans, longitudinal, transverse, seif, star, and complex dunes (Figure 2.3).

2.7 Soil Type

Apart from fluvial, aeolian, and colluvial sediments, present day soils veneer of associated parental materials (both consolidated and unconsolidated) constitute an important unit of quaternary stratigraphy. In Shekhawati region, these soils are of lateritic (red) and arid type's characteristics of residual rocky peneplains and residual aeolian plains respectively.

2.8 Generalized chronological sequence of geomorphic units with Geology

Period	Geology	Geomorphic Unit (Landform)	Climate
3rd peneplantation in progress (under aeolian environment)			
Recent	Younger Aeolian Sediments	15. Sand heap, Sand ripple 14. Playa 13. Wadi	Present arid to semiarid
Sub-Recent to Recent	Younger Alluvium	12. Present (Active) dune 11. Present river channel and braid	Warm sub-humid
Sub-Recent (Early Holocene)	Older Aeolian Sediments	10. Present flood plain 09. Backswap 08. Stable dune 07. Pediplain 06. Pediment 05. Bornhardt (Comprising of basement rocks)	Arid
2nd peneplantation (under aeolian environment)			
Upper Pleistocene to Early Holocene	Older Alluvium	04. Left out channel 03. Older flood plain 02. Terrace	Humid and Cold
Pleistocene (?)	Sandstone/Siltstone, Conglomerate (Indicate the ancient river bed, do not form any land forms)		
1st peneplantation (under fluviatile environment)			
Precambrian	Delhi Supergroup of rocks and post Delhi intrusive	01. Hill ranges, discontinuous hills, outcrops	

2.9 Drainage System of Rajasthan

The present drainage system of Rajasthan is supported by Banas, Banganga, Chambal, Ghaggar, Kali Sind, Luni, Mahi, and Parbati rivers.

Banas literally means, "hope of the forest". Rises in the Aravalli range (25°31`N and 73°28` E) in Udaipur, about 4.8 kms from the fort of Kumbhalgarh and continues through the territories of Udaipur, Jaipur, Bundi, Tonk and Karauli falls into the Chambal (25°55` N and 76°44` E) at the holy sangam Rameshwar.

Banganga or Utangan literally mean as 'Arrow River'. A river of north India, rises in Jaipur near Bairat (27°24`N and 76°10`E). It flows generally in an easternly direction through Jaipur, Bharatpur and Dholipur and Agra and after a course of about 376 kms joins the Jumna 16 kms east of Fatehabad in Agra.

Chambal identified as 'Charmwati' by Sanskrit writers is a river of Central India and Rajasthan and one of the chief tributaries of the Jumna. It rises in Indore (22°27` N and 75°31` E), 14 kms south-west of Mhow, in the Janapav hills 2019 feet above the msl. It flows in the northern slopes through Gwalior, Indore and Sitamau and skrits of Jhalawara entering Rajasthan at Chaurasgarh 312 kms from its source. It receives many tributaries in Central India, the chief being the Chambal and Sipra, both of which rise in the Vindhayan Mountain. In Rajasthan the Chambal breaks through a scrap of the Patar plateau, the bed getting narrower and after a winding course of 48 km it receives the Bamani at Bhainsrorgarh. Continuing northeast, Kali Sind joins the river from the south and the Mej from the west while lower down where the frontiers of Jaipur, Kota, and Gwalior meet, the Parbati flows into it. The Chambal forms the boundary between Jaipur, Karauli, and Dholipur on the one side and the Gwalior on the other.

Ghaggar, a river of north India rises on the slopes of the Himalayas (30°4`N and 77°14` E), passing within 4.8 kms of Ambala, touching Patalia enters Hissar and finally losses itself near Hanumangarh. The river was an affluent of the Indus, and the dry bed of the old channel being still traceable.

Kali Sind is frequently referred to in the Sanskrit literature and is mentioned by Abul Fazl as one of the principal river of Malwa. Sarangpur and Gagraun are the principal places on its banks. It is probable that the river derives its name from the prevalence of black (Kali) basalt in its bed. The river is a tributary of Chambal, which forms the part of Central India and Rajasthan. It rises in the Vindhayas (22°36`N and 76°25`E) at the village of Barjhiri. It flows for about 288 kms through Gwalior, Dewas, Narsinghgarh and Indore in Central India. After that, it travels through Kota and Jhalawar in Rajasthan piercing the Mukandwara hills near Gagraum and falls into the Chambal 360 kms from it sources near the village of Piparia in Kota (25°32`N and 76°19`E).

Luni, literally means "Salt River", the Lonavari or Lovanavari of the Sanskrit writers. Luni river of Rajasthan rises in the hills south west of Ajmer (26°25`N and 74°34`E) and is first known as Sagarmati. After passing Govingarh it is joined by the Sarsuti, which has its sources in the scared lake of Pushkar and from this points the river is known as Luni. It at once, enters Jodhpur territory and after a course of about 320 kms generally within the southwestern direction is finally lost in the marshy ground at the head of Rann of Kutch (24°40` N and 71°15`E).

Mahi is the "Mophis" of Ptolemy and "Mais" of the Periplus. Mahi, a river of western India, rises in the Amjhera of Gwalior 1850 feet above the sea level (22°52`N and 75°5`E) and flows for about 160 kms through the southwestern corner of Central India passing through Gwalior, Dhar, Jhabua, Ratlam and Sailana. It then enters Rajasthan and flows in the northern direction, intersecting eastern half of Banaswara, till it reaches Udaipur then forms boundary between Dungarpur and Banaswara. It then passes into Gujarat.

Parbati is a tributary of the Chambal draining part of Central India and Rajasthan. Rising in the Vindhyas (22°52`N and 76°39`E) at the village of Makgardha, it flows in a northern direction either through or along the borders of Bhopal, Gwalior, Narsinghgarh and Rajgarh in Central India and Tonk and Kota in Rajasthan. After the course of 352 kms it joins the Chambal at Palighat in the northeastern corner of Kota (25°51`N and 76°37` E).

2.10 Drainage System of Research Area

The present drainage system in Shekhawati region is supported by Kantli, Rupangarh, Dohan, Dongar, Sabi (Sahibi), Chandrawati, Udaipur-Lohargarh ki Nadi, Madhobini Nadi, Sota Nadi, Dhudhala Nadi, Bandi Nadi, Sukh Nadi (Singhana Nadi), and Khariya Nadi (Figure – 2.4).

2.11 General Parameters of Minor Rivers / Streams of Research Area

Mendha originates approximately 2 km northeast of the village Aspura Bara (27°26′ N - 75°46′ E) in Sikar district. It flows in southwestern direction for about 100 kms.

Rupangarh originates in the Aravalli hills south of Kishangarh around Naulacha-Muhami village. Initially it flows in northern direction and then in northeastern direction for about 70 kms.

Udaipur-Lohargarh ki Nadi originates in the Raghunathgarh hills near Salaraimataji/Sakraimataji village (27°37′ N - 75°22′ E). Intially it flows in northeastern direction and then in northwestern direction for about 55 kms.

Sota Nadi originates in Garhtaknet village (27°27′ N - 75°48′ E) of Sikar district. It flows in northeastern direction for about 35 kms.

Sabi Nadi originates in the Saiwar hills (27°25′ N - 75°54′ E), a reserved forest in Sikar and Jaipur distiicts. The nadi flows in southeastern direction for about 20 kms.

Khariya Nadi originates in the Surani hills (27°29′ N - 75°48′ E) of Jaipur district. It flows in southwestern direction for about 10 kms.

Madhobini Nadi originates in Amarsar village (27°22′ N - 75°50′ E) of Jaipur district and flows in southeastern direction for about 25 kms.

Dhudala originates in the confluence of Khariya and Madhobini Nadis (27°16′ N - 75°59′ E). It flows in southeastern direction for about 10 kms.

Doha Nadi: originates in Mankri village (27°48′ N – 75°48′ E) of Sikar district and flows in northeastern direction for about 17 kms.

Dongar Nadi originates in Babai village (27°54′ N – 75°46′ E) of Jhunjhunu district. It flows in northesastern direction for about 13 kms.

Chandrawati originates in the confluence of Dohan and Dongar Nadis (27°55′ N – 75°54′ E) and flows in northeastern direction for about 25 kms.

2.12 Characteristic of Kantli River Channel in Research Area

Kantli River, accordingly to the oral tradition "Is the river, which cuts its own courses". Kantli is Ephemeral River flowing in northern direction, originating from the hills near Gidhalaya village in Sikar district and finally gets dried up and disappear in the dunal tracts near Navrangpura of Rajgarh village in Churu district. The total length of Kantli River is about 135 kms, flows through the hilly terrain in the upper reaches, cutting across the Aravalli ranges for about 52 kms upto Mainpura in Jhunjhunu district and then flows through the vast aeolian dune field in the north. The total catchment area is about 1746 sq km.

Channel Segment	Flow Direction	Channel Type	Channel Length
Gidhalya-Thikriya	NNW	Rectilinear to sinuous with local rectangular to angulate bends at Thikriya, Kanwal, etc.	17Kms
Thikiriya-Chaukri	NW	Rectilinear to sinuous	05 Kms
Chaukri-Guhala	N	Rectilinear to sinuous	05 Kms
Guhala-Sunari	NE	Sinuous to straight with local meandering at Pachlagi and Papra.	12 Kms
Sunari-Sanwal ka Bas	NW	Rectilinear to sinuous, rectangular at Mainpura, braided at Chawara, Kair, and Sanwal ka Bas.	30 Kms
Sanwal ka bas-Islampur	N	Sinuous to rectilinear, braided	22 Kms
Islampur-Lamba Gothra	N	Rectilinear to rectangular	13 Kms
Lamba Gothra-Tigiwas	NW	Sinuous, bend with local meanders at Dhattarwala, Dhandharia, Ranjitpura. Straight channel at few places.	20 Kms
Tigiwas-Naurangura	N	Rectilineat	04 Kms

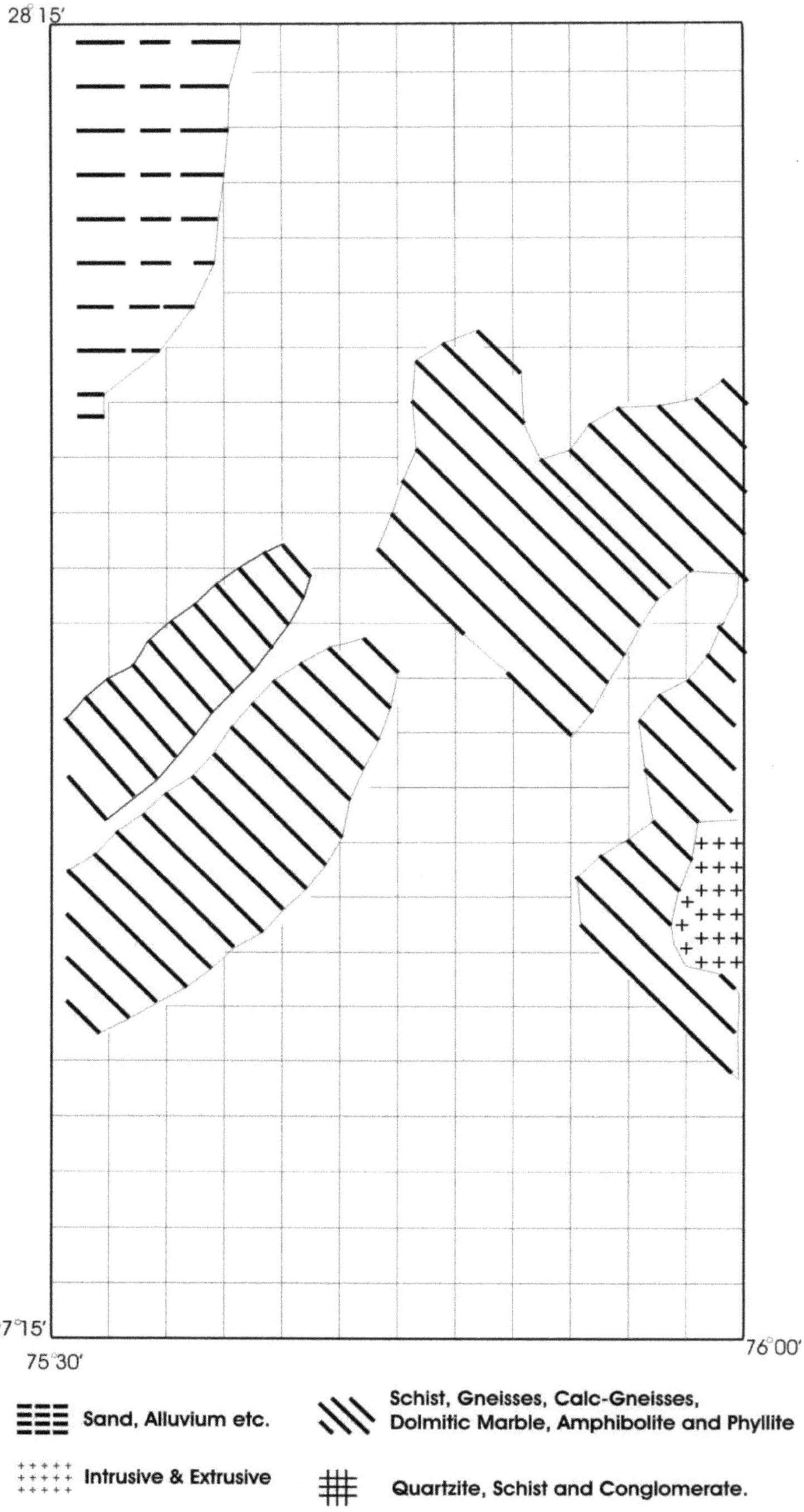

Figure 2.1 – Geological Formation in Research Area

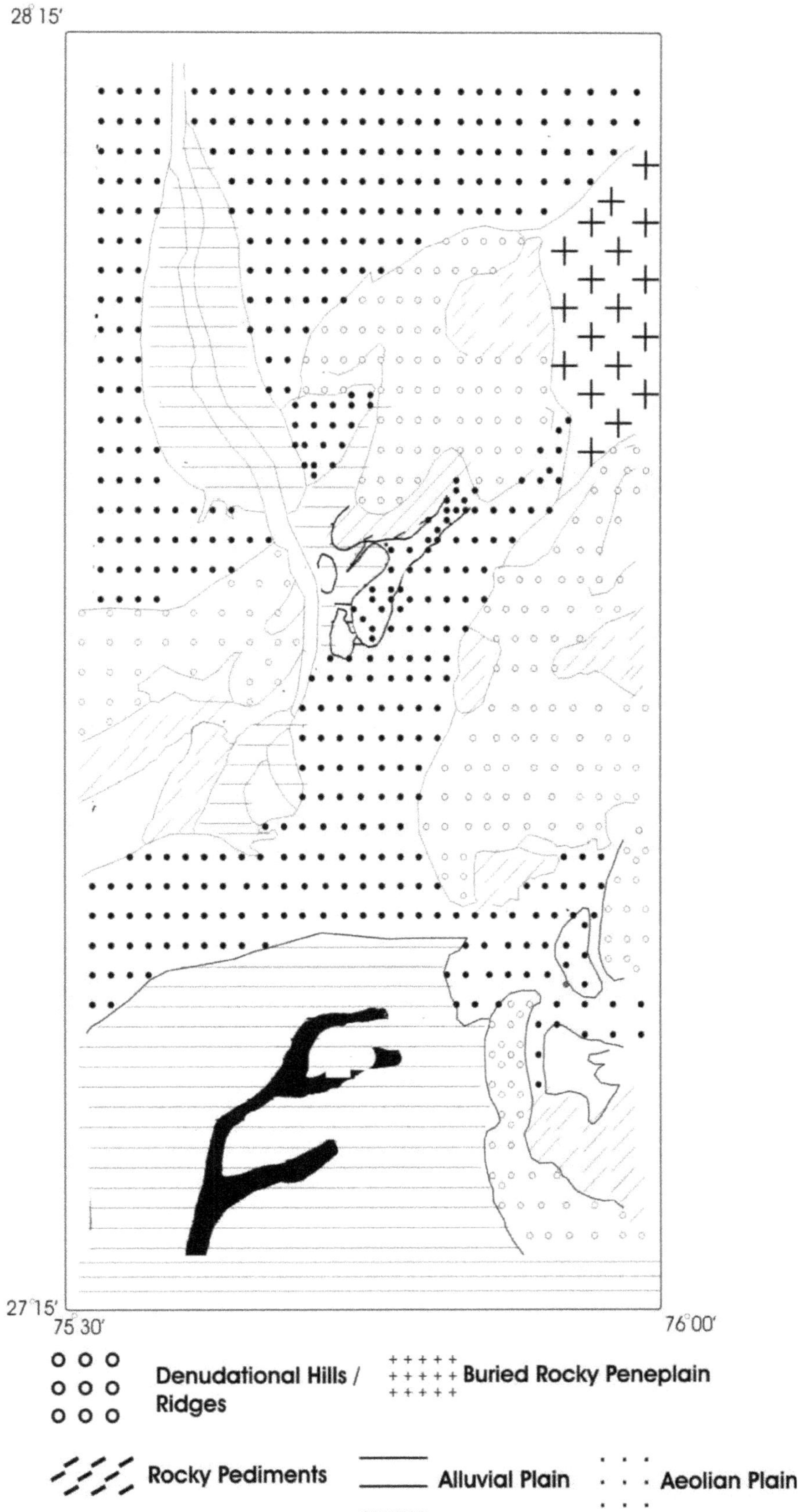

Figure 2.2 – Geomorphological Features in Research Area

Figure 2.3 – Dune Types in Research Area

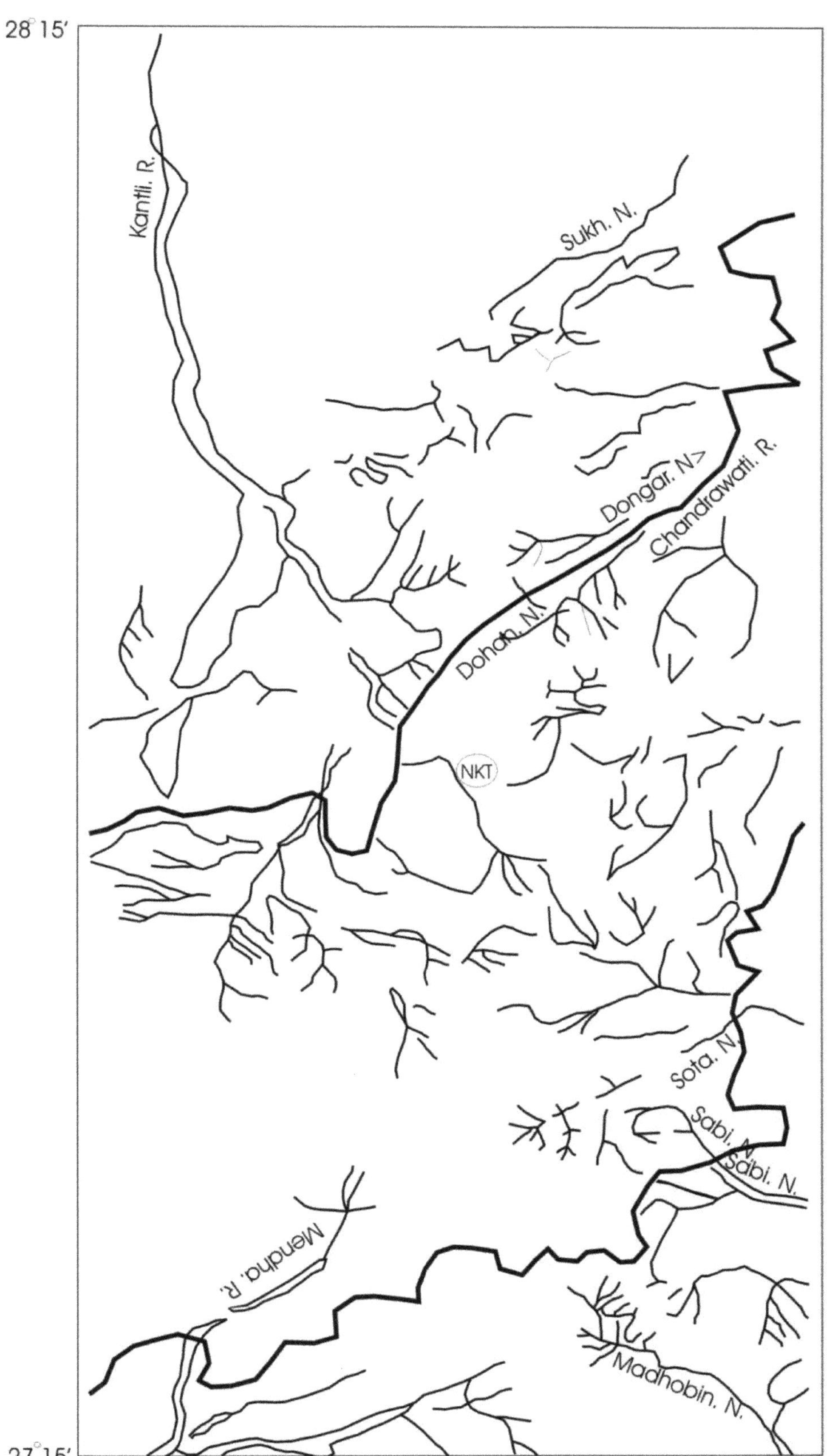

Figure 2.4 – Drainage Systems in Research Area

CHAPTER 3 - EXPLORATION

Overall, it gives a brief account of material culture discovered during exploration. It discusses the location of sites and their nature along with their geomorphic setting. The periodization of sites is base on comparative study of ceramics, lithic assemblages and metal processing indicators.

3.1 Palaeolithic Research in Rajasthan

Rajasthan saw the beginning of Palaeolithic research in 1950s. Antecedent to this period, only three discoveries were made by Carlleyle in 1871, Hackett in 1870s and Seton Karr in 1928, which are now only of historical importance because exact location and context of those artifacts are unknown (Stein, 1989).

From 1950s onwards, Palaeolithic research in Rajasthan has revealed numerous sites that provide us with a broader and better picture of this culture. In 1950s, Deshpande (IAR 1953-54:37 and IAR 1954-55: 58) discovered Lower Palaeolithic pebble tools, flakes, handaxes and cleavers from Gambhiri and Berach Valley near Chittorgarh. Later following the footsteps of M. N. Deshpande, Rao located six Palaeolithic sites from Bamani and Ruparel, the tributaries of Chambal and Breach Valley (IAR 1956-57: 5). In the meanwhile, in 1953 Sankalia also discovered Palaeolithic tools in a gravel deposit from the river Banas at Nathdwara in Udaipur District (IAR 1953-54: 37). In 1955, Soundara Rajan found handaxes and cleavers near Bhangarh in Alwar District from Sanwan River, a tributary of Banganga (IAR 1956-57: 8). Again, in 1956-57, Rao discovered Lower and Middle Palaeolithic tools from Chamli River, a tributary of Chambal River near Badoli and later also excavated at Sonita, Navaghat and Bhainsrorgarh all on the Chambal River (IAR 1956-57: 5). In 1956-57 Archaeological Survey of India (IAR 1956-57: 5-8), unearthed Lower Palaeolithic tools in Beawar on the Wagan River in Chittorgarh district. In 1958-61 Misra carried out Prehistoric research in eastern and western Rajasthan, in Chambal, Banas and all its tributaries in south Rajasthan, Banganga and Sanwan in north Rajasthan and Luni and all its tributaries in western Rajasthan (Misra, 1966). Allchin, Goudie and Hegde (1978) carried out Prehistory and Paleography research in Great Indian Desert from Jaipur to Jaisalmer and from Bikaner to Bhiwani in east–west orientation and from Mt Abu to Delhi along the western side of the Aravalli and from Pali to Bikaner in north–south orientation.

3.2 Explored Palaeolithic Sites in Research Area

Palaeolithic research carried out by Central and State government bodies viz. Archaeological Survey of India, Geological Survey of India and Rajasthan State Archaeological Department have surveyed the region along the River and its tributaries in Rajasthan. Such research has brought to light number of sites in the southern Rajasthan whereas central, western and northern Rajasthan was shown to be devoid of any Palaeolithic sites. Research Institute of Deccan College under the supervision of Sankalia (1956) and later followed by Misra (1967) carried out their research on the same footings as previously carried out by the government agencies and extended the horizon of Palaeolithic findings to eastern and western Rajasthan. The foreign collaborative work of Allchin, Goudie and Hegde (1978) has further extended the horizon in Great Indian Desert. In spite of all these collective works by different agencies, the present research area was shown devoid of any Palaeolithic findings.

In the present survey, 3 Lower Palaeolithic sites were reported from the research area. The findings of these sites has given lead for the future research in the central and northeastern part of Rajasthan known as Shekhawati region whereas Middle Palaeolithic and Upper Palaeolithic sites are still not discernible in the research area.

1. **Laluda Ki Dhani (LKD):** A Lower Palaeolithic secondary site located on a foothill at Laluda Ki Dhani (27°55'258"N – 75°49'308"E), situated 1 Km west of village Papurna in Khetri taluka of Jhunjhunun (hereafter JJN) district. The village can be approached through a metal road and is nearly 27 Kms north of Nim Ka Thana (hereafter NKT) Khetri Maudh. The explorations in this area revealed one handaxe and two scrapers in quartzite raw material (Figure – 3.1). On the basis of typo-technological studies these tools can be dated to Lower Palaeolithic Period.

2. **Payga Kundalia (PKD):** Another Lower Palaeolithic secondary site (27°50'000"N – 75°58'440"E) located 1 Km west of the village Sohanpura in NKT taluka of Sikar (hereafter SKR) district. This village can be approached through a metal road and is situated 31 Kms northwest from NKT (Khetri Maudh). The site is situated near an ephemeral stream that drains the village from north to south and the landform is dominated by small hillocks. The survey in the foothills (Figure – 3.2) of this area discovered one handaxe, one cleaver, two scrapers and three hammer stones. All these tools are in quartzite raw material. Based on typo-technology comparisons these tools are assign to the Lower Palaeolithic Period. This multi-cultural site revealed Mesolithic Rock Shelters with paintings, Chalcolithic and Early Historic potteries.

3. **Pir Baba Ki Bani (PBB):** (27°57'948"N – 75°52'771"E) is secondary Lower Palaeolithic site that is located 2.5 Km southwest of the Tonda Village in Kherti taluka of JJN District. This village can be approached through a metal road and situated 37 Kms northwest from NKT (Khetri Maudh). The area is drained by Paneta *nalla* that is tributary of Chandrawati River. This nalla flows across the village and shows an open shrub area. The exploration for approximately 800 m stretch of the *nalla* revealed one handaxe, one cleaver and four scrapers in quartzite (Figure – 3.3). Again the Typo-technology of the tools suggests the Lower Palaeolithic Period for this site.

(a) (b) (c)

Figure 3.1 – Laluda ki Dhani (a) handaxe; (b – c) scrapers

(a)

Figure 3.2 – Payga Kundalia(a) general view(b) handaxe; cleaver; scraper & hammer stones

Figure 3.3 – Pir Baba ki Bani (a) handaxe; (b) cleaver; (c) scrapers

3.3 Mesolithic Research in Rajasthan

The Mesolithic research in Rajasthan became known from the work of Rao in 1950s with the finding of Microlithic sites from Kadmali, Berach and Gambhiri valleys in Chittorgarh district (IAR 1956-57: 5). During the same period, Thapar discovered Microlithic sites from Kakoni on the Parwani in Kota district (IAR 1956-57: 5-8). Later on Soundara Rajan in 1955 discovered more Microlithic sites from Jodhpur district (IAR 1956-57: 8). As a routine village-to-village survey, Archaeological Survey of India in 1956-57 discovered Microlithic sites on the Berach River in Chittorgarh and Udaipur districts and some near Bhainsrorgarh on Chambal River. Similarly, State Archaeological Department of Rajasthan in 1958-63 discovered Microlithic sites from Khainsawar near Nagore in Jodhpur, Bhilwara, Chittorgarh, Ajmer and Tonk districts. In 1971 and 1977, Misra excavated two major Mesolithic sites in Rajasthan. Tilwara (Misra, 1971) located in the eastern fringes of the desert in Barmer district of Marwar region and Bagor (Misra, 1967), which sits over the Kothari River in Udaipur district of Mewar region. The findings at Bagor have given new vista for better understanding of Mesolithic culture in Rajasthan. Hooja (1988), has listed 146 Mesolithic sites in Rajasthan out of which 67 has been reported from Udaipur, Chittorgarh and Bhilwara district in southeastern Rajasthan; 75 sites were reported from Ajmer, Jaipur, Sikar and Jhunjhunun districts in central and northeastern Rajasthan; Barmer, Pali, and Jaisalmer districts in western Rajasthan revealed 5 sites and in Bikaner in north Rajasthan only one site has been reported.

3.4 Explored Mesolithic Sites in Research Area

Generally, lithic assemblages are taken to predate the use of metal. Therefore, the Mesolithic of Rajasthan is believed to be necessarily earlier than the Chalcolithic. However, there is some, as yet insufficient, evidence to suggest that sometime the technological transition from Lithic to Chalcolithic was not as total as one had previously been led to believe, nor sudden and universal. However, the Mesolithic site of Bagor and Ganeshwar indicates that microlithic using hunter-gatherers inhabited the region even when Chalcolithic stage has been firmly establish in parts of these regions. This implies that some degree of hunter-gathering lifestyle continued to be in existence even when a more advanced technological skill had come into use. In this direction, present survey on Mesolithic research was carried out to find sites showing single culture having microliths with or without pottery and multi-cultural site having microliths with other cultural materials. In the light of these assumptions, 12 Mesolithic sites reported from the research area have given a single culture with microliths and lithic debitages and multi-cultural sites with Chalcolithic and Early historic pottery and few sites near to the mining areas.

1. **Beed Ki Dhani (BKD):** A multi-cultural site (27°41'N – 75°54'E) located 1 km northeast of Dariba village in NKT taluka of SKR district. The village is situated 23 Kms southeast of NKT (Khetri Maudh) drained by perennial streams with open scrub area having few patches of agricultural field around the site. The nature of the site points towards it being a seasonal settlement. The exploration revealed multi-cultural assemblage of microlithic core (Figure – 3.4) and Chalcolithic pottery. The site measures around 405 Sq m and located in the vicinity of an ancient mining area. Few pottery sherds were also collected from this area.

2. **Bhadwari (BWR):** A multi-cultural site (27°34'166″N – 75°39'182″E) located 0.2 Kms southwest of Bhadwari village in NKT taluka of SRK district. The village is approached through metal road and that is situated 28.5 Kms southwest of NKT (Khetri Maudh). The area is drained by a tributary of Kantli River flowing in west-east orientation of the village. The landform shows series of sand dunes that rise up to 6 - 7 m height from the surrounding area. The ephemeral nature of the site perhaps indicates it being as a seasonal settlement measure 375 Sq m and located on one such sand dune. The exploration revealed microlithic blades and lithic debitages found in a single cluster (Figure – 3.5). Along with microliths, a multi-cultural assemblage of the site reveals Chalcolithic and Early Historic pottery.

3. **Biharipur (BRP):** A multi-cultural site (27°53'186″N – 75°53'710″E) located 2 Kms south of the village Biharipur in NKT taluka of SKR district. The site is identified as a seasonal settlement measures around 525 Sq m located on the sand dune that rise up to 10 m height from the surrounding area. The area is drained by tributary of Chandrawati River flowing in north-south direction across the village, which is around 41 Kms west from NKT (Khetri Maudh). The exploration revealed microlithic core and a fragment of shell bangle (Figure – 3.6).

4. **Bhudoli (BDL):** A single culture site (27°41'865″N – 75°46'727″E) located 1 Km north of Bhudoli village in NKT taluka of SKR district. The village is nearly 5 Kms south of NKT (Khetri Maudh) and drained by ephemeral streams on the northern, southern and eastern side of the village. The landform around the site shows sand dune ridges that rise up to 4 - 5 m height from the surrounding area. The ephemeral nature of the site perhaps point towards it being a seasonal settlement, which measures only 27 Sq m and located on one of the sand dune revealing microlithic blades and lithic debitage collected in clusters of 2–3 m area.

5. **Gidhali / Nali No. 1 (GDL):** A single culture site (27°47'448"N – 76°03'696"E) located 1 Km north of the village Haripura in NKT taluka of SKR district. The village is around 35 Kms west of NKT (Khetri Maudh). The landform around the site shows series of sand dunes that rise up to 8 – 10 m height from the village, which is on the verge of vandalism by the

present construction worker. The nature of the site point towards it being a seasonal settlement, which measures 239 Sq m and located on one such sand dune revealing microlithic blades, cores and lithic debitage (Figure – 3.7) with stray findings of terra cotta figurine.

6. **Gidhali / Nali No. 2 (GDL):** A multi-cultural site (27°47'435"N – 76°03'630"E) located on the adjacent sand dune from GDL No. 1. The site perhaps point towards it being a seasonal settlement measures 346 Sq m and located on one such sand dune that rise up to 6.5 m heights revealing microlithic cores and lithic debitage (Figure – 3.8). A multi-cultural assemblage of the site also revealed Chalcolithic and Early Historic pottery with stray findings of iron piece.

7. **(Juna Tila), (GNR) No. 3:** A single culture site (27°40'531"N – 75°48'745"E) located 2 km northwest of the village Ganeshwar in NKT taluka of SKR district. The village is around 12 Kms southeast of NKT (Khetri Maudh) situated on the slope of the hillock. The village is drained by Chandrabhaga nadi, a tributary of Kantli River on the west and Raotala nalla on the southeast of the village. The ephemeral nature of the site perhaps points towards it being a seasonal settlement measures only 74 Sq m revealing microlithic blades and lithic debitages collected from a single cluster in the sand dune that rise to 12m height from the surrounding peneplain area (Figure – 3.9).

8. **Kakrana (KRN):** A multi-cultural site (27°53'476"N – 75°39'458"E) located 0.5 km north of the village Kakrana in NKT taluka of SKR district. The village is around 22 Kms northwest of NKT (Khetri Maudh) drained by Kantli River flowing south to north on the western side of the village. The landform shows number of sand dunes around the site. The ephemeral nature of the site point towards it being a seasonal settlement, which measures 312 Sq m revealing microlithic blades and lithic debitages collected in a single cluster from stabilize sand dune. A multi-cultural assemblage of the site also reveals Chalcolithic pottery with stray findings of terra cotta figurine and clay lumps along with the ancient mining area..

9. **Mataji No Timbo (MNT):** A multi-cultural site (27°38'160"N – 75°41'655"E) located 2 Kms south of the village Govindpura in NKT taluka of SKR district. The village is around 18 Kms southeast of NKT (Khetri Maudh) drained by Kantli River flowing east to west on the southern side of the village. The landform around the site shows series of sand dunes that rise up to 4 - 5m height from the surrounding area. The site has identified as a seasonal settlement, which measures 429 Sq m located on one such stabilize sand dune revealing microlithic cores and lithic debitages collected in a single cluster. A multi-cultural nature of the site revealed Chalcolithic and Early Historic pottery.

10. **Payga Kundalia (PKD):** A multi-cultural site as mentioned earlier in the Palaeolithic section. The Mesolithic period is identified with three rock shelters (Figure – 3.10) out of which two of them shows painting on it, which has been presumed to be from Mesolithic period (personal communication with Sharmaji, Kot Putali). The painting is in ochre color depicting hunting scenes, group dance and other group rituals (personal communication with Madan Lal Meena, Nim Ka Thana). There are series of cup marks observed on the stone floors, which indicate that it has been used for preparing colors. The exploration has also revealed Chalcolithic and Early Historic pottery.

11. **Sadmata Ka Maula (SKM):** A multi-cultural site (27°42'142"N – 75°49'926"E) located on the southeast periphery of the village Ganwari in NKT taluka of SKR district. The village is nearly 8 Kms southeast of NKT (Khetri Maudh) drained by a tributary of Kantli River flowing south to north and north to west on the eastern side of the village. The ephemeral nature of the site perhaps point towards it being a seasonal settlement, which measures 1988 Sq m revealing microlithic core and lithic debitages collected in clusters from the slope of the hillock. The landform around the site shows series of hillocks. Along with the microliths, site reveals Chalcolithic pottery.

12. **Thoi:** A multi-culture site (27°32'595"N – 75°45'196"E) located 0.2 Km on the eastern periphery of the village **Thoi** in NKT taluka of SKR district. The village is situated on the foothills, which is around 32 Kms southwest of NKT (Khetri Maudh. The area is drained by a tributary of Kantli River flowing south to north on the western side of the village. The landform shows series of sand dunes around the site. The ephemeral nature of the site perhaps point towards it being a seasonal settlement (Figure – 3.11), which has revealed microlithic cores, blades and lithic debitages collected in patches from sand dune ridges. Along with microliths, the site also revealed Early Historic pottery.

3.5 Chalcolithic Research in Rajasthan

Chalcolithic research in Rajasthan imparts with preliminary investigation by Tessitor at Kalibangan in the year 1918 to 1921 (Tod, 1894). After two decades, Stein carried out exploration in Saraswati valley as an archaeological tour along the Ghaggar – Hakara River (Stein, 1942) and later Gupta published Stein's work in the year 1989. Ghosh in 1950 to 1953 extended the work of Stein in Ghaggar River and discovered Harappan and Early Historic sites along the dry bed of the river (Stein, 1989). Misra in 1963-64 conducted exploration in south Rajasthan and brought to light the vista for Prehistoric and Proto historic sites in southern Rajasthan (Misra, 1967). Agrawala and Kumar brought to light Shekhawati region of Rajasthan with the findings of new sites Ganeshwar and Jodhpura and give new evidence regarding regional Chalcolithic culture of India (Agrawala

Figure 3.4 – Beed ki Dhani (core)

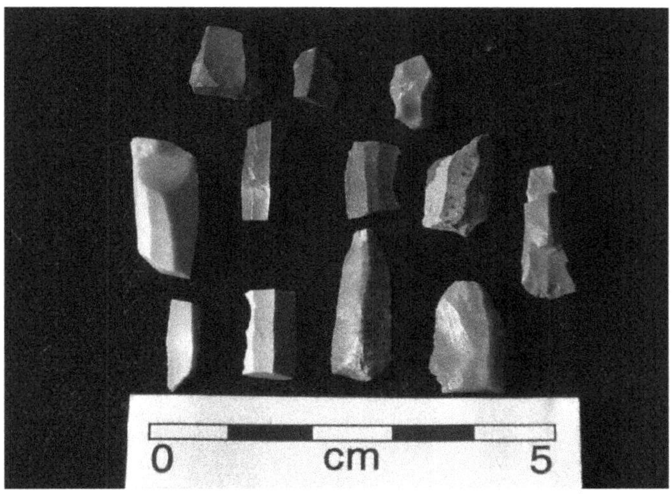

Figure 3.5 – Bhadwari (blades)

Figure 3.6 – Biharipur (core and shell bangles)

(a)

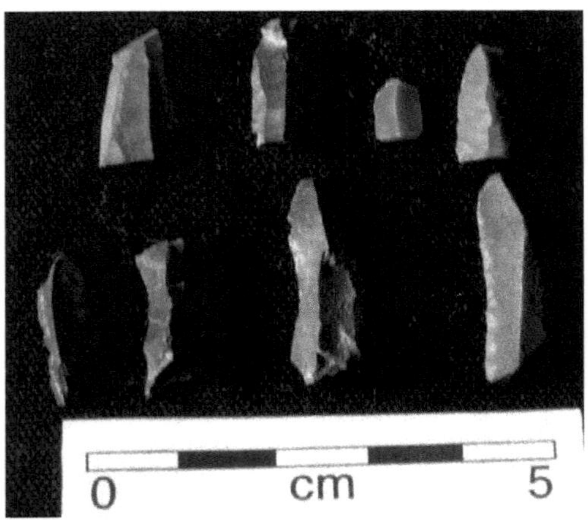

(b)

(c)

Figure 3.7 – Gidhali No 1 (a) general view (b) blades (c) cores

(b)

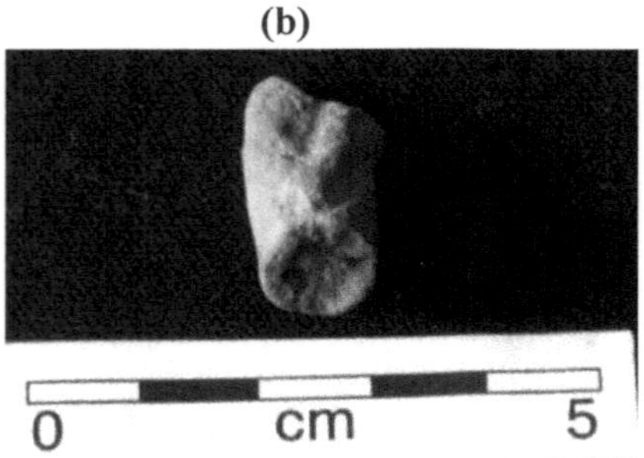

(c)

(a)

Figure 3.8 – Gidhali No 2 (a) general view (b) cores and (c)terra cotta figurine)

(a)

(b)

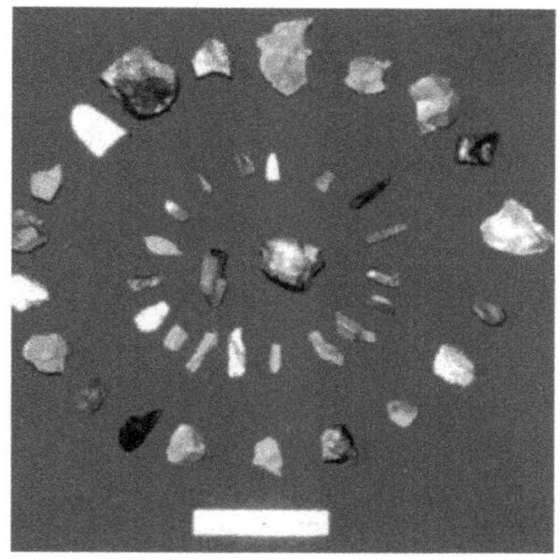

Figure 3.9 – Juna Tila at Ganeshwar (a) general view (b) cores, blades & flakes

(a)

(b)

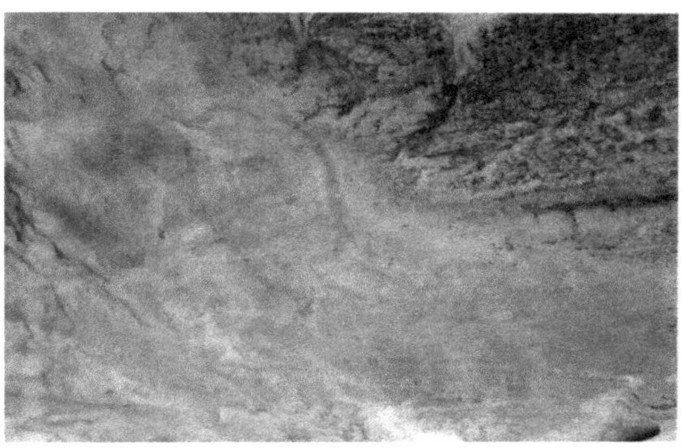

(c)

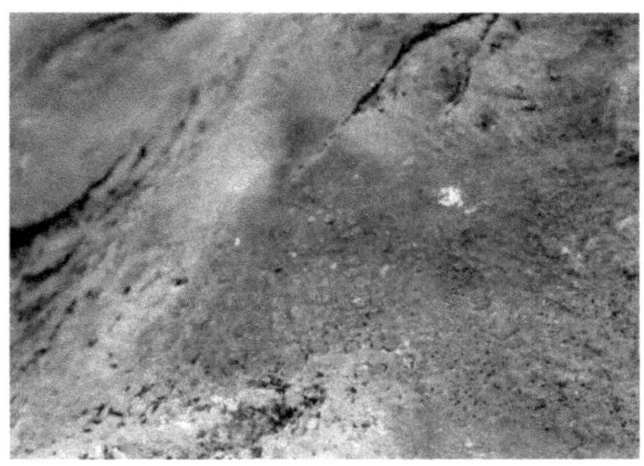

(d)

Figure 3.10 – Payga Kundalia (a – b) general view of rock shelters; (c – d) rock paintings

Figure 3.11 – Thoi - general view

and Kumar, 1982). Further routine village-to-village survey conducted by Archaeological Survey of India and State Archaeological department and exploration carried by Hooja for her Ph.D. dissertation in southeastern, central and northeastern region of Rajasthan has added new sites in Rajasthan (Hooja, 1988).

3.6 Explored Chalcolithic (Ganeshwar Cultural) Sites in Research Area

For the first time in the history of Indian Archaeology, a rich find of Chalcolithic copper objects has been found at a single site of Ganeshwar (Agrawala, 1981b; Agrawala and Kumar, 1982). These findings of Chalcolithic Copper Objects at Ganeshwar, which has raised many important queries: the site is near to the mines and in association with "OCP" from beginning to the end? Copper objects such as thin blades, arrowheads and fishhook show similarity with the Harappans copper objects? Presence of round terracotta cakes is also equally important for the inter-relationship with the Harappans? Findings of microliths along with the copper objects suggest that they have developed metallurgical technology while still being in foraging stage? They have exploited and successfully worked on the nearby mining areas as early as 3rd millennium BCE.

Ganeshwar is the key site in Ganeshwar–Jodhpura Cultural Complex was excavated by RC Agrawala and Vijay Kumar in the year 1979 to 1982. During the excavation thousands of copper objects in association with "OCP" was found. The stratigraphic position of the cultural material shows Mesolithic, followed by "OCP", unpainted Black & Red ware, PGW and Sunga–Kushana level. The earliest date for "OCP" is dated around 3000 – 2800 BCE (Agrawala, 1978, 1981b, and 1984; Agrawala and Kumar, 1982). Whereas Jodhpura in Jaipur district of Rajasthan reveals stratigraphic position of the cultural material showing "OCP", Black & Red ware and PGW. At both the sites Black & Red ware precedes PGW, which is similar to Noh at Bharatpur district of Rajasthan and Atranjikhera at Etawah district of Uttar Pradesh. The earliest date at Jodhpura for the upper level of "OCP" is dated around 2500 to 2200 BCE (Agrawala & Kumar, 1982; Agrawala, 1984).

In the present survey, 33 Ganeshwar Cultural sites were reported. In the central and northeastern region of Rajasthan that is, Ganeshwar culture sites is known from a typical kind of pottery called as "OCP" (Ganeshwar), which emanate as an archaeological entity in its earlier stages of development. The term "OCP" for Ganeshwar pottery was a misnomer, which has also been misunderstood. Ganeshwar pottery is define as red slipped painted pottery with profuse incised design, executed with the help of a sharp, comb like instrument (Agrawala and Kumar: 1982). Due to the sketchy availability of information on Ganeshwar pottery, hitherto nothing much has been said or understood about the culture as a whole. In the present study, an attempt has been made to examine the explored pottery with the aim to present corpus of Ganeshwar ceramics types.

Thus, Ganeshwar culture emerges as an amalgamation of Early Chalcolithic culture from northern Rajasthan and Semi-Nomadic Copper using culture from southeastern Rajasthan. The Ganeshwar culture is marked by incised Red ware, Black painted Red ware and Red Slipped ware. The sites of Bagor and Gilund from southeast Rajasthan and Kalibangan from north Rajasthan show resemblance in ceramic tradition with the Ganeshwar incised ware. The incised pottery from Bagor phase 2 (Misra 1973b, Fig: 22-23), incised pottery from Gilund in Chalcolithic level (IAR 1959-60, plate: XLIV) and fabric D from Kalibangan (Bala 1997, plate: 13.11), show the resemblance in design pattern with the Ganeshwar incised red ware. Hence, Ganeshwar culture marked the influence from both the Chalcolithic Culture from northern Rajasthan and Semi-Nomadic Copper using culture from southeastern Rajasthan to form an individual entity in the Shekhawati region of Rajasthan that is known as Ganeshwar Culture.

1. **Bavadi Ki Jhamin (BKJ):** A single culture site (27°42'055"N - 75°50'582"E) located 0.1 Km east of the village Bhitarli Ganwari in NKT taluka of SKR district. There are ephemeral streams running on western and eastern side of the village. The village is nearly 10 Kms southeast from NKT (Khetri Maudh). The landform around the site shows series of hillocks with agriculture fields in the peneplain. On the basis of ephemeral nature of the site, it appears to be a temporary settlement. The site is small and measures 270 Sq m only. The pottery recovered from the site identify as Ganeshwar ware compare with the fabrics and shapes, which shows affinities with Ganeshwar pottery recovered during exploration. The Ganeshwar pottery identified as red ware pots, bowls, basins and dishes except one undiagnostic sherd in gray ware shows vertical grooves on the exterior (Figure – 3.12).

2. **Beed Ki Dhani (BKD):** A multi-cultural site as mentioned earlier in Mesolithic section appears to be a seasonal settlement. The exploration revealed few diagnostic sherds of Chalcolithic period identified as Ganeshwar ware. The Ganeshwar pottery reported from the site includes red ware pots and bowl (Figure – 3.13).

3. **Bhadwari (BWR):** Beside microliths, the exploration revealed only 9 diagnostic sherds. Out of these 5 sherds have been identified as Ganeshwar ware (Figure – 3.14) on the basis of fabric and shapes except red ware bowl with incised design on neck and body, which appears to be important feature of the Ganeshwar pottery. Remaining 4 sherds identified as Early Historic pottery.

4. **Biharipur (BRP):** A multi-cultural site (27°53'186"N – 75°53'710"E) located 2 Kms south of the village Biharipur in NKT taluka of SKR district. The site is identified as a seasonal settlement measures around

525 Sq m located on the sand dune that rise up to 10 m height from the surrounding area. The area is drained by tributary of Chandrawati River flowing in north-south direction across the village, which is around 41 Kms west from NKT (Khetri Maudh). The exploration revealed diagnostic and undiagnostic sherds, identified as Chalcolithic and Early Historic pottery. The Chalcolithic pottery (Figure – 3.15) shows affinity with Ganeshwar ware in fabric, shapes and incised design.

5. **Gadhka (GDK) (Kunda ki Dhani):** A multi - cultural site (27°46'916"N – 75°52'417"E) located 2 Kms north of the village Mahawa near Narsinghji ki Mandir in NKT taluka of SKR district. The area is drained by an ephemeral streams running on the southern side of the village. The land features shows hills on the north and agricultural fields on the south. The site measures around 800 Sq m and only 2 diagnostic sherds mark Chalcolithic pottery in red ware (Figure – 3.16). The typical feature of the pottery observed while rubbing the surface it turns powdery.

Ganeshwar: Ganeshwar is very important and is a key site for Ganeshwar – Jodhpura culture. R.C. Agarwala and Vijay Kumar excavated the site from 1979 to 1982, but unfortunately, the final excavation report has not been published as yet. The site is situated in Ganeshwar village in NKT taluka of SKR district. The village can be approached from NKT by metal road. The site is nearly 12 Kms southeast of NKT (Khetri Maudh). The village is situated on the slope of a hill and the area is drained by Chandrabhaga nadi (a tributary of Kantli River) on the west and Raotala nala on the southeast. The cultural material is collected from various locations that have been given numbers. Localities 1, 2, 3, and 4 are situated on the northeast and peneplains between the hillock on the east and the sand dune ridge on the west of the village. These mounds are badly disturbed by the road construction and agricultural activities. Though pottery was recovered from various localities, but it appears to be one single mound, except the locality number 5 and 6 that is situated on the foothills of the hillock and locality number 3 is situated on the ridge of the sand dune (Figure – 3.17 to 3.25).

6. **Ganeshwar (GNR) No. 1:** A single culture site (27°40'368″N – 75°49'012″E) located 1 Km northwest of the village Ganeshwar. The site measures around 77 Sq m comprise Chalcolithic pottery.

7. **Ganeshwar (GNR) No. 2:** A single culture site (27°40'447″N – 75°48'931″E) located 1.5 Km northwest of the village Ganeshwar. The site is spread in an area of 4840 Sq m revealing Chalcolithic and Early Historic pottery with the stray findings of terra cotta ring stand and stone hammer.

8. **Ganeshwar (GNR) No. 4:** A single culture site (27°40'447″N – 75°48'931″E) located 2 Km northwest of the village Ganeshwar. The area measures around 992 Sq m revealed Chalcolithic and Early Historic pottery. Besides pottery terra cotta mould, lump of sand stone and mace-head in quartzite found during the exploration.

9. **Ganeshwar (GNR) No. 5:** A multi-cultural site (27°39'938″N – 75°49'499″E) located 0.5 Km east of the village Ganeshwar on the slope of hill. The site measures around 1978 Sq m comprise Chalcolithic pottery.
10. **Gidhali/Nali No. 2 (GDL):** A multi-cultural site having rich findings of microliths in terms of blades and cores. The exploration has also revealed Chalcolithic and Early Historic pottery. The Chalcolithic pottery shows resemblance with Ganeshwar ware (Figure – 3.26) in fabric shapes and incised design.

11. **Gogajiwali Dongri (GWD):** A multi-cultural site (27°40'766"N – 75°37'927"E) located 0.5 Km southwest of the village Guhala in Udaipurwati (hereafter UPW) of JJN district. The site is drained by Kantli River flowing from south to northeast on the northern side of the village, which is situated 20 Kms southwest from NKT (Khetri Maudh). The ephemeral nature of the site identified as a seasonal settlement, which measures around 369 Sq m. The exploration revealed Chalcolithic pottery, which shows affinity with Ganeshwar ware in fabric and shapes (Figure – 3.27). Beside this, site has also revealed Early Historic pottery.

12. **Gujjaro Ki Dhani (GKD):** A multi-cultural site (27°46'101"N – 75°59'672") located 0.5 Km east of the village Bewa Patan in NKT taluka of SKR district. The site is drained by Krishnawati River on the eastern side of the village, which is around 26 Kms northwest from NKT (Khetri Maudh). The site rise up to 3m height from the surrounding area form by agriculture field. This is a very large site measures around 4318 Sq m. The exploration revealed Chalcolithic pottery (Figure – 3.28) shows affinity with Ganeshwar ware in fabric and shapes. The site also revealed Early Historic pottery.

13. **Jasi Ka Bas (JKB):** A single culture site (27°37'997"N – 75°41'233"E) located 2 Kms south of the village Jasi Ka Bas in NKT taluka of SKR district. The site is drained by tributary of Kantli River flowing in east-west direction on the southern side of the village, which is around 17 Kms southwest from NKT (Khetri Maudh). There are series of sand dunes in a large area and the site is located on one such dune that rises up to 4 m height from the surrounding area. The site measure around 857 Sq m and appears to be a seasonal settlement. The exploration revealed Chalcolithic pottery (Figure – 3.29), which shows resemblance with Ganeshwar ware. The typical feature of the pottery observed here while rubbing the surface it turns powdery. Along with this pottery, site

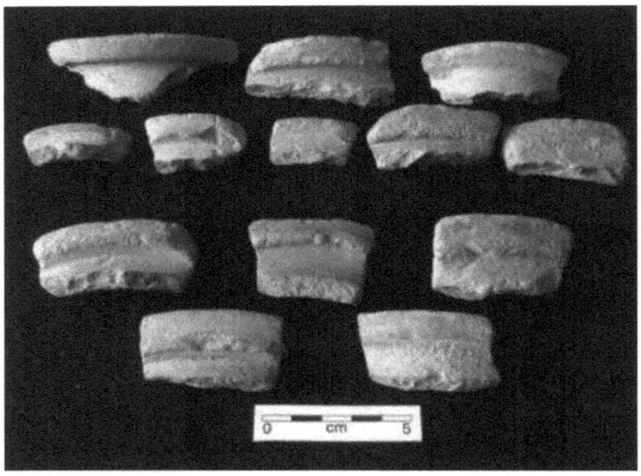

(a)

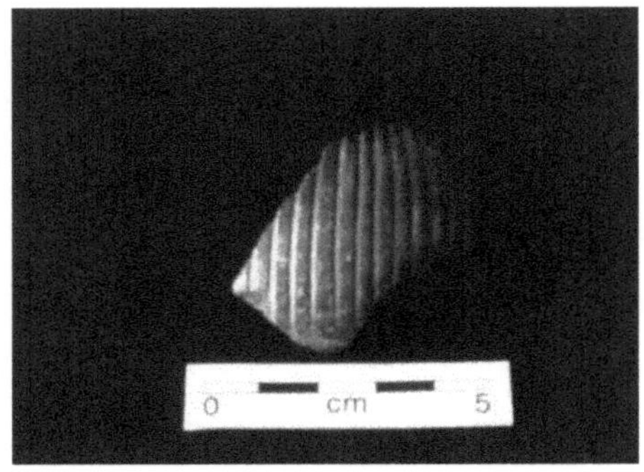

(b)

(c)

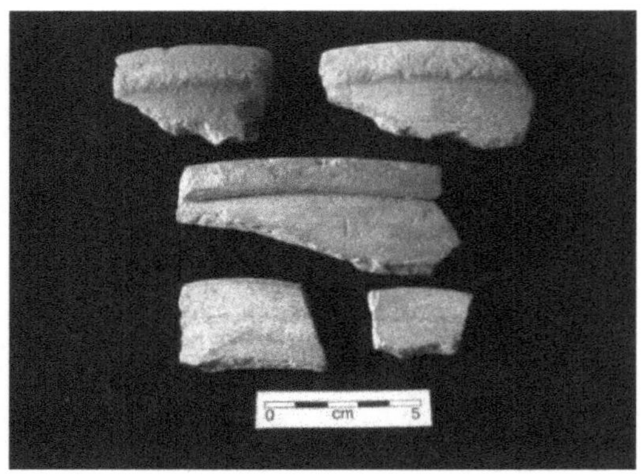

(d)

Figure 3.12 – Bavadi ki Jhamin (a) pots; (b) bowl with vertical grooves; (c) bowls; (d) basins

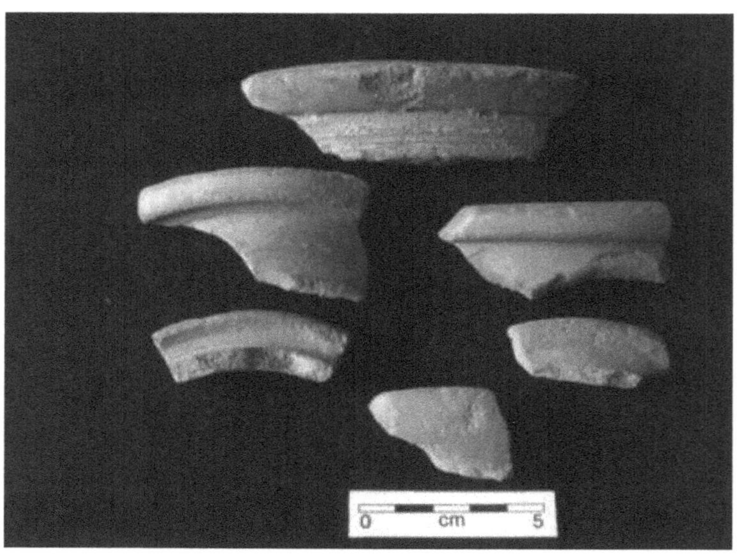

Figure 3.13 – Beed ki Dhani (pots & bowls)

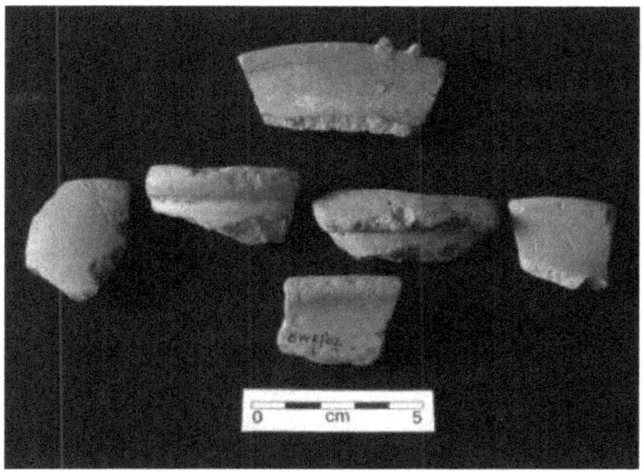

(a)

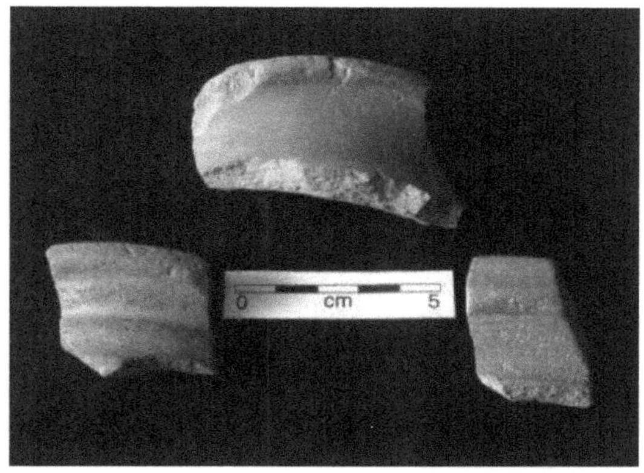

(b)

Figure 3.14 – Bhadwari (a) bowls; (b) pots; (c) incised bowl

Figure 3.15 – Biharipur (bowls)

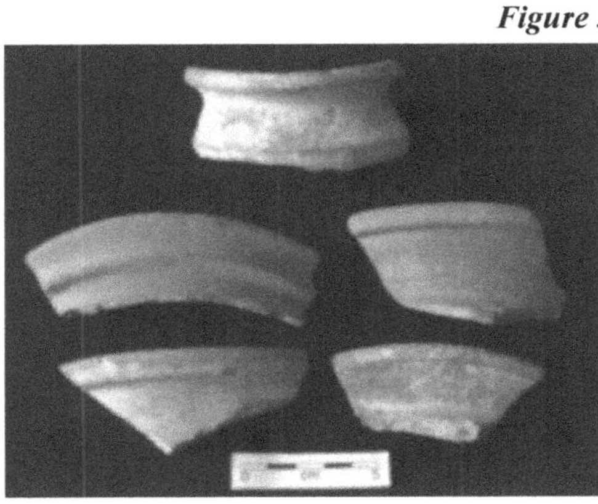

(a)

(b)

Figure 3.16 – Gadhka (a) pots; (b) basins

revealed Chalcolithic blades, core and lithic debitage with stray findings of terra cotta bead.

14. **Jodhpura (JPR):** A multi-cultural site (27°47'383"N – 75°40'342"E) located on the periphery of the village Jodhpura in Khetri taluka of JJN district. The village is nearly 13 Kms northwest from NKT (Khetri Maudh). The site is drain by Kantli River on the northern and eastern side of the village. The landform around the site shows open scrub area on the northern and western side and remaining area is surrounded by hills. The site measures around 500 Sq m and appears to be a seasonal settlement. The Chalcolithic pottery shows affinity with Ganeshwar ware in fabric and shapes with few undiagnostic sherd show reserved slip technique on red over yellow slip and incised horizontal grooves (Figure – 3.30).

15. **Jogi Badh (JBD):** A multi-cultural site (27°39'505"N – 75°43'506"E) located 1 Km east of the village Bhagot in NKT taluka of SKR district. The village is around 22 Kms southwest from NKT (Khetri Maudh). The area is drained by ephemeral stream running across northern and southern side of the village. The landform around the site shows sand dunes rises up to 5 – 6 m height from the surrounding area. The site measures 6860 Sq m, which appears to be larger area revealing Chalcolithic and Early Historic pottery. Here the Chalcolithic pottery shows resemblances with Ganeshwar ware in fabric, shapes and incised design. Along with Chalcolithic pottery, there is few sherds shows powdery nature of the surface while rubbing it (Figure – 3.31).

16. **Kakadiyo (KDY):** A multi-cultural site (27°37'929"N – 75°40'016"E) located 1 Km east of the village Thikriya in NKT taluka of SKR district. The village is around 18.5 Kms southwest from NKT (Khetri Maudh). The area is drained by Kantli River flowing on the western and southern side of the village. The site has identified as seasonal settlement measures around 576 Sq m located on the agriculture field that rise up to 3 m height from the surrounding area. The cultural material revealed Chalcolithic pottery (Figure – 3.32), which shows resemblance with Ganeshwar ware in fabric and shapes.

17. **Kakrana (KRN):** Beside microliths, the site has also revealed Chalcolithic pottery, which shows affinity with Ganeshwar ware in fabric and shapes (Figure – 3.33). There are few sherds, which form powdery while rubbing the surface. The site is near to the ancient copper mining area locally known as "Kakrana Pahar".

18. **Kanwat (KWT):** A single culture site (27°35'560"N – 75°41'279"E) located 2 Km west of the village Kanwat in NKT taluka of SKR district. The village is around 25 Kms southwest from NKT (Khetri Maudh). The area is drained by a tributary of Kantli River flowing on the eastern side of the village. The landform shows sand dunes around the hillock, locally known as "Kanwat Pahar". The site measures around 575 Sq m located on one such dune, which appears to be a temporary settlement. The exploration revealed Chalcolithic pottery (Figure – 3.34), which shows affinity with Ganeshwar ware in fabric.

19. **Kharagbinjpur (KBP):** A single culture site (27°43'156"N – 75°53'319"E) located 0.5 Km northwest of the village Baleshwar in NKT taluka of SKR district. A deserted village as a pilgrim place situated 14 Kms northwest from NKT (Khetri Maudh). Perennial streams running down from the hills drain the area. The dense and open scrub areas with deserted look give picturesque view to the place. The site measures around 1000 Sq m due to the ephemeral nature of the site, it is appears as a temporary settlement. The site has identified as Ganeshwar culture site on the basis of the pottery (Figure – 3.35), which shows resemblances with Ganeshwar ware in fabric.

20. **Kho:** A single culture site (27°46'000"N – 75°33'570"E) located 1 Km northwest of the village Kho in NKT taluka of SKR district. The village is around 20 Kms northwest from NKT (Khetri Maudh). Ephemeral streams running across the village drain the site, which shows hills and open scrub area. The site appears as a temporary settlement revealed few diagnostic sherds identified as Ganeshwar ware.

21. **Kot:** A multi-cultural site (27°47'850"N – 75°42'072"E) located 1 Km south of the village Sunari in Khetri taluka of JJN district. The village is situated around 15 Kms northwest from NKT (Khetri Maudh). The site is located on the bed of Kantli River, which flows from south to north in the research area. The site measures around 68,400 Sq m having thick habitation deposit, which is visible in the section of the mound. The exploration revealed Chalcolithic and Early Historic materials. On examining the Chalcolithic pottery (Figure – 3.36), it shows major affinities with the Ganeshwar ware.

22. **Mataji No Tibo (MNT):** Beside microliths, the site has revealed Chalcolithic pottery, which shows affinity with Ganeshwar ware in fabric, shapes and incised design. There are sherds, which turns in powdery form while rubbing the surface.

23. **Meeno Ki Dhani (MKD):** A single culture site (27°42'067"N – 75°50'903"E) located 1 Km northeast of the village Bhitarli Ganwari in NKT taluka of SKR district. The village is around 10 Kms southeast from NKT (Khetri Maudh and drained by ephemeral streams running across the village. The landform around the site shows hills and open scrub area with agriculture field in patches. The site is located on one of such field, which measures around 282 Sq m and identified as temporary settlement. On basis of the pottery, it is identified as Ganeshwar site, which shows affinity with Ganeshwar ware in fabric and shapes (Figure – 3.37).

Figure 3.17 – Ganeshwar (general view of the localities)

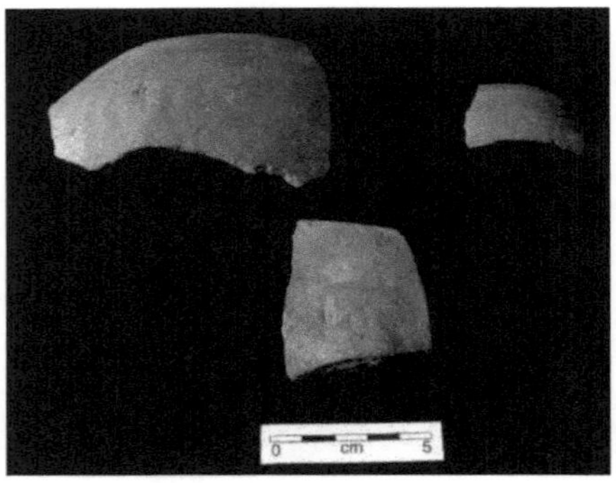

(a)

(b)

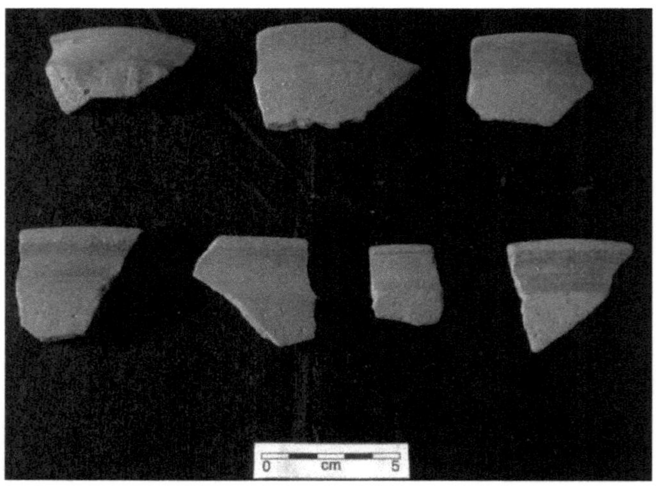

(c)

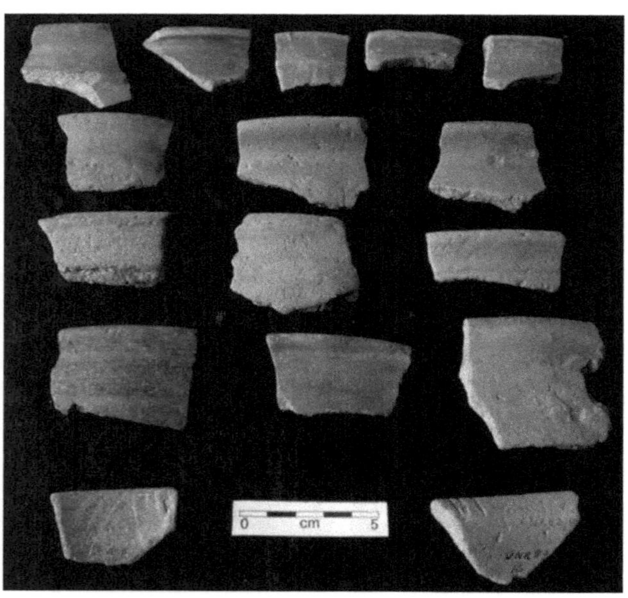

(d)

Figure 3.18 – Ganeshwar Bowls (a) Type A (1); (b) Type B (1); (c) Type B (2); (d) Type B (3)

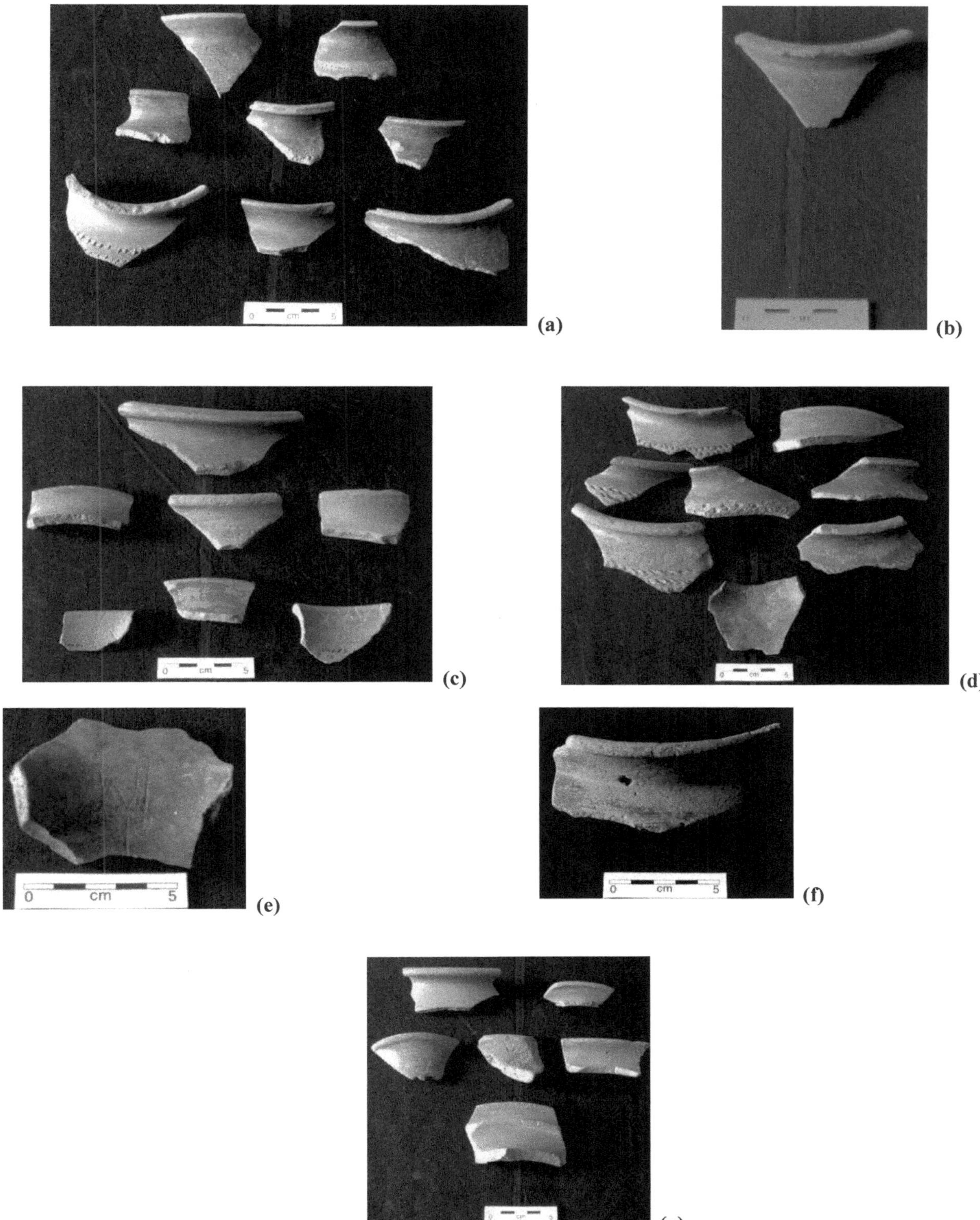

Figure 3.19 – Ganeshwar Pots (a) Type A (1); (b) Type A (3); (c) Type B (1); (d) Type B (2); (e) Type B (2); (f) Type B (3); (g) Type B (4)

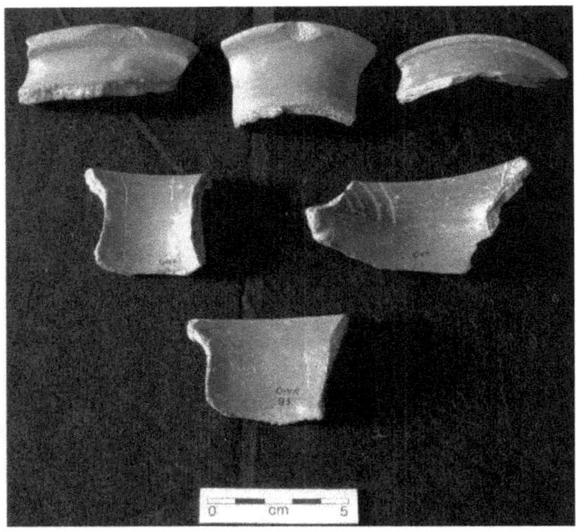

Figure 3.20 – Ganeshwar (Jars)

(a)

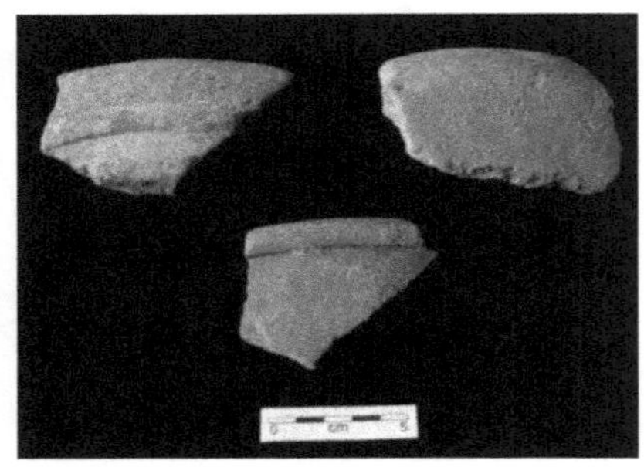

(b)

Figure 3.21 – Ganeshwar (Basins) (a) Type A (1); (b) Type B (1)

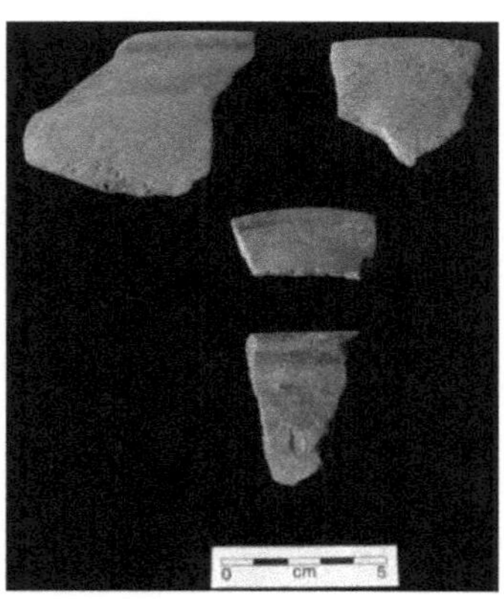

Figure 3.22 – Ganeshwar (Dishes)

Figure 3.23 – Ganeshwar (Handi)

(a)

(b)

(c)

(d)

(e)

(f)

(g)

(h)

Figure 3.24 – Ganeshwar (a – h) incised designs

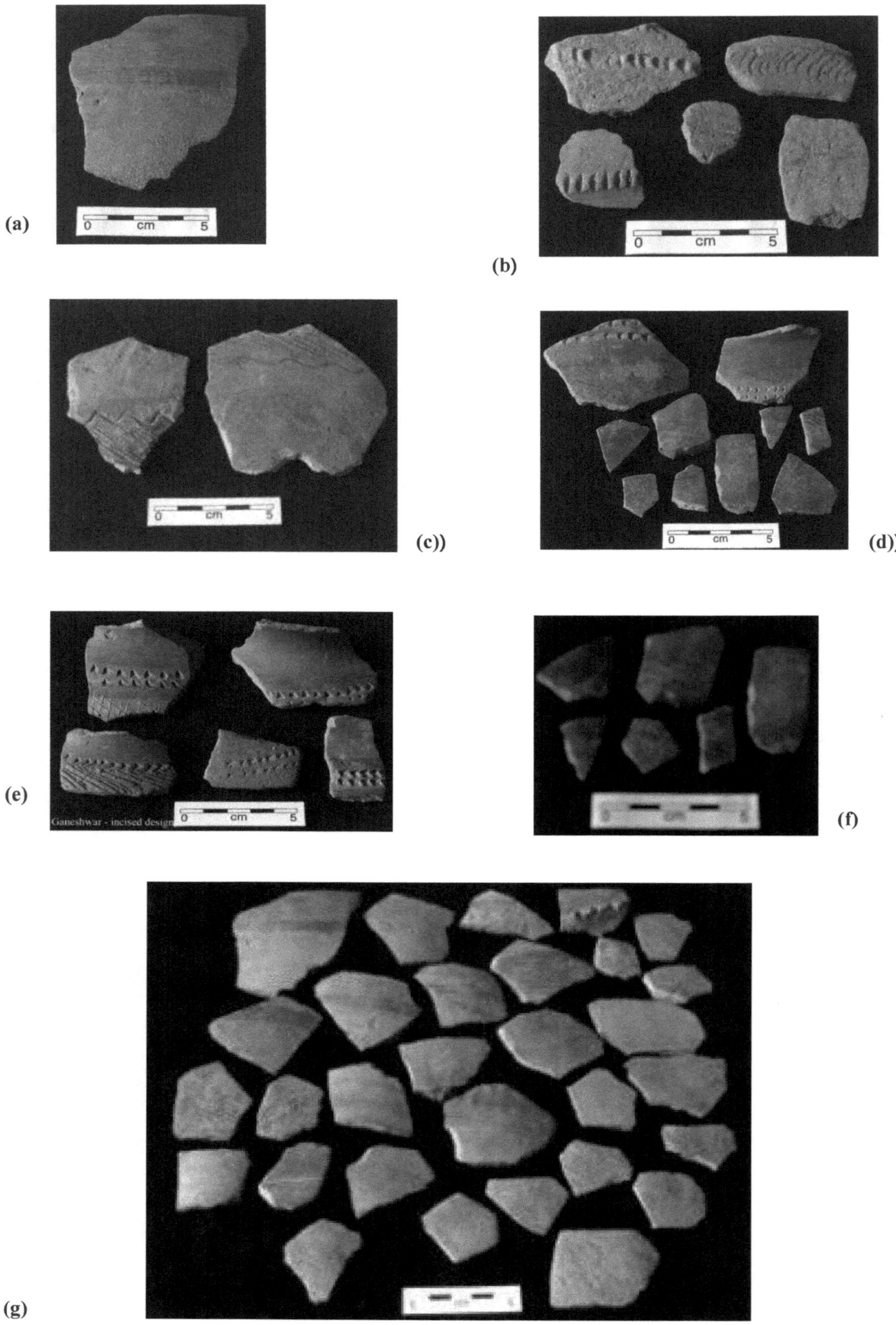

Figure 3.25 – Ganeshwar (a – g) incised & painted sherds

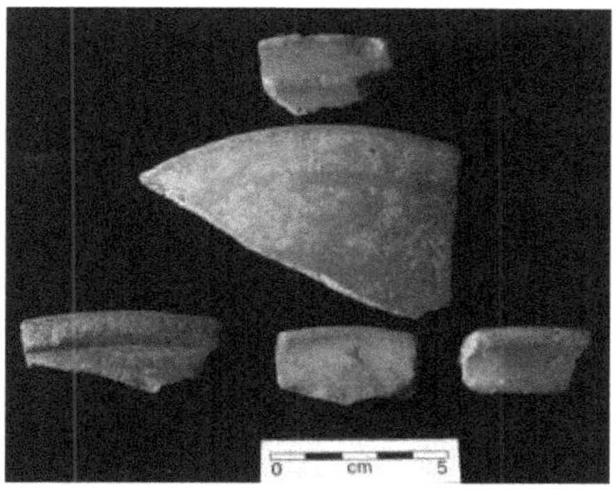

(a)

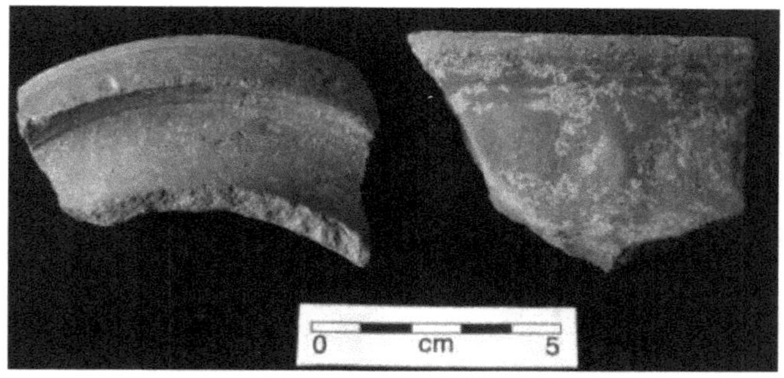

(b)

Figure 3.26 – Gidhali No 2 (a) bowls; (b) pots

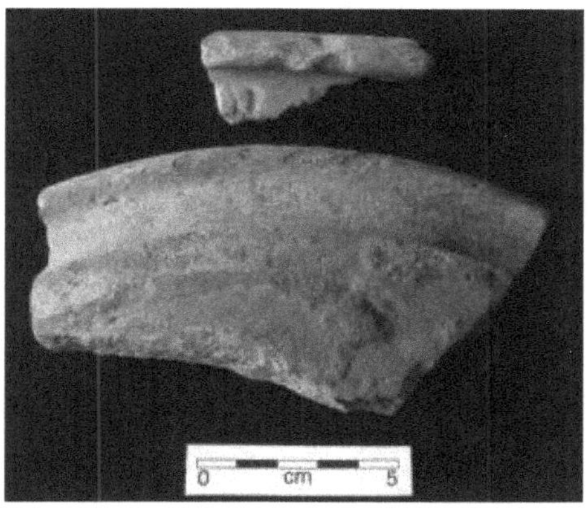

(a)

(b)

(c)

Figure 3.27 – Gogajiwali Dongri (a) basins; (b) bowls; (c) pots

(a)

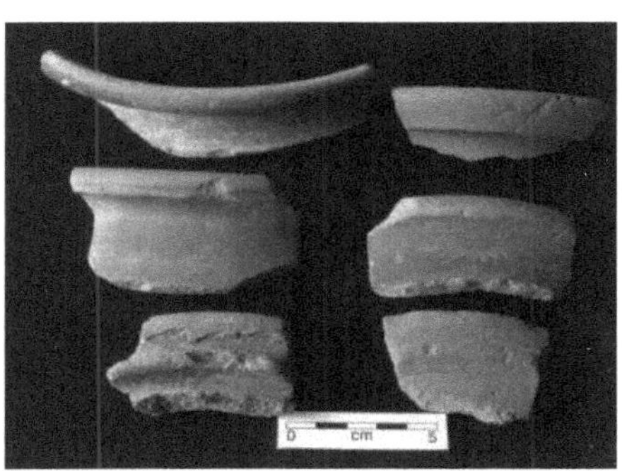

(b)

Figure 3.28 – Gujjaro ki Dhani (a) pots; (b) bowls

(a)

(b)

(c)

(d)

(e)

(f)

Figure 3.29 – Jasi ka Bas (a - b) bowls; (c) dish; (d) pots; (e) incised sherds; (f) lithic core & blades

24. **Murajawali:** At Murajawali two localities has been identified and named as MWL No. 1 (27°43'954"N – 75°52'155"E) and MWL No. 2 (27°44'143"N – 75°52'192"E), which is at a distance of 200 m apart from each other. The site is located 1 Km southwest of Gagharewala Balaji Mandir in the village Nimod in NKT taluka of SKR district. The village is around 13 Kms east of NKT (Khetri Maudh) and drained by a seasonal nalla on the southern side of the village. The site is located in an open scrub area and measures around 66 Sq m and 300 Sq m as temporary settlements. The site has been identified as metal processing activity area, which has revealed few diagnostic sherds resemblance with Ganeshwar ware in fabric (Figure – 3.38).

25. **Nalot (NLT):** A multi - cultural site (27°32'N – 75°40'E) located 1.5 Km northwest of the village Nalot in SMP taluka of SKR district. The village is situated 36 Kms southwest from NKT (Khetri Maudh) and drained by ephemeral streams on the northern side of the village. The site is located in an agriculture field measures around 751 Sq m with open scrub area around it. On basis of ephemeral nature of the site, it has identified as seasonal settlement with Chalcolithic pottery (Figure – 3.39) shows affinity with Ganeshwar ware in fabric, shapes and incised design.

26. **Narshingpuri (NSP):** A single culture site (27°40'635"N – 75°36'363"E) located 0.5 Km south of the village Narshingpuri in UPW taluka of JJN district. The village is situated 22 Kms southwest from NKT (Khetri Maudh). The area is drained by a tributary of Kantli River on the northern and southern side of the village. The site is located on a stabilize sand dune measures around 7630 Sq m that rise up to 5.5 m height from the surrounding area. On basis of the material culture, it is identified as a Chalcolithic site reveals Chalcolithic blades having crescent guiding ridge technique and copper arrowhead and antimony rod (Figure – 3.40).

27. **Payga Kundalia (PKD):** It is another important key site represents almost all the cultural phases except Early Historic phase I in the research area. Here the Chalcolithic phase is identified with pottery (Figure – 3.41), which shows affinity with Ganeshwar ware in fabric, shapes and incised design.

28. **Salodara (SDR):** A single culture site (27°52'563"N – 75°57'413"E) located 1.5 Kms north of the village Salodara in NKT taluka of SKR district. The village is situated 34 Kms northwest from NKT (Khetri Maudh). The landform shows hills and agriculture field around the site. The exploration revealed Chalcolithic pottery (Figure – 3.42) collected from the slope of the hill shows resemblance with Ganeshwar ware in fabric and shapes.

29. **Sadmata Ka Maula (SKM):** Beside Chalcolithic pottery, the site has revealed microliths, which has mention earlier in the previous section. There are around 27 diagnostic sherds in red ware, which shows resemblance with Ganeshwar related Chalcolithic site (Figure – 3.43).

30. **Shimarala (SRL):** A single culture site (27°32'N – 75°42'E) located 0.3 Km west of the village Shimarala in SMP taluka of SKR district. The village is situated 29.5 Kms southwest from NKT (Khetri Maudh). The area is drained by a tributary of the Kantli River flowing on the eastern side of the village. The site is located on a sand dune, which measures around 532 Sq m as a temporary settlement. The exploration revealed lithic debitage of Chalcolithic period.

31. **Swamiyo Ka Kuha (SKK):** A multi-cultural site (27°41'413"N – 75°34'374"E) located 2 Km northeast of the village Kotri in UPW taluka of JJN district. The village is situated 26 Kms southeast from NKT (Khetri Maudh). The area is drained by tributary of Kantli River flowing west-east direction on the northern side of the village. The landform around the site is of sand dune that rises up to 7m height from the surrounding area. The exploration revealed Chalcolithic and Early Historic pottery in an area about 1332 Sq m. The Chalcolithic pottery (Figure – 3.44) shows resemblance in fabric and shapes with Ganeshwar ware except red ware bowl with incised geometric design on the neck, which show its typical pottery to notice.

32. **Tapekeshwar (TKW):** A multi-cultural site (27°39'770"N – 75°52'029"E) located 2 Km west of the village Jatala in NKT taluka of SKR district. The village is situated 37 Kms southwest from NKT (Khetri Maudh) drained by Kasawati Nadi near the village. The landform is of dense and open scrub area around the site. The site is situated on the hill behind Tapekeshwar Mandir. Information gathered on personal communication with the priest of the temple that during repair work of the temple large number of sherds and bricks were found. On examining the bricks it is dateable to Sunga – Kushana period. The pottery collected from the area has revealed few Chalcolithic and Early Historic pottery. The Chalcolithic pottery (Figure – 3.45) shows resemblance with Ganeshwar ware in fabric showing powdery nature of the surface while rubbing the surface.

33. **Thaneshwar / Ganwari Banda (TNW):** A single culture site (27°41'532"N – 75°51'659"E) located 0.5 Kms east of the Thaneshwar Mandir in village Bhitarli Ganwari in NKT taluka of SKR district. The village is situated 10 Kms southeast from NKT (Khetri Maudh) and drained by Thana ki Nalla. The site has identified as metal processing activity area measures around 732 Sq m as a temporary settlement. The pottery shows affinity with Ganeshwar ware in fabric and shapes (Figure – 3.46).

(a)

(b)

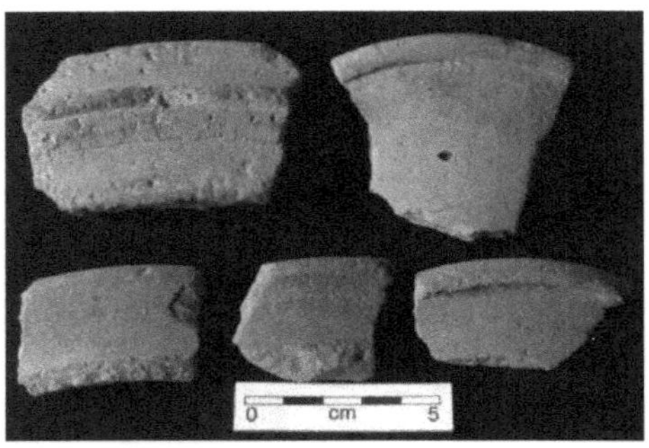

(c)

(d)

Figure 3.30 – Jodhpura (a) basins; (b) bowls; (c) dishes; (d) pots

Figure 3.31 – Jogi Badh (bowls)

(a)

(b)

Figure 3.32 – Kakadiyo (a) basins; (b) bowls

(a)

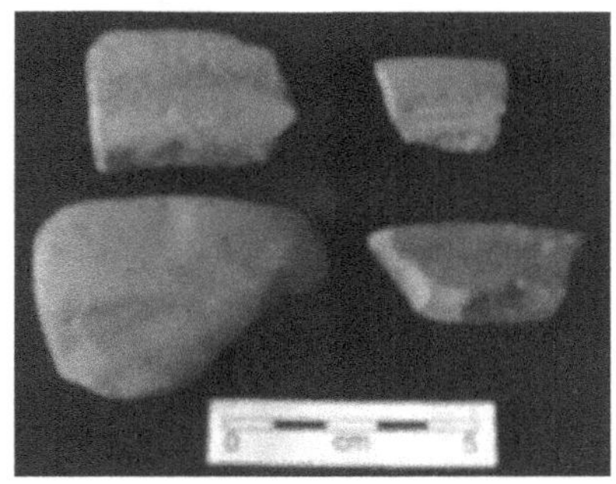

(b)

Figure 3.33 – Kakrana (a) bowls; (b) pots

Figure 3.34 – Kanwat (bowls & pots)

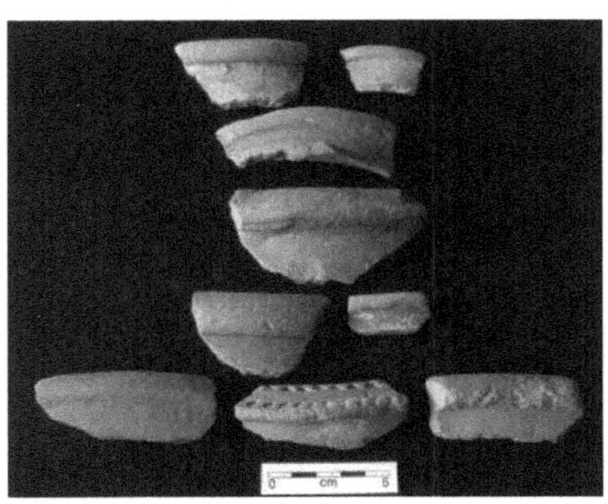

Figure 3.35 – Kharagbinjpur (pot, basin & bowls)

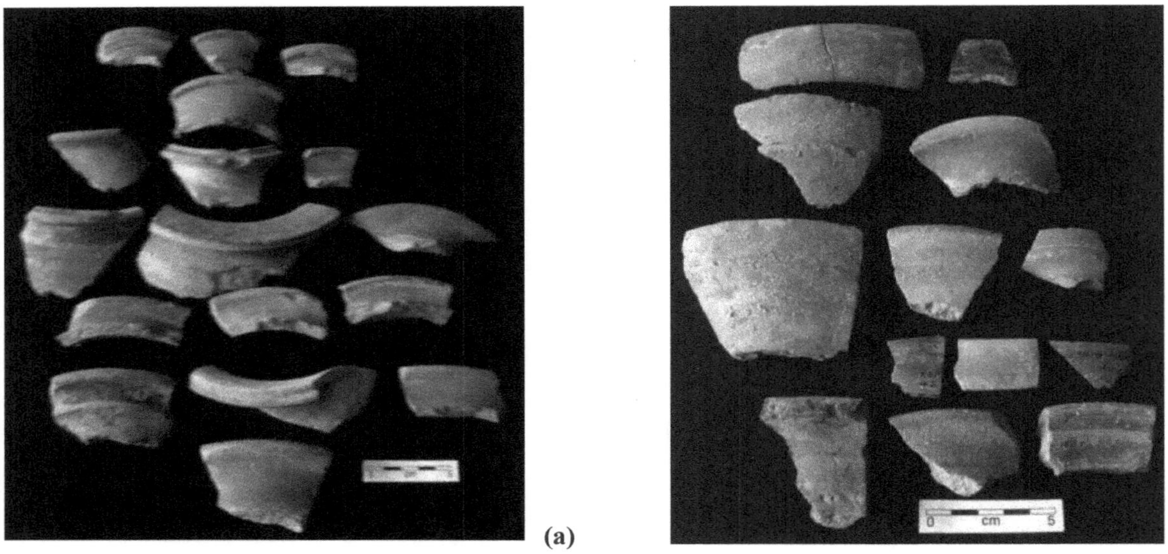

(a) (b)

Figure 3.36 – Kot (a - b) bowls

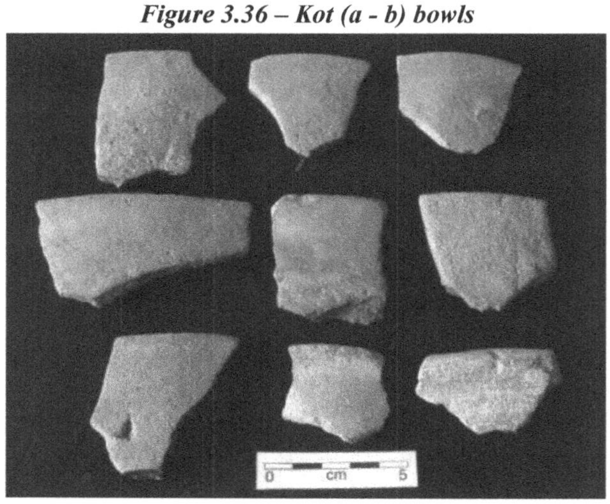

(a)

(b)

Figure 3.37 – Meeno ki Dhani (a) bowls; (b) pots

Figure 3.38 – Murajawali (pots)

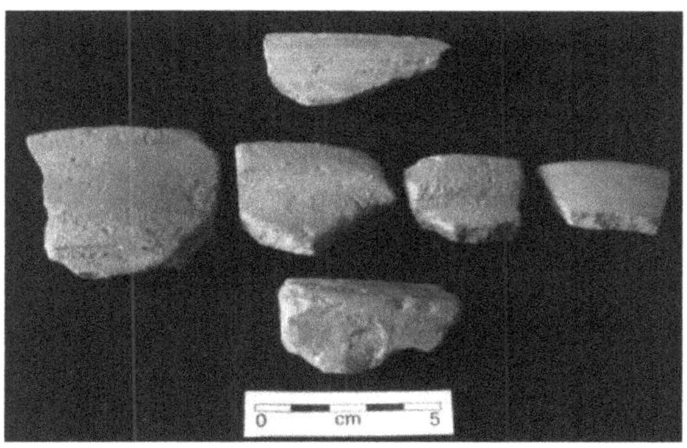

(a)

(b)

Figure 3.39 – Nalot (a) bowls; (b) pots

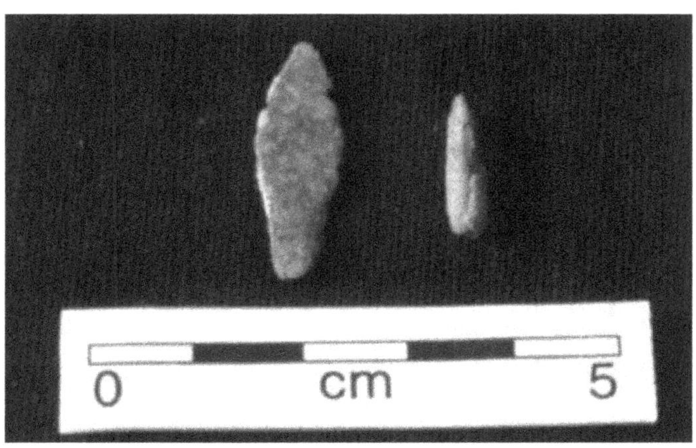

Figure 3.40 – Narshingpuri (copper arrowhead & antimony rod)

(a)

(b)

(c)

(d)

Figure 3.41 – Payga Kundalia (a) basins; (b) bowls; (c) pots; (d) incised sherds

(a)

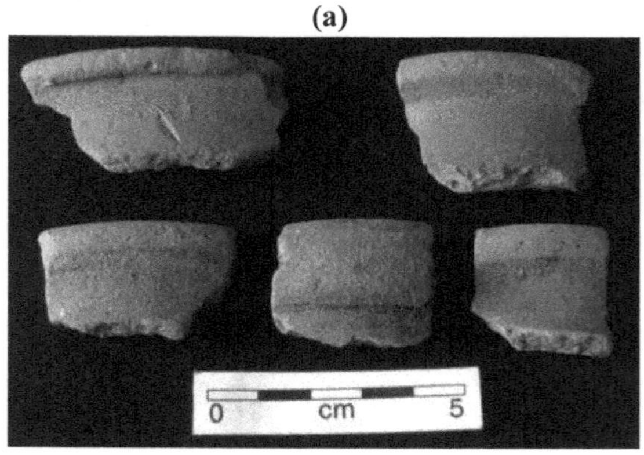

(b)

Figure 3.42 – Salodara (a) bowls; (b) pots

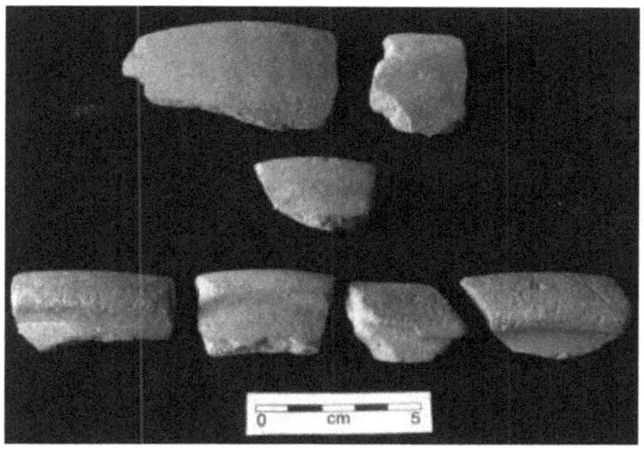

(a)

(b)

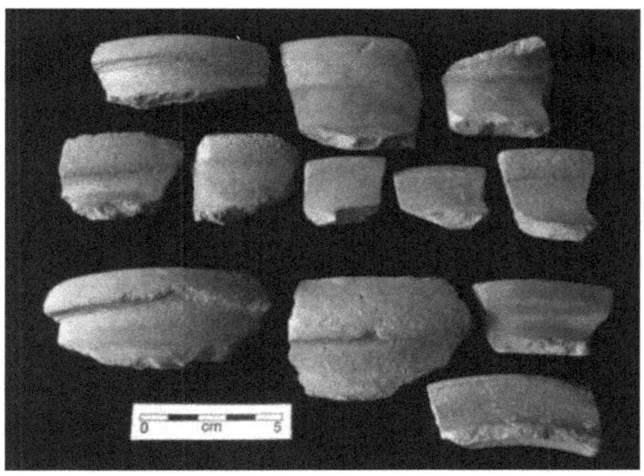

(c)

Figure 3.43 – Sadmata ka Maula (a) dishes; (b) bowls; (c) pots

(a)

(b)

(c)

(d)

Figure 3.44 – Swamiyo ka Kuha (a) basins; (b) bowls; (c) incised bowl; (d) pots

Figure 3.45 – Tapekeshwar (basins)

(a)

(b)

(c)

(d)

Figure 3.46 – Thaneshwar (a) basins; (b) bowls; (c) pots; (d) incised designs

3.7 Early Historic Research in Rajasthan

Periodic work carried out on several sites in Rajasthan. In 1940-41, Stein surveyed Bikaner and Bahawalpur states, mainly in search of prehistoric sites in the arid beds of Saraswati and Drishadvati (Ghaggar) (Stein, 1989:1-40). Although the chief contribution of his survey lies in a definite identification of Harappan settlements in the region, it has also revealed the existence of the PGW and a new ceramic tradition of the early century of the Common Era, approximately contemporaneous with Kushana and early Gupta periods. This tradition designated as 'Rang Mahal culture or Rang Mahal ware from the type-site near Suratgarh in Ganganagar district (Stein, 1989: 58-98 and Hanna, 1959).

3.8 Early Historic Period in Rajasthan

After Painted Gray Ware Culture, all inhabitants in northern Rajasthan seem to have deserted perhaps due to the desiccation of Saraswati River. Then in the early centuries of the Commom Era, saw the rise of flourishing Rang Mahal Culture in northern Rajasthan. The scenario of eastern Rajasthan during Early Historic period seems a continuation of Painted Gray Ware Culture as exemplified at Noh and Bairat as at many sites in northern India.

An Early Historic site of Noh, Bairat, Rairh, Jodhpura, Nagar, Sambhar, Nagari, Rang Mahal and Ahar from Rajasthan forms four stages, which furnishes some ideas of the material culture of the Early Historic period.

Stage No. 1: From 6th century – 3rd century BCE confined two sites Noh and Bairat both in eastern Rajasthan, shows the overlap of PGW and NBPW appears to represent the cultural traits of this period. Pottery types marked as black and red ware and PGW.

Stage No. 2: From 3rd century – 2nd century BCE confined sites viz. Bairat, Noh, Rairh etc. Pottery types marked as NBPW including Plain Grey ware forms main ceramic assemblages. Artefacts marked as uninscribed cast coins; human and animal terra cotta figurines; copper and iron implements; beads of semiprecious stone, glass, ivory and terra cotta; shell bangles and bone tools; a seal with an inscription in Brahmi from Noh and many polished Chunar sandstone pillars from Bairat. Coinages marked as PMC coins, Indo-Greek coins, Mitra coins and Janapada coins of Sibis and Malavas.

Stage No. 3: From 2nd century BCE - 1st century CE confined sites viz. Rairh, Sambhar, Nagar and Noh. Pottery types marked as red ware and occasionally used gray ware in plain spouted jars and knobbed lids. Seals with Brahmi inscription from Ahar and Sambhar.

Stage No. 4: From 1st century – 3rd century CE, confined sites viz. Rairh, Nagar, Noh, Rang Mahal and Nagari. Pottery types marked as red ware, red polished ware and Rang Mahal Ware consists of varieties of black painted designs including those of birds and animals and incised ones. Variety of sprinklers and spouted vessels of plain and decorated type besides votive tanks and ritualistic bowls and glaze pottery. Artefacts such as hand made human and animal figurines prepared out of single or double moulds; beads of agate, lapis lazuli, paste, terra cotta, faience and glass; mirror and chariots of bronze from Rairh; soap stone toilet casket; legged and rotary querns; terra cotta rottles; wheels and gamesman; bone, terra cotta, and stone discs; bangles of shells, copper and terra cotta; ear ornaments, skin rubbers, and copper objects and iron implements. Coinages marked as Malavas coins, Kushanas coins and Murundas coins. Bronze seal with Brahmi legends from Rang Mahal.

3.9 Explored - Early Historic - I Sites in Research Area

Early Historic I represent unpainted PGW, unpainted black and red ware and NBPW. In the present survey two Early Historic - I sites were reported from the research area.

1. **Kakadiyo (KDY):** Mention earlier in Chalcolithic section as a multi-cultural site. The exploration revealed diagnostic plain gray ware sherds, which marked the Early Historic I level in the research area. Beside Early Historic I pottery, the site also shows affinity with Rang Mahal ware.

2. **Kot:** A multi-cultural site having thick habitational deposit revealing Chalcolithic and Early Historic phases. Here Early Historic I phase is marked by pottery, which shows gray ware bowls, Black and Red ware and Black Slip ware sherds.

3.10 Explored - Early Historic - II Sites in Research Area

Early Historic II represents Rang Mahal ware sites. In the present survey 21 Early Historic - II sites were reported from the research area.

1. **Bhadwari (BWD):** As mention previous in Mesolithic and Chalcolithic section as a multi-cultural site. The site has revealed diagnostic sherds identified as Early Historic II pottery (Figure – 3.47) only on the basis of its fabric with 2 diagnostic plain gray ware sherds.

2. **Bhagegawala Tila (BWT):** A single culture site (27°40'637"N – 75°45'096"E) located 2.5 Kms southeast of the village **Bhudoli** in NKT of SKR district. The village is situated 5 Kms south from NKT (Khetri Maudh) and drained by Chandrabhaga Nadi on the east. The site is located on a sand dune that rises up to 4.5 m height from the surrounding area and measures 2889 Sq m. The site revealed red ware sherd identified as Early Historic pottery (Figure – 3.48) on basis of its ware. There is 1 diagnostic plain gray ware sherd.

3. **Biharipur (BRP):** A multi-cultural site revealed Mesolithic and Chalcolithic material. Beside this, site

has also revealed Rang Mahal ware comprises diagnostic sherds in red ware and undiagnostic rusticated and painted sherds except brown ware with thumb nail impression and dark gray ware (Figure – 3.49).

4. **Gadhka (GDK) (Kunda ki Dhani):** A multi - cultural site (27°46'916"N – 75°52'417"E) located 2 Kms north of the village **Mahawa** near Narsinghji ki Mandir in NKT taluka of SKR district. Ephemeral streams on the southern side of the site drain the area. The land features shows hills on the north and agricultural fields on the south. On basis of the pottery, it has been identified as Early Historic site, which measures around 800 Sq m. The pottery is marked by red ware with undiagnostic painted sherd having black color on red slip (Figure – 3.50).

5. **Ganeshwar (GNR) No. 5:** Along with Chalcolithic pottery, GNR No. 5 has revealed sculpture head in black schist stone and painted sherds from Early Historic II (Figure – 3.51).

6. **Gidhali/Nali No. 2 (GDL):** As mention previous in Mesolithic and Chalcolithic section as a multi-culture site having rich findings of microlithic blades, cores and lithic debitage. The site has also revealed undiagnostic sherds having horizontal and wavy applied bands, external ledge on the body and vertical applied bands of sandy texture on the body, which has been identified as Rang Mahal ware (Figure – 3.52).

7. **Gogajiwali Dongri (GWD):** A multi-cultural site. The Early Historic II pottery (Figure – 3.53) identified as Rang Mahal ware shows undiagnostic sherds having rustication on the body with applied bands of wet design, rustication on the body with applied horizontal bands of sandy texture, black color bands on the exterior surface and thumb impression on the interior surface. There is one undiagnostic sherd in plain gray ware.

8. **Gujjaro Ki Dhani (GKD):** A multi-cultural site revealed Chalcolithic and Early Historic II pottery. The Early Historic pottery (Figure – 3.54) identified as undiagnostic painted sherds having black color horizontal bands on exterior surface and reserve slipwares having black color bands over red surface. Besides a spout, lamp and few diagnostic sherds in red ware were recovered from the site.

9. **Jodhpura (JPR):** A multi-cultural site having Early Historic II pottery identified as undiagnostic painted sherds having red slip with black color band and incised leave pattern on the exterior surface. There are few diagnostic sherds in red ware (Figure – 3.55).

10. **Jogi Badh (JBD):** Along with Chalcolithic, the site revealed Early Historic II pottery, which shows rustication on the body with appliqué bands of wet design and horizontal applied bands of sandy texture with few red ware sherds (Figure – 3.56).

11. **Kakadiyo (KDY):** A multi-cultural site. The Early Historic pottery identified as Rang Mahal ware shows undiagnostic sherds having rustication on the body with horizontal bands of sandy texture and incised horizontal lines (Figure – 3.57).

12. **Kali Khankaria (KKR):** A single culture site (27°50'286"N – 75°56'760"E) located 1 Kms north of **Imloha** village in NKT taluka of SKR district. The village is around 32 Kms west from NKT (Khetri Maudh) and drained by perennial streams. The land feature shows open scrub area around the site. The site has been identified as Early Historic II on basis of the pottery classified as red ware pots and bowls (Figure – 3.58).

13. **Kot:** A unique site in the research area located on the bed of the Kantli River revealed Early Historic II pottery identified as Rang Mahal ware. The pottery shows undiagnostic sherds having incised stamp marks, horizontal incised lines, nail-headed incised notches and appliqué pattern (Figure – 3.59). The important findings of terra cotta sealing having five rows of Brahmi script along with terra cotta disc and anvil and fragments of shell – glass – iron bangles.

14. **Mataji No Tibo (MNT):** A multi-cultural site having Mesolithic, Chalcolithic and Early Historic pottery. Early Historic pottery identified as diagnostic and undiagnostic sherds with few sherds shows oblique lines with incised design between them, horizontal incised lines and incised circular mark with stray findings of terra cotta figurine collected from the stabilize sand dune.

15. **Mokulwas (MKW):** A single culture site (27°43'226"N – 75°54'228"E) located 1 Km south of the village **Baleshwar** in NKT taluka of SKR district. The village is nearly 14 Kms east from NKT (Khetri Maudh) and drained by perennial streams running down from the hillock. The landform shows dense and open scrub area around the site. The site measures around 6660 Sq m having Early Historic pottery identified as red ware shows resemblance in fabric with Rang Mahal ware (Figure – 3.60).

16. **Nalot (NLT):** A multi cultural site. The Early Historic phase revealed 2 diagnostic sherds with typical Rang Mahal ware showing floral and notch design (Figure – 3.61).

17. **Payga Kundalia (PKD):** As mention earlier as a key site having all the cultural phases in the research area. Early Historic pottery (Figure – 3.62) is marked by red ware along with this; it shows stray findings of glass bangle fragments and arakanut bead collected from the peneplain area.

18. **Sainiyo ki Dhani (SKD):** A single culture site (27°46'139"N – 75°59'604"E) located 0.5 Kms east of the primary school in the village **Bewa Patan** in NKT taluka of SKR district. The village is around 26

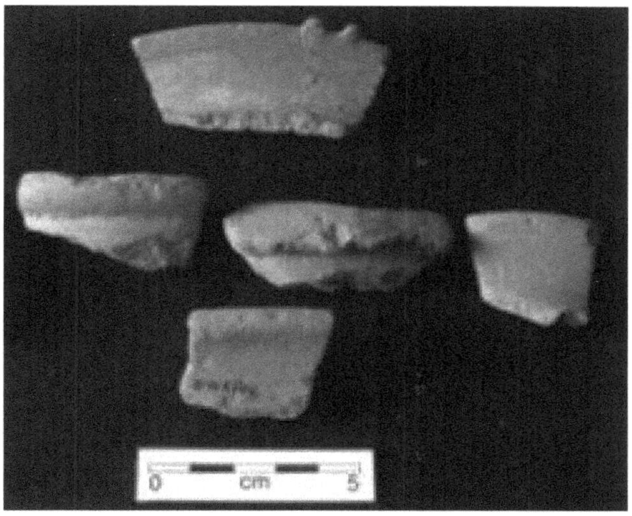

Figure 3.47 – Bhadwari (bowls)

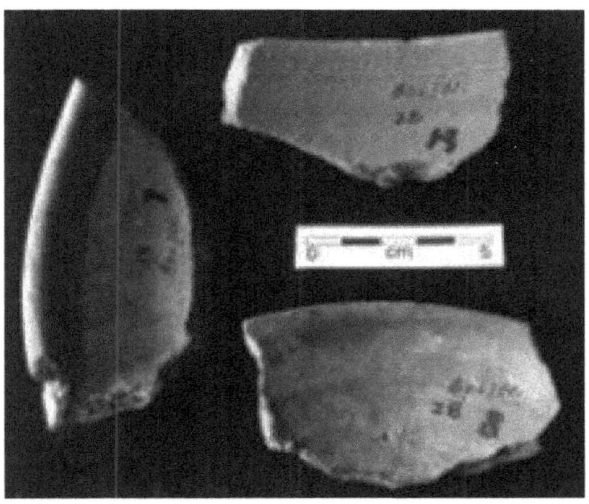

(a)

(b)

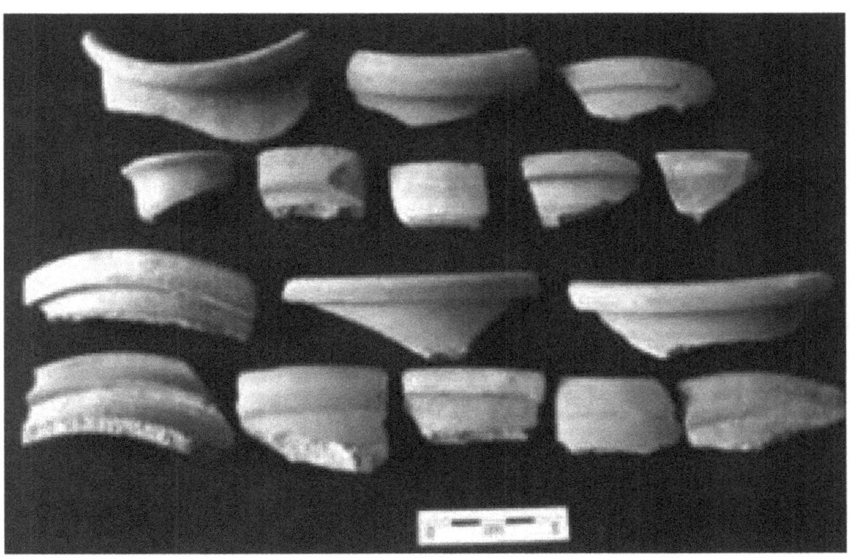

Figure 3.48 – Bhagegawala Tila (a) dishes; (b) bowls; (c) pots

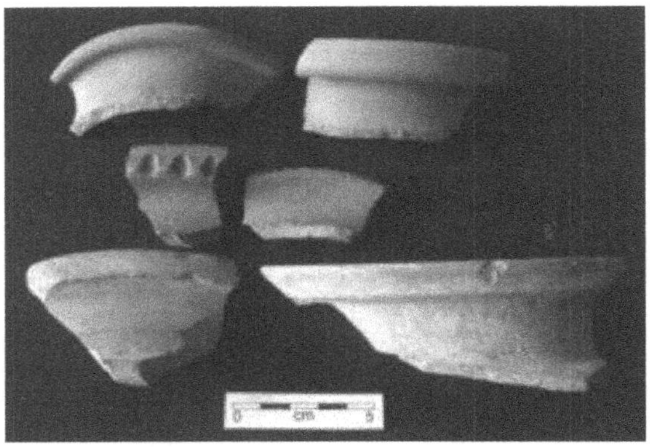

(a)

(b)

Figure 3.49 – Biharipur (a) pots; (b) rusticated & painted sherds

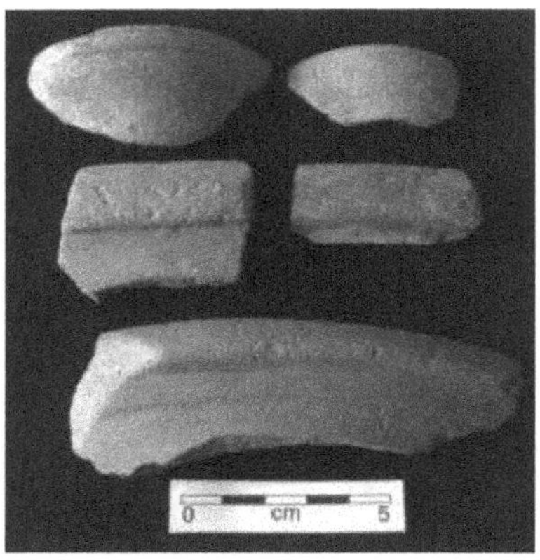

(a)

(b)

Figure 3.50 – Gadhka (a) bowls; (b) painted sherds

Figure 3.51 – Ganeshwar 5 (a) painted sherds; (b) black schist stone sculpture

Kms northwest from NKT (Khetri Maudh) and drained by Krishnawati River on the eastern side of the village. The landform shows agricultural field and the site is situated on one of such field that rise upto 3 m height. The site measures around 9379 Sq m revealed Early Historic pottery, which shows typical Rang Mahal sturdy ware and red ware pots and bowls (Figure – 3.63).

19. **Tapekeshwar (TKW):** A multi-cultural site. Along with Chalcolithic pottery, the site revealed undiagnostic sherds identified as Early Historic pottery, which shows rustication on the body with applied bands of sandy texture (Figure – 3.64).

20. **Thoi:** A multi-culture site. Along with microliths, the site also revealed Early Historic pottery identified as red ware pots and bowls.

21. **Tonda (Rampura - Tonda):** A single culture site (27°50'160"N – 75°53'675"E) located 3 Kms north of **Mehada** village in Khetri taluka of JJN district at the distance of 42 Kms northwest from NKT (Khetri Maudh). The site is situated within the reserved forest area, which measures around 1500 Sq m and drained by a tributary of Chandrawati River on the northern and southern side. The site revealed red ware sherds identified as Early Historic pottery with few painted sherds (black on red surface) which is typical feature observed in Early Historic II pottery (Figure – 3.65).

3.11 Earliest Evidence of Metal

Metal mining is practiced in India since time immemorial. The mining for gold, silver, copper, lead, zinc, iron metals and alloys are mentioned in ancient religious literatures. The finding of metal objects from the archaeological sites of Harappan tradition is well known. Kautilya in his Arthashastra described the metal mining and metallurgical activities on different metallic ores. In Rajasthan metal mining is being conducted since ancient past, the testimony of which are numerous mine sites, mine waste dumps, slag heaps and remains of furnaces or retorts. Colonel James Todd (1894) indicating the importance of mining in Rajasthan. He mentioned working mines of silver-tin in Zawar, copper in Dariba and lead near Gwalior and he further mentions that from Zawar and Dariba, Mewar Kingdom was getting revenue of ₹ 300,000/- annually. Study of the old metal mines, which are popularly known as old workings, started in Rajasthan by Geological Survey of India during British period. The references of mining and metallurgy exists in several writings i.e. Kautilya's Arthashastra, Tavernier, Alberuni, Buchana-Hamilton and Ain-i-Akbari.

The earliest evidence of copper–bronze technology comes from Harappan sites of Harappa, Mohenjodaro and Chanudaro. There are few more other sites in Baluchistan and from unstratified copper hoards (Sankalia, 1970). Beside this, there are several Chalcolithic sites have been excavated in different parts of Indian subcontinent, each yielding copper-bronze objects. The use of copper as a form of ornament has traced back to the early levels at Mehrgarh, where there is evidence for a single copper bead from the Neolithic levels (period IB), circa 6000 BCE (Jarrige, 1982). Whereas Sialk in Iran shows the development of copper tools made from cold hammering to castings in open moulds, casting in closed moulds, smelting and casting by cire perdue method in successive periods (Coghlan and Constantini, 1951). Mundigak in Afghanistan shows development in tool typology, as well as the transition from copper to bronze (Casal, 1961).

In 1881, V Ball in Economic Geology of India has mention that copper ore mining in Khetri and other districts of Rajasthan especially in Dariba and Singhanan had been carried out in large scale till the end of 19th century. In 1991, SR Rao mentions that the Harappans themselves did not mine the ores or even extract copper, which was done by others at the mines at Rajasthan, Afghanistan and other areas. It may be more apt to assume that the Harappan metal workers smelted copper near the mining areas. Lynn Willies, 1992 carried out preliminary investigation at Khetri copper belt in Rajasthan. He came with result that most of existing copper mines sustain the ample of evidence of old workings, which are calibrate C 14 dates around 14th – 19th century and he came out with the observation that the ancient mines can be dated further.

Another body of data that has yet to be fully understood is the work on the Ganeshwar-Jodhpura culture of Rajasthan (Agrawala, 1984). The site of Ganeshwar is around 35 kms from Khetri copper belt. No radiocarbon dates are known from the site and the ceramics allow only general association with the "so called" Ochre Colored Pottery culture (Kenoyer, 1991; Shaffer, 1991). After his survey in 1979, Agrawala reported over thousand copper objects, including 400 arrowheads, 50 fishhooks, 60 flat celts and numerous other objects (Agrawala, 1984).

Copper

Copper occurs in several mineral forms, such as native, sulphide, oxide and complex minerals. Each requires a different metallurgical treatment to obtain copper. The complex ores contain copper with lead, zinc, gold and silver minerals.

The type of copper ore and its tenor generally determine the method of treatment. The lowest grade ores are simple and easily treated. Native copper deposits run as low as 0.4 %. Sulphide ores run as low as 0.55 % or less and high-grade ores may range from 5 to 30%. Most oxidized ores may range from 0.4 to 10 %. Ores carrying 4 % or more copper are generally smelted directly to avoid concentration losses; low-grade ores of about 0.6 % copper are either leached or concentrated and the concentrated are smelted (Jensen and Bateman, 1981).

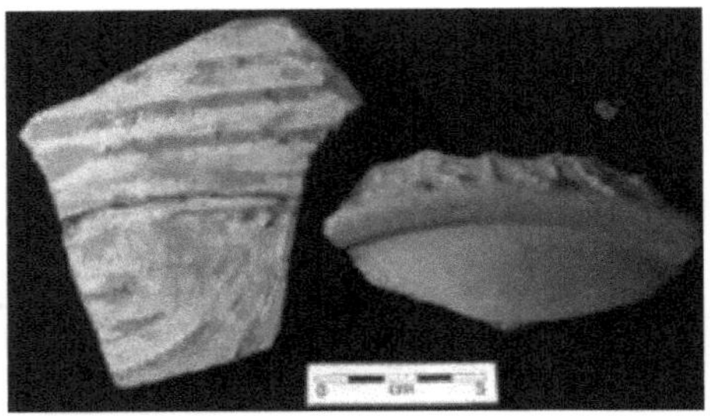

(a)

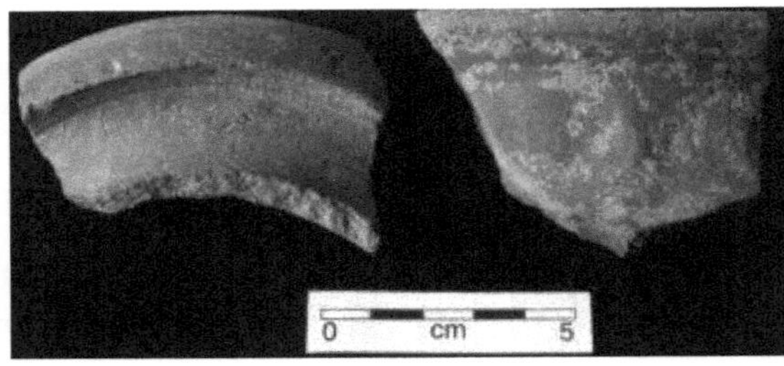

(b)

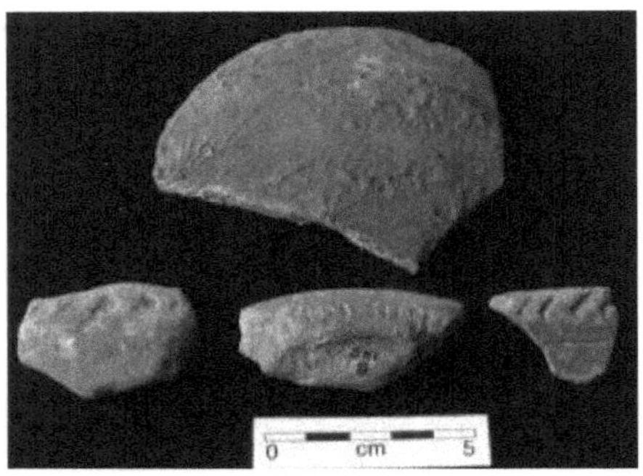

(c)

Figure 3.52 – Gidhali No 2 (a) rusticated sherds; (b) pots; (c) rusticated & incised sherds

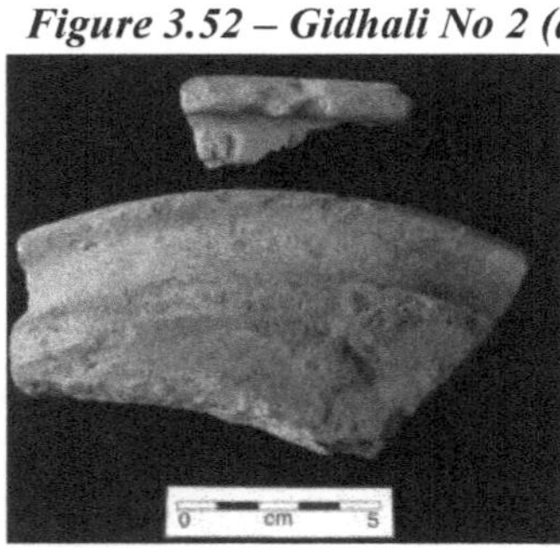

(a)

(b)

(c)

Figure 3.53 – Gogajiwali Dongri (b) bowls; (c) incised & painted sherds

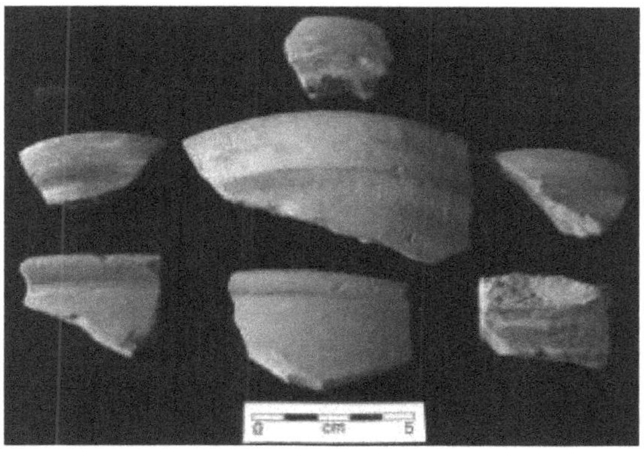

(a)

(b)

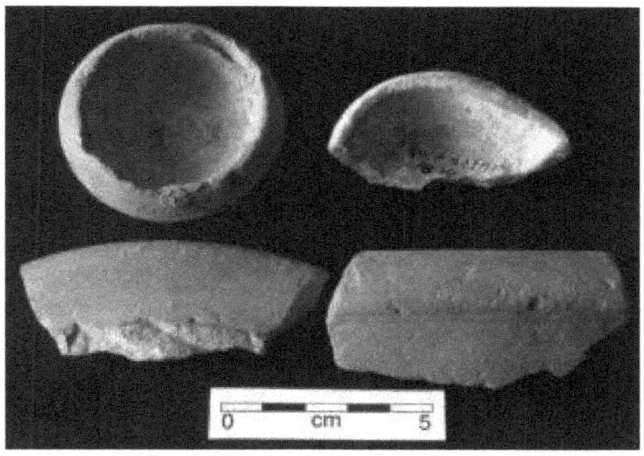

(c)

Figure 3.54 – Gujjaro ki Dhani (a) bowls; (b) painted sherds; (c) dishes & lamps

(a)

(b)

Figure 3.55 – Jodhpura (a) pots; (b) incised & painted sherds

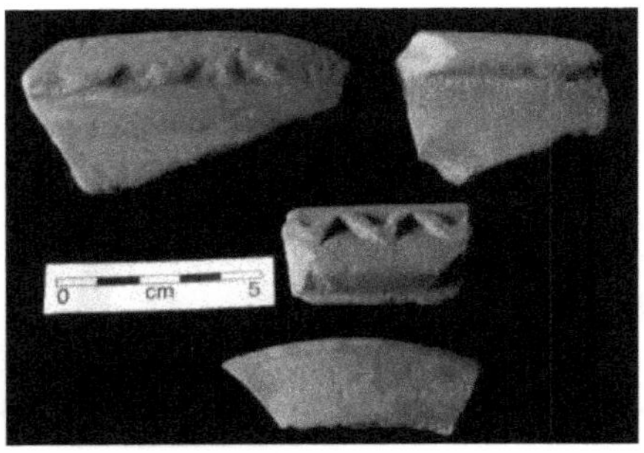

(a)

(b)

(c)

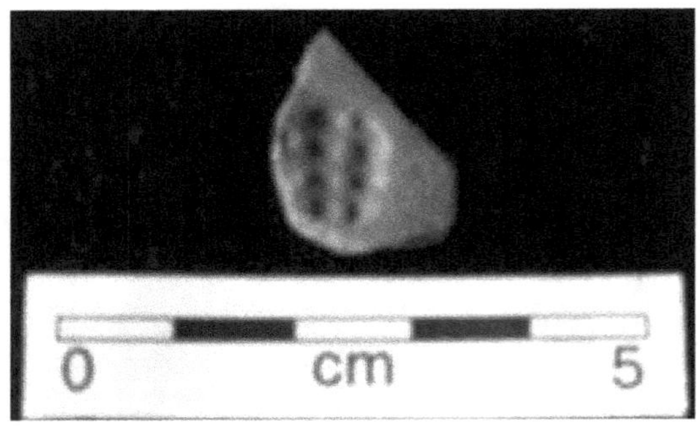

(d)

Figure 3.56 – Jogi Badh (a) dishes; (b) incised & painted sherds; (c) pots; (d) stamped design

Figure 3.57 – Kakadiyo (pots)

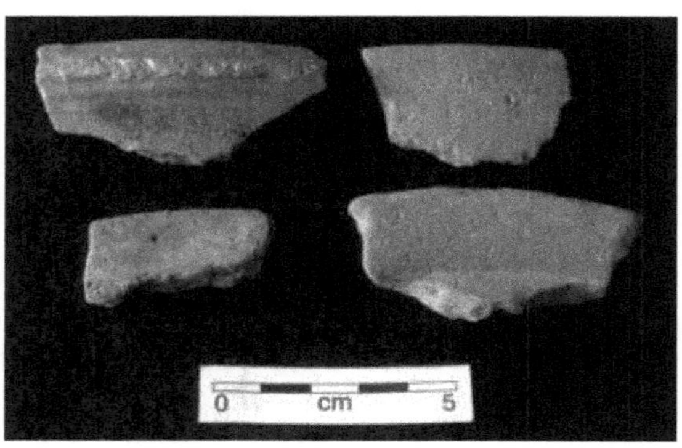

(a)

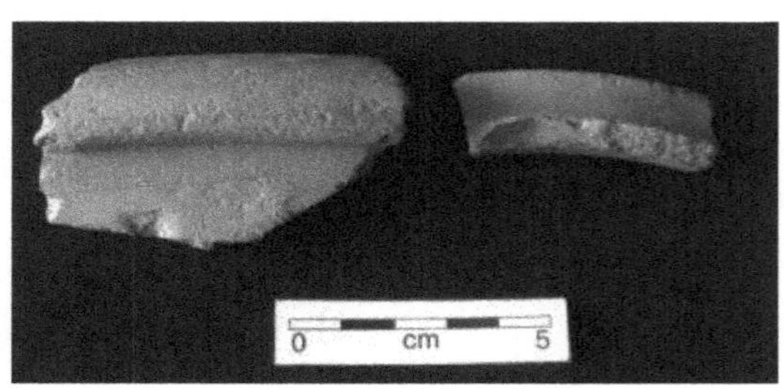

(b)

Figure 3.58 – Kali Khankaria (a) bowls; (b) pots

Figure 3.59 – Kot (a) plain gray ware bowls; (b) bowl, (c) basins; (d) rusticated sherds; (e) black & red ware sherds;

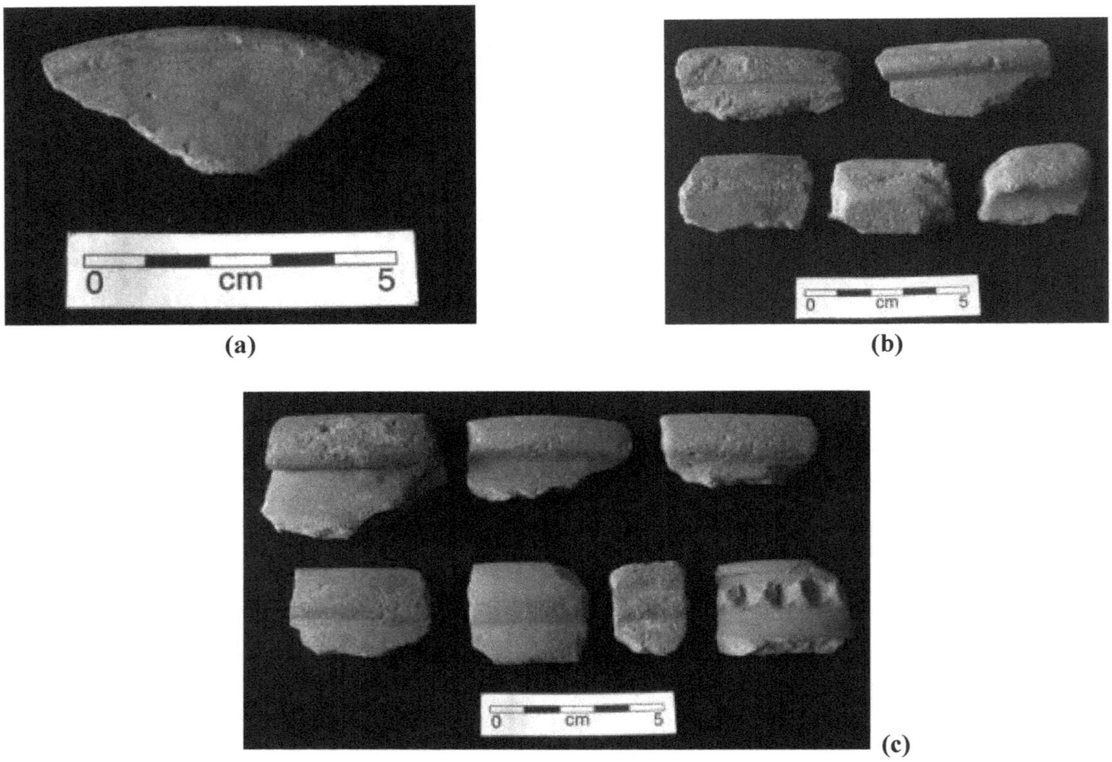

(a) (b) (c)

Figure 3.60 – Mokulwas (a) dish; (b) bowls; (c) pots

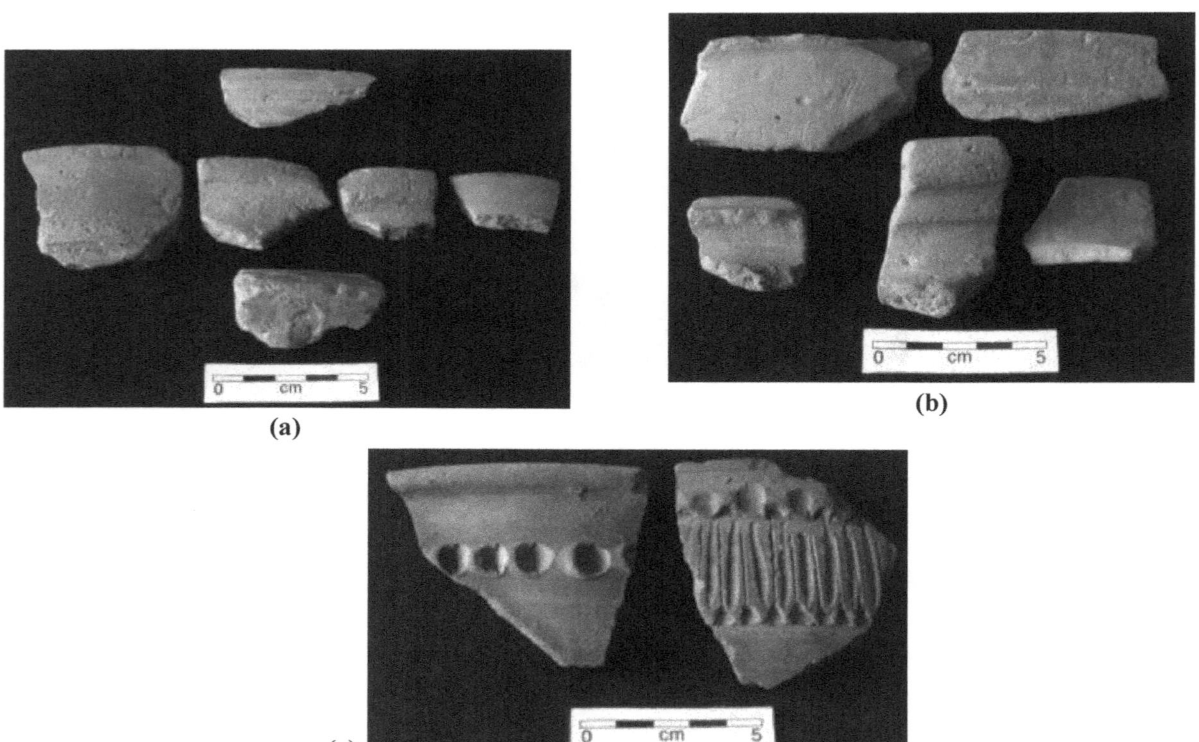

(a) (b) (c)

Figure 3.61 – Nalot (a) bowls; (b) basins; (c) bowl & basin

(a)

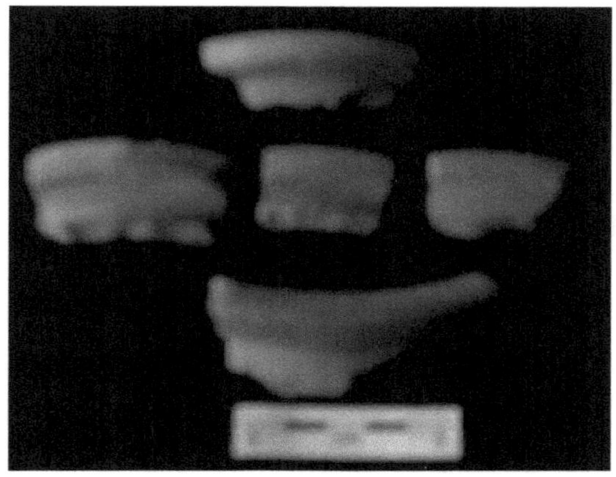

(b)

Figure 3.62 – Payga Kundalia (a) basins; (b) pots

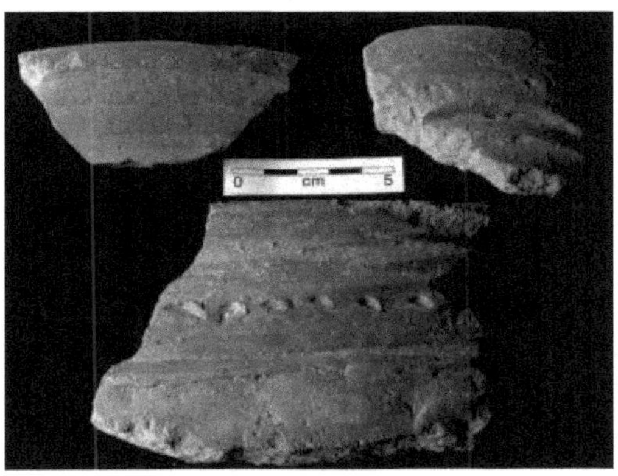

(a)

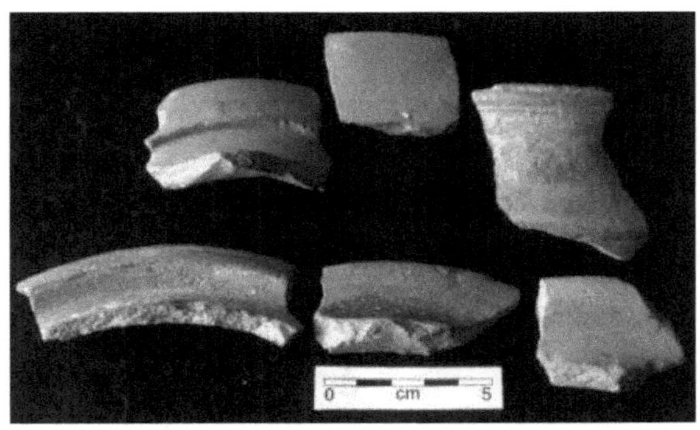

(b)

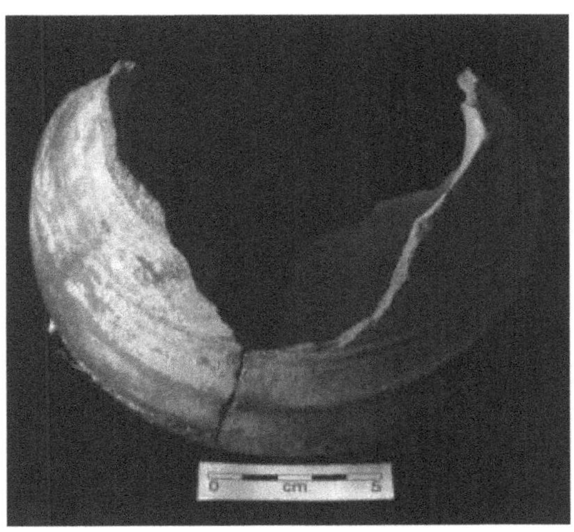

(c)

Figure 3.63 – Sainiyo ki Dhani (a) turf; (b) pots & bowls; (c) pot

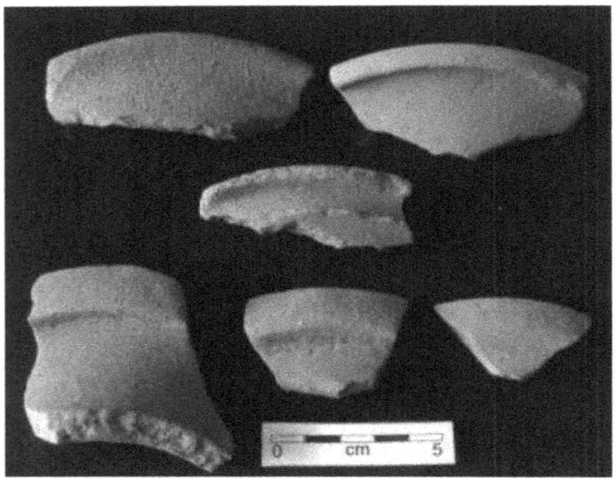

(a)

(b)

Figure 3.64 – Tapekeshwar (a) bowls; (b)rusticated sherds

(a)

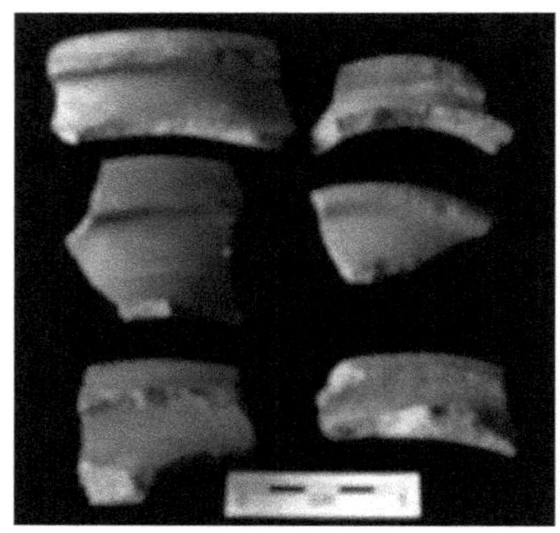

(b)

(c)

(d)

Figure 3.65 – Tonda (a) bowls; (b)pots; (c) painted sherds; (d) general view of the site

Table Shows Copper Mineral Ores

Minerals	Composition	% of Cu
Native: Copper	Cu	100.00
Sulphides:		
Chalcopyrite	$CuFeS_2$	34.50
Bornite	Cu_5FeS_4	63.30
Chalcocite	Cu_2S	79.80
Covellite	Cus	66.40
Enargite	Cu_3AsS_4	48.30
Tetrahedrite	$Cu_8Sb_2S_7$	52.10
Tennantite	$Cu_8As_2S_7$	57.00
Oxidized:		
Cuprite	Cu_2O	88.80
Tenorite	CuO	79.80
Malachite	$CuCO_3Cu(OH)_2$	57.30
Azurite	$2CuCO_3Cu(OH)2$	55.10
Chrysocolla	$CuSiO3..2H_2O$	36.00
Antlerite	$Cu_3SO_4(OH)_4$	54.00
Brochantite	$Cu_4SO_4(OH)_6$	56.20
Atacamite	$CuCl_2.3Cu(OH)_2$	59.40

Copper is one of the very essential minerals and considerable alloy of copper is used such as bronze and brass. The bronzes are copper-tin-zinc (88 % Cu, 10 % Sn, 2% Zn) and the brasses are copper-zinc alloys (55 to 99% Cu). There is also a nickel, aluminium and steel alloy of copper; minor special alloys utilize argon, beryllium, cadmium, chromium, cobalt, iron, lead, magnesium, manganese and silicon.

The chief gangue minerals of copper ores are rock matrix, quartz, calcite, dolomite, siderite, rhodochrosite, barite and zeolite (Jensen and Bateman, 1981).

Table Shows Common Gangue Minerals

Class	Name	Composition
Oxides	Quartz	SiO_2
	Other silica	SiO_2
	Bauxite, etc	$Al_2O_3.2H_2O$
	Limonite	$Fe_2O_3.H_2O$
Carbonates	Calcite	$CaCO_3$
	Dolomite	$(Ca,Mg)Co_3$
	Siderite	$FeCO_3$
	Rhodochrosite	$MnCO_3$
Sulphates	Barite	$BaSO_4$
	Gypsum	$CaSO_4{+}2H_2O$
Silicates	Feldspar	
	Garnet	
	Rhodonite	$MnSiO_3$
	Chlorite	
	Clay minerals	
Miscellaneous	Rock matter	
	Fluorite	CaF_2
	Apatite	$(CaF)Ca_4(PO_4)_3$
	Pyrite	FeS_2
	Marcasite	FeS_2
	Pyrrhotite	Fe1-xS
	Arsenopyrite	FeAsS

Iron

Iron is the second most abundant metal in the earth crust; the character of its natural compounds prevented its use as early as some other metals. It was known by 4000 BCE and the Egyptian Pharaohs regarded it more highly than gold, but this probably was the rare meteoritic iron. Apparently by 1200 BCE iron was manufactured but was still rare and its industrial use did not commence before 800 BCE, which marks the start of Iron Age. Steel came into use about 800 years later and the blast furnace in the 14th century CE (Hegde, 1991).

Around first half of third millennium BCE iron objects, which were not deliberately smelted were recovered from Tell Asmar, Chagar Bazaar and Mari in Mesopotamia and Alaca Huyuk in Anatalio (Coghlan and Constantini, 1956). During Bronze Age melting copper, under certain circumstances produce malleable iron in the furnace. It was very likely that these stray iron objects were produced from this unexpected extraction of the metal. Some of them may be produced from naturally occurring meteoritic iron. It is not yet quite clear where exactly the first deliberate smelting of iron started. Archaeological records indicate that from 2000 to 1200 BCE iron objects in West Asian sites gradually increased. Almost a similar slow development in the production of iron objects took place in India.

The earliest iron objects in India found at Atranjikhera and Hallur, dated around 11th - 9th century BCE. Hastinapur, Alamgirpur, Kausambi and Atranjikhera sites within Ganga - Yamuna Doab dated around 7th century BCE. Taxila, Rupar, Hastinapur, Sonkh, Kausambi, Rajghat and Ujjain dated around 6th - 5th century BCE. By the 4th century BCE iron technology spread all over the country and industry reached its mature stage, which is reflected in archaeological records (Ghosh, 1989).

Of, the iron ore minerals, magnetite is the richest but of minor quantity; hematite is the mainstay of the iron industry; limonite and siderite are of importance (Jensen and Bateman, 1981).

Table Shows Iron Ore Minerals

Mineral	Composition	% of Fe	Commercial Classification
Magnetite	$FeO.Fe_2O_3$	72.40	Magnetic or black ores
Haematite	Fe_2O_3	70.00	Red ore
Limonite	$Fe_2O_3.nH_2O$	59.00 or 63.00	Brown ore
Siderite	$FeCO_3$	48.20	Spathic, black band, clay iron stone

Other iron bearing minerals, such as pyrite FeS_2, pyrohotite Fe1-xS, marcasite FeS, and chamosite $Fe_2Al_2S1O_5(OH)_4$ do not occur in significant amounts in high grade large deposits and are not considered as potential iron reserves in the foreseeable future. Common impurities in iron ores are silica, calcium carbonate, phosphorus, manganese (especially in hematite), sulphur, alumina, water and titanium. The making of a usable product of iron involves two steps: first the reduction of the iron ore to pig iron; and second the treatment of the pig iron to make cast iron, wrought iron or steel.

3.12 Ancient Mining and Metal Processing Activity in Research Area

Mining is surface examination of a mineralised area normally gives an idea of the mineralisation. The old working gives considerable significances of the primary minerals worked out and the nature, size, shape and disposition of ores. It also gives an idea of cut-off grade of the ancient miners and also the mining conditions which may be expected to be encountered in exploratory mining.

There are numerous old working sites spread over the entire terrain from *Khetri* in the north to *Zawar* in the south and *Shakkargarh* in the east to *Basantgarh* in the west. The earliest evidence of old working (C-14 dates of wooden ladder) found at Rajpura Dariba mines, dated back to 2500 years ago (Brewis, 1991). There are no written records available about ancient mining activity, but there are profuse surface indications in the form of mine dumps, slag heaps, retorts, blow pipes etc., which attest to the mining and metallurgical knowledge of the ancient miners (Roy et.al.1998).

The six centres of copper ore deposits in India are: Aravalli region in Rajasthan and north Gujarat, Chhota Nagpur in Bihar, Garhwal and Almora in Uttaranchal, Jabalpur in Madhya Pradesh, Kurnool and Agnigundala in Andhra Pradesh and Chitaldurg in Mysore (Hegde, 1965). There are around 165 copper minerals, out of which only seven are known from India (Hegde, 1965):

1. Chalcopyrite or copper pyrite ($Cu_2SFe_2S_3$), having brass yellow color and metallic luster.
2. Chalcocite (Cu_2S), having grayish black color.
3. Bornite (Cu_3FeSO_4), having pinkish brown color.
4. Tetrahedrite ($4Cu_2SSb_2S_3$), (Sulphantimonite) having steel grey color.
5. Covellite (CuS), having indigo blue color.
6. Malachite ($CuCO_3$, $Cu(OH)_2$), having bright green color.
7. Azurite ($2CuCO_3Cu(OH)_2$), having azure blue colors.

Among these seven, **Chalcopyrite** is commonly observed copper mineral. A pure chalcopyrite contains 30.5% Copper, 30.6% Iron, and 39% Sulphur. In the Aravalli's, chalcopyrite ore deposit are observed near Khetri, Babai, Singhana, Akhwali, Dariba, Dev Bari, Delwara, Kotri, etc. In Aravalli region at Khetri, Babai, Dariba, and Singhana, chalcopyrite belt are continuous for about 25 km. All these copper ore deposits are delineated with ancient mining (Hegde, 1965).

Based on physical characters and use, the ancient metal mines are classified into three main categories:

The first category: Open Mines are open in nature and occur as irregular pits to trenches, and show shallow to moderate depth. Some of these might be trial pits. Generally these are filled-up by rubble, soil or vegetal material.

The second category: Shallow Mines occurs as vertical shafts with circular openings, narrow in diameter but deep. These are mostly situated close to the mine debris and seen in cluster of more than one. These are sub-classified as lined with stone mortar and lime, or un-lined. The lined shafts were used for haulage of ore or de-watering the mine when working below the water table. The un-lined shafts were meant for mine ventilation.

The third category: Deep Mines are inclines or tunnel-like openings meant for entering into underground mine. The ancient metal mines can be classified based on the metal excavated: Copper mines, Multi-Metal mines (lead–zinc–copper), Iron mines, Gold mines and Silver mines.

Ancient mining and metallurgical research was carried out to probe its relation with the archaeological sites. In the present study 19 ancient mining-areas and 14 metal-processing activity areas have been identified. Hence, this region gives the evidence of exploitation of metals by the inhabitants in different periods.

3.13 Explored Ancient Mining in Research Area

1. **Akawali Area** located in "Akawali Pahar" in Babai village, which is around 29 Kms northwest from NKT (Khetri Maudh) in Khetri taluk of JJN district. The area is divided into five blocks: north block, central block, south block, Kalota south block and eastern block (after GSI). The ancient miners have extensively worked in this area. The old workings comprise small pits, deep shafts, stopes and deep mine. There are four localities identified in and around Akawali Pahar, which has been discussed below as:
2. **Akawali Mine** (27″55'688″N - 75°45'349″E) located in Akawali Pahar. The exploration revealed 9 ancient mining areas in two clusters, which comprises pit mines only. In these clusters, one cluster comprises 5 mines while the second cluster comprises 4 mines.
3. **Akawali Mine No. 1** (27°55'756″N - 75°45'513″E), which is also located in Akawali Pahar. The mine has identified as ancient mining area comprises **small pits**, which was extensively work by the ancient miners.
4. **Akawali Mine No. 2** (27°55'760″N - 75°45'520″E) located 0.2 km on the top of the Akawali Pahar. The mine identified as **deep mine**, which was also extensively work by the ancient miners.
5. **Suradhi (SRD)** (27°55'961"N – 75°46'625"E) located 2 km northwest Babai village. The mine identified as **open mine**.
6. **Samiyo ki Dhani** located in Baleshwar village in NKT taluka of SKR district. The village is nearly 13 Kms east from NKT (Khetri Maudh). The exploration revealed open mine comprises ancient copper mining area.

7. **Beed Ki Dhani** (BKD) a multi-cultural site revealed Mesolithic and Chalcolithic material along with ancient mining comprises open mine, which is around 3 m wide and 15 m deep. The cultural materials were collected from the vicinity of the ancient mining area.
8. **Bavan Kuhi** (BVK) (27°44'405"N – 75°29'294"E) located 0.5 km west of the village Dhanota in UPW taluka of JJN district. The village is situated around 40 Kms west from NKT (Khetri Maudh). The area drained by ephemeral streams and the landform shows agricultural land on the northern side of the site. The site has revealed ancient mine area comprises **open pit mine** having 52 open pits along with the drainage channels, which was used to extract water from the pits and narrow shafts lead into big stopes. The mining area is locally known as Bavan Kuhi (fifty two wells). Some of the stopes are more than 35m wide. Dumps containing around 0.2 million tonnes of slags lying at places in the Dhanota village (after GSI).
9. **Kakrana** (KRN) a multi-cultural site revealed Mesolithic and Chalcolithic materials. Along with this cultural material, the site has shows open mine, which indicate mining area used by the ancient miners.
10. **Khetri Mine No. 1 and 2** (28°04'267"N – 75°48'287"E) located 3 km south towards the village Kharkhara in Khetri taluka of JJN district. The village is around 34 Kms northwest from NKT (Khetri Maudh). The Khetri mine areas are the part of Khetri copper complex, which extends 3 km by 1 km. The surface manifestations of the sulphides mineralization in this area are quite prominent and present in the form of gossans, old workings and huge heaps of metal slags.
11. **Kho** is a single culture site revealed Chalcolithic pottery along with the ancient mining area comprises open mine for copper. The area also gives the evidence for soap stone mine within the vicinity.
12. **Chandmari Mine** (28"00'N – 75°46'E) located 2.2 km south from Kolihan-Chandmari mine. The site is around 27 Kms north from NKT (Khetri Maudh). The ancient mining area comprises **open mine** and **deep mine** (500m length, 200m wide and 148m deep). The mine was initially used as an open mine and later on forms deep mine. At present Hindustan Copper Limited (hence after HCL) is drilling out the deep mine having the depth around 434 feet. At the footsteps of the open mine, there are opening for old working on its wall, which shows heaps of mine spoil, gossan bands and malachite staining.
13. **Kolihan Mine No. 486** (28°00'934"N – 75°46'215"E) located 2 km west of the village Kolihan in Khetri taluka of JJN district. The village is around 27 Kms north from NKT (Khetri Maudh). The ancient mining area comprises **deep mine**. Kolihan section lies in the northern part of Khetri Copper Belt (hence after KCB). The site falls in KCB, undertaken by HCL.
14. **Kolihan - Chandmari** (28°00'421"N – 75°46'344"E), located between Kolihan and Chandmari, which is 2 km south from Kolihan mine. The mine is around 29 Kms north from NKT (Khetri Maudh). The ancient mining area comprises **open pit mine**. The intervening section between Chandmari pit in the south and Kolihan mines in the north. The old workings with dumps of mine spoil show that the ancient miners worked in this area. The old workings are mostly located either at hanging wall or footwall side of gossans exposures. These workings seem to extend to a depth of 15-30 m.
15. **Abrak ki Khan** (ABK) (27°49'428"N - 75°52'231"E), located 2 Kms east of the village Mavanda Kala in NKT taluka of JJN district. The village is nearly 13 Kms northeast from NKT (Khetri Maudh). The ancient mining area comprises **open mine** having quartz-mica schist situated on the hillock, which was extensively worked by the ancient miners.
16. **Gura - Ponkh** located in village Ponk in UPW taluka of JJN district. The village is situated 28 Kms northwest from NKT (Khetri Maudh). The area drain by ephemeral streams running across the village and the landform shows hillock and open scrub area around the site. The site has revealed ancient soapstone mine comprises **open mine**.
17. **Ponkh ki Dhani** (27°49'109"N – 75°35'937"E) located in the village Ponk in UPW taluka of JJN district. The village is round 28 Kms northwest from NKT (Khetri Maudh). The ancient mining area comprises **open mine**.
18. **Sandhawala Kuha** (SWK) (27°48'N – 75°35'"E) located 0.5 Kms southwest of the village Ponk in UPW taluka of JJN district. The village is around 28 Kms northwest from NKT (Khetri Maudh). The site revealed ancient mining area comprises **open mine**.
19. **Sawalpur** (SWP) (27°38'N – 75°05'E) located 2 km east at the vicinity of the village Sawalpur in NKT taluka of SKR district. The village is around 30 Kms southeast from NKT (Khetri Maudh). The area drain by Buchara Bandh located 3 km on the north and 2.5 km on the south of the village. The ancient mining area comprises **open pit mines** having quartzite-mica minerals.

3.14 Explored Metal Processing Activity in Research Area

1. **Agri Ki Dhani** (AKD) (27°39"441"N - 75°50'779"E) located on the eastern periphery of the village Agri Ki Dhani in NKT taluka of SKR district. The village is nearly 15 Kms southeast from NKT (Khetri Maudh). The area drain by seasonal Bani or Khad ki nalla on the northern side of the slope and the landform shows hillock around the site. The site identified as a metal processing activity area comprises the findings of vitrified clay lumps, metallurgical slags, terracotta tuyeres and unidentified sherds colleted from the slope of the hillock. There are also patches of slag deposits on the slope of the hillock.
2. **Baleshwar** (BLW) (27°43'316"N – 75°53'418"E) located 1 km southwest of the village Baleshwar in NKT taluka of SKR district. The village is also nearly 13 Kms east from NKT (Khetri Maudh). The area

drain by perennial streams running down from the hillock and the landform shows dense and open scrub area. The site identified as a metal processing activity area comprises metallurgical slags spread on slope of the hillock.

3. **Kharagbinjpur** (KBP) is a single culture site. Along with Chalcolithic material the site also identified as a metal processing activity area with comprises the findings of vitrified clay lumps and metallurgical slags collected from the slope of the hillock. The site is located on the foothills in the open scrub area of about 1 Km along with the structural remains and slag deposits. The huge slag deposit scattered on the surface indicates that the area was intensively used for metal producing.
4. **Thaneshwar / Ganwari Banda** (TNW) is a single culture site. Along with Chalcolithic material the site also revealed metal processing activity area with the findings of vitrified clay lumps, terra cotta tuyeres, stone muller, circular cut stone and lithic debitage collected from the foothills in the open scrub area. The important feature noticed is the insitu terra cotta tuyeres.
5. **Ganeshwar** (GNR) No. 1 is a single culture site comprising chalcolithic material with indication of metal processing activity area comprises the findings of vitrified clay lumps, charred wood and metallurgical slags lying scattered on the surface.
6. **Ganeshwar** (GNR) No. 6 is a metal processing activity area (27°40'175"N – 75°49'577"E) located 1.5 km east of the village Ganeshwar in NKT taluka of SKR district. The village is nearly 12 Kms southeast from NKT (Khetri Maudh). The area drain by Chandrabhaga nadi (tributary of Kantli River) on the west and Raotala nalla on the southeast of the village and the landform shows dense and open scrub area on hillocks on the east and sand dune ridges on the west of the site. The site comprises findings of metallurgical slags on the surface of the hillock. There are stone alignments in circular fashion on the surface, presume to be used during metal processing activities.
7. **Shivnaryan ki Dhani** (27°49'049"N – 75°53'541"E) located 5 km south of the village Jhilo in NKT taluka of SKR district. The village is around 38 Kms northwest from NKT (Khetri Maudh). The area drain by a tributary of Chandrawati River and the landform shows open scrub area around the site. The findings of terra cotta tuyeres in different sizes collected from open scrub area on the slope of the hillock shows as a macro metal processing activity area.
8. **Murajawali** (MWL) No. 1 is a single culture site. Along with Chalcolithic material the site has also identified, as a metal processing activity area comprises with the findings of vitrified clay lumps and terra cotta tuyeres collected from the foothills.
9. **Murajawali** (MWL) No. 2 (27°44'143"N – 75°52'192"E) located 200m northwest of MRW No. 1 or 1 km southwest of Gagharewala Balaji Mandir in the village Nimod in NKT taluka of SKR district. The village is nearly 13 Kms east from NKT (Khetri Maudh). The area drain by seasonal nalla flowing on the southern side of the village and the landform shows hillock and open scrub area around the site. The site identified as a metal processing activity area measures around 300 Sqm comprises with the findings of vitrified clay lumps and terra cotta tuyeres collected from the foothills.
10. **Naka ki Dhani** (NKD) (27°49'105"N – 75°35'938"E) located 5 km southwest of the village Ponk in UPW taluka of JJN district. The village is around 28 Kms northwest from NKT (Khetri Maudh). The site identified as metal processing activity area measures around 300 Sqm comprises with the findings of vitrified clay lumps and terra cotta tuyeres collected from the foothills.
11. **Pitrawali Dongri** (27°49'877"N – 75°35'686"E) located 6 km southwest of the village Ponk in UPW taluka of JJN district. The village is around 28 Kms northwest from NKT (Khetri Maudh). The site identified as a metal processing activity area comprises the findings of vitrified clay lumps and terra cotta tuyeres collected from the foothills. The site shows metal producing indicator in form terra cotta tuyeres lying scattered on the surface with patches of fired activity. Around the village, there are number of ancient mining area viz. Sat Kui (near Basadi ki Dhani), Lohi Khan, Ponk ki Bawadi, and Sandhawala Kuha.
12. **Salodara** (SDR) is a single culture site. Along with Chalcolithic material the site also shows metal processing activity area comprises the findings of vitrified clay lumps and metallurgical slags along with the drainage channel.
13. **Singhana** (SGN) (28°07'N – 75°49'E) located 1 km west of the village Singhana in Khetri taluka of JJN district. The village is approximately 40 Kms north from NKT (Khetri Maudh). The site identified as a metal processing activity area comprises of huge metal slag deposits scattered all around the village. Some of the village lanes and houses were constructed over the slag deposits. This shows that the area was intensively used for metal processing or dumping ground for the ancient metallurgists. The present village inhabits over the slag heaps. There are three slag heaps in the village. The site falls in the Khetri Copper Belt (KBC), which is around 80 km copper belt from Singhana to Raghunathgarh. The area forms the northern most continuation of KCB, under cover of blown sand/alluvium and strike of about 11 km. Pb – isotope analysis carried out on slags has given very interesting result.
14. **Kot** is a multi-cultural site revealed Chalcolithic, Early Historic – I and Early Historic – II materials. Along with this, it also identified, as metal processing activity area comprises tuyeres and vitrified clay lumps.

CHAPTER 4: RESULT OF ANALYSIS

The result obtained from typological lithic classification of Palaeoliths and Mesolithic is present in a comparative format to highlight the correlation aspects. And the result obtained from typological classification of Chalcolithic and Early Historic pottery is presented after investigating relationship between vessel forms, fabric and function and highlighting the data to assist in a better understanding of the culture.

.

4.1 Lithic Classification of Palaeolithic Sites

In the present survey, 3 Lower Palaeolithic secondary sites were reported. The findings of these sites have given lead for future research, whereas Middle Palaeolithic and Upper Palaeolithic sites are still not discernible in the Shekhawati region.

Typologically, Lower Palaeolithic tools found from the sites have been classified into chopping tool (1), handaxes (3), cleavers (2) and scrapers (7). The raw materials used for making these tools are sand stone (1), quartz (3) and quartzite (7).

1. **Laluda ki Dhani** the site comprises handaxe (1) and scrapers (2) (Figure – 4.1). Handaxe is in ovate shape made from reddish brown sand stone measuring 16.1x8.9x5.2 cms. Scrapers are in oval made from gray quartzite measuring 11x7.2x4.7 and 9.2x9.6x2.3 Cms.
2. **Payga Kundalia** tools from this site comprise chopping tool, handaxe and cleaver made from milky quartz. Chopping tool is in irregular shape measuring 8.5x9.6x7.7 Cms. Handaxe is in narrow elongated shape measuring 14x7x5 Cms. Cleaver is in trapezoid shape measuring 8.6x9x4.3 Cms. This site also reveals scraper that is oval and made from reddish brown quartzite, which measures 6.6x5.4x3.7 Cms (Figure 4.2).

3. **Pir Baba ki Bani** also comprises handaxe (1), cleaver (1) and scrapers (4). Handaxe is in oval measuring 16.1x9.1x5.5 Cms and cleaver is in oblong shape measuring 14.1x9.5x5 Cms. Both these tools are made from reddish brown quartzite. Whereas scrapers are in oval shape made from reddish brown and gray quartzite that measures 9x9.8x4.4, 7.6x7.3x3.3, 9.5x8.3x3.3 and 9x6.5x2.9 Cms (Figure 4.3).

4.2 Lithic Classification of Mesolithic Sites

Mesolithic research was carried out to find sites showing single culture site having microliths with or without pottery and multi–cultural site having microliths with other cultural materials. In the light of these assumptions 12 Mesolithic sites were reported from the research area, which has given single culture sites with microliths, multi–cultural sites with Chalcolithic and Early Historic materials and few sites near to the mining area.

Typologically, Microliths are classified into cores, blades, backed blades, core trimming, flakes and chunks. Cores are classified into four different shapes prismatic (6), cylindrical (1), rectangular (1) and irregular (10). Out of these 18 cores, 13 cores are made from chert, 4 from jasper and 1 from chalcedony. Whereas blades are classified into raw material, dimension and section. Out of 36 blades, 18 blades are made from chert, 7 from jasper, 5 from chalcedony, 4 from carnelian and 2 from agate. These blades show two types of sections: triangular (21) and trapezium (15). Along with above said microliths, there are 6 core trimmings in chert and jasper and four backed blades in chert. Further lithic debitage has been sorted out in flakes and chunks. Out of this, 166 flakes sorted out in different raw materials identified as chert (120), jasper (39), chalcedony (5) and agate (2). Similarly, 115 chunks are sorted out in chert (64), jasper (33), chalcedony (11), agate (6) and carnelian (1), (Figure – 4.4 to 4.18).

4.3 Pottery Classification of Chalcolithic and Early Historic Sites

Pottery classification is grouping of similar entities. The similarity of entities within groups reflects something inherently significant in their nature. Groups are usually based on certain common features of material, technique and style and their significance is interpreted culturally. Whereas the identification of wares is based on large number of attributes, which includes function, decoration, paste composition, texture, color, firing, forms and geographical location. The fabric description is based on texture of the sand, which has been graded from sandpaper: sandpaper No. 100: Fine, sandpaper No. 060: Medium and sandpaper No. 050: Coarse. A ceramic group in which all this attributes remain constant is categories as a ware.

The examination of color reading has been done by using Munsell Color Chart that is perhaps most useful when considering the variation of firing conditions in a large sample of sherds of similar composition.

Vessel Forms

The titles Jar, Pot, Bowl and Dish are use as convenient and universally recognized labels for the four major groupings of pottery. In most general terms, these represent a continuum from tall, thin vessels (Jars) through globular to squat vessels that range from restricted to open (Pots and Bowls) to extremely open vessels (Dishes) (Dales and Kenoyer, 1986). A fifth group, Basins, which has thick, deep and squat dishes. A sixth group, Handi's is a cooking vessel.

Jars: Jar is a vessel that is taller than it is broad – i.e., its Internal Height (IH) > its Maximum Body Diameter (MBD).

Pots: Pot body forms are globular with IH = MBD or squat forms with IH < MBD, other forms with THT <

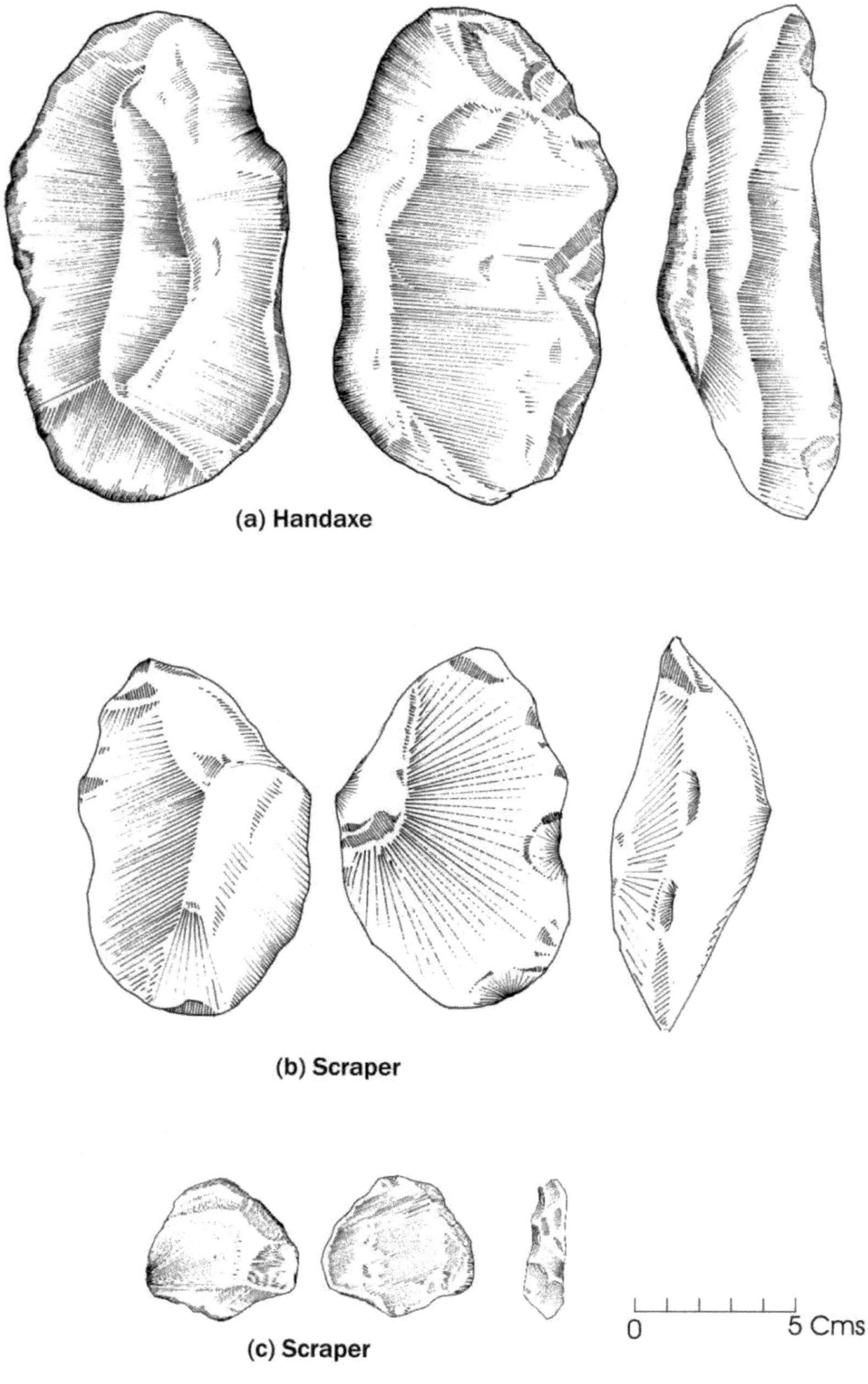

Figure 4.1 – Laluda ki Dhani (a) handaxe; (b-c) scrapers

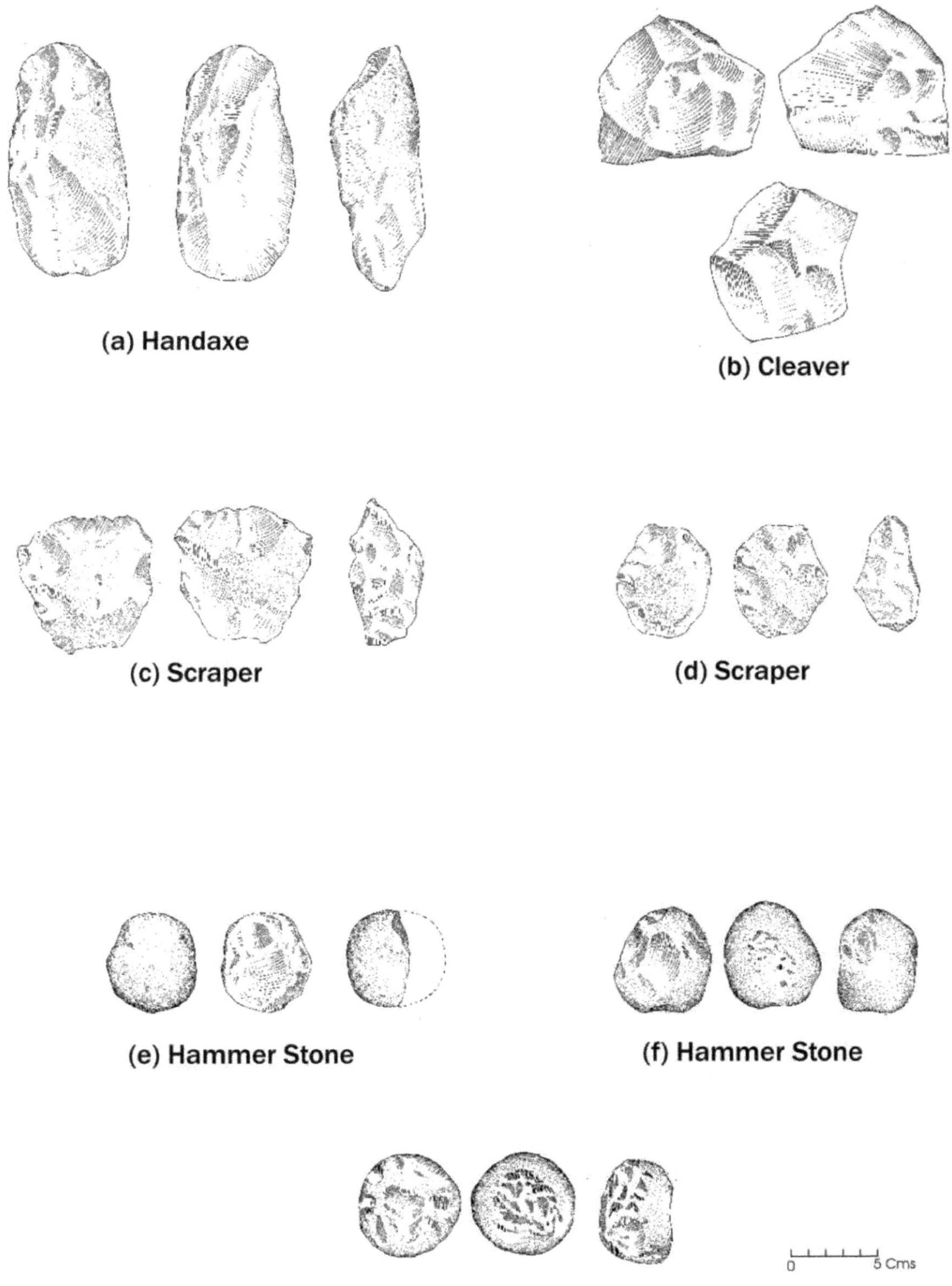

Figure 4.2 – Payga Kundalia (a) handaxe; (b) cleaver; (c-d) scrapers; (e-g) hammer stones

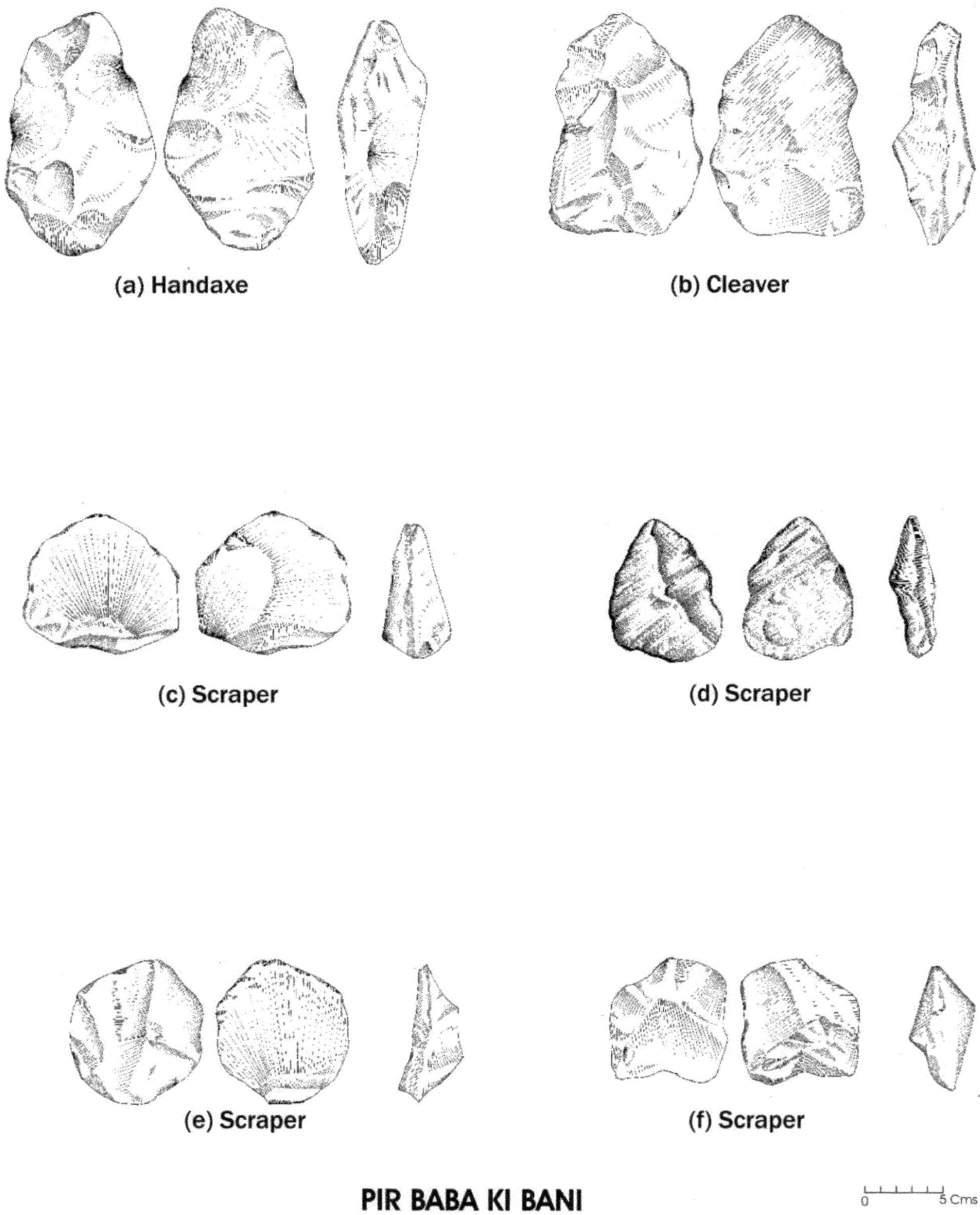

Figure 4.3 – Pir Baba ki Dhani (a) handaxe; (b) cleaver; (c-f) scrapers

Lithic Classification of Mesolithic Sites											
	Site name	Cores				Blades				Flakes	Chunks
		Material	Shape	Measurment	Remarks	Material	Measurment	Section	Remarks		
1	Beed Ki Dhani Total (1 no)	Chert	Prismatic	3x3.2x3.5							
2	Biharipur Total (5 nos) (Fig. 4.4)	Chert	Prismatic	3.2x3.1x3.2		-	-	-	-	-	Chert (2) Jasper (1) Chalcedony (1)
3	Bhadwari Total 37 nos) (Fig. 4.5)	-	-	-	-	Chert	1.3x0.7x0.3	Trapezium	Broken	Chert (13)	Chalcedony (6)
						Chert	1.1x0.7x0.2	Trapezium	Broken	Jasper (2)	Chert (2)
						Chert	0.8x0.9x0.1	Triangular	Broken	Chalcedony (2)	Agate (2)
						Chert	1.3x0.4x0.2	Triangular	Backed Blade		
						Jasper	1.5x0.8x0.3	Trapezium	Broken		
						Jasper	1.5x0.8x0.2	Trapezium	Broken		
						Carnelian	2.1x0.8x0.3	Triangular	Broken		
						Carnelian	0.9x0.6x0.1	Triangular	Broken		
						Chalcedony	1.6x1.0x0.3	Triangular	Broken		
						Chalcedony	1.7x1.1x0.2	Trapezium	Broken		
4	Bhudoli Total (21 nos) (Fig. 4.6)	-	-	-		Chalcedony	1.9x1.0x0.2	Triangular	Broken	Chert (5)	Jasper (4)
						Chalcedony	0.8x0.4x0.1	Triangular	Broken	Jasper (2)	Chert (3)
						Agate	1.6x1x0.3	Trapezium	Broken	Agate (1)	Carnelian (1)
						Agate	0.7x0.6x0.1	Traingular	Broken		Agate (1)
5	Ganeshwar 3 Total (40 nos)	-	-	-	-	Chert	1.3x0.7x0.1	Triangular	Broken	Chert (3)	Chert (13)
						Chert	2.5x0.8x0.2	Triangular	Broken	Agate (1)	Jasper (5)
						Chert	2.0x0.7x0.1	Triangular	Broken	Jasper (3)	Chalcedony (3)
						Chert	2x1.2x0.2	Triangular	Broken		
						Chert	2.8x1.2x0.2	Triangular	Broken		
						Chert	2.2x0.7x0.1	Triangular	Broken		
						Chert	1.2x1x0.1	Trapezium	Broken		
						Chert	2x1.5x0.3	Trapezium	Broken		
						Jasper	2.3x1.2x0.2	Trapezium	Broken		
						Jasper	1.3x0.7x0.1	Triangular	Broken		

	Site name	Cores				Blades				Flakes	Chunks
		Material	Shape	Measurement	Remarks	Material	Measurement	Section	Remarks		
							Jasper	1.5x1x0.2	Triangular	Broken	
						Chalcedony	1.8x0.8x0.2	Trapezium	Broken		
6	Gidhali # 1 Total (82 nos) (Fig. 4.7; 4.8; 4.9; 4.10)	Jasper	Irregular	2.8x2.5x1.5		Chert	2x0.8x0.1	Trapezium	Broken	Cher (25)	Chert (18)
		Jasper	Irregular	2.5x1.6x1.3		Chert	1.8x0.9x0.2	Trapezium	Broken	Jasper (8)	Jasper (9)
		Chert	Irregular	3x2.5x1.7		Chert	1.5x2.9x2.2	Trapezium	Broken	Chalcedony (3)	Agate (2)
		Chert	Irregular	1.5x2.9x2.2		Chert	1.1x0.7x0.2	Traingular	Broken		Chalcedony (1)
		Chert	Prismatic	2x1.7x1.5		Jasper	1.6x0.7x0.2	Trapezium	Broken		
		Chert	Irregular	1.5x2.5x1.6		Carnelian	0.8x0.5x0.1	Traingular	Broken		
		Chert	Irregular	2.6x1.5x0.7		Chert	2.5x0.8x0.2	-	Backed Blade		
						Chert	2x1x0.4	-	Backed Blade		
						Jasper	1.8x0.6x0.2	-	Backed Blade		
7	Gidhali # 2 Total (90 nos) Fig. 4.11; 4.12; 4.13; 4.14)	Chert	Prismatic	2.5x2.5x1.1						Chert (44)	Chert (21)
		Chert	Prismatic	2.7x2.7x1.7						Jasper (10)	Jasper (6)
		Chert	Prismatic	1.5x1.7x1.1							
		Jasper	Cylindrical	3x1.7x0.7							
		Jasper	Irregular	2.5x2.7x0.7							
		Chalced ony	Irregular	2.6x2x1.3							
		Chert	-	-	Core Trimming						
		Jasper	-	-	Core Trimming						
		Jasper	-	-	Core Trimming						
8	Kakrana Total 8 nos) (Fig. 4.15)	-	-	-	-	Jasper	1.2x0.5x0.1	Trapezium	Broken	Chert (4)	Chert (2)
											Jasper (1)
9	Mata Jino Timbo Total (19 nos) (Fig. 4.16)	Chert	Irregular	3.4x2.6x2.1	Flaked Irregularly					Chert (10)	Jasper (2)

	Site name	Cores				Blades				Flakes	Chunks
		Chert	-	-	Core Trimming					Jasper (4)	
		Chert	-	-	Core Trimming						
10	Sadh Mata Ka Maula Total (21 nos) (Fig. 4.17)	Chert	Irregular	4x4.5x3.5	Flake Irregularly from all sides	Chert	0.8x0.8x0.2	Triangular	Broken	Jasper (2) Chert (10)	Chert (2) Jasper (1)
		Chert	Rectangular	2.5x1.7x0.6	Flake Core	Chert	1.1x0.9x0.2	Triangular	Broken		
						Chert	0.6x0.6x0.1	Triangular	Broken		
						Carneli an	0.9x0.6x0.2	Triangular	Broken		
11	Thoi Total (21 nos) (Fig. 4.18)	Jasper	-	-	Core Trimming	-	-	-	-	Chert (6)	Jasper (4)
										Jasper (8)	Chert (1) Agate (1)

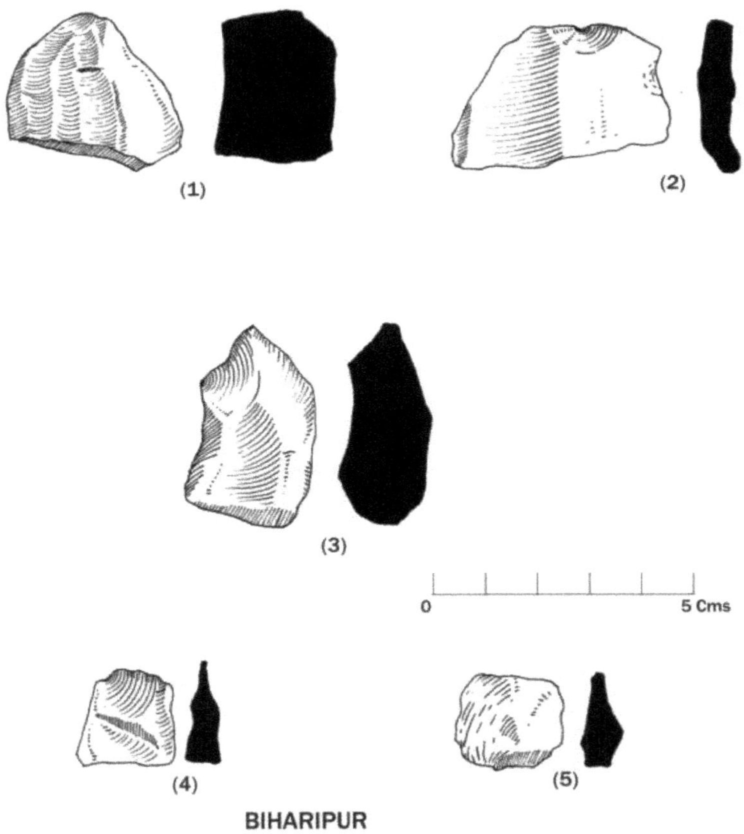

Figure 4.4 – Biharipur (1) core; (2-5) chunks

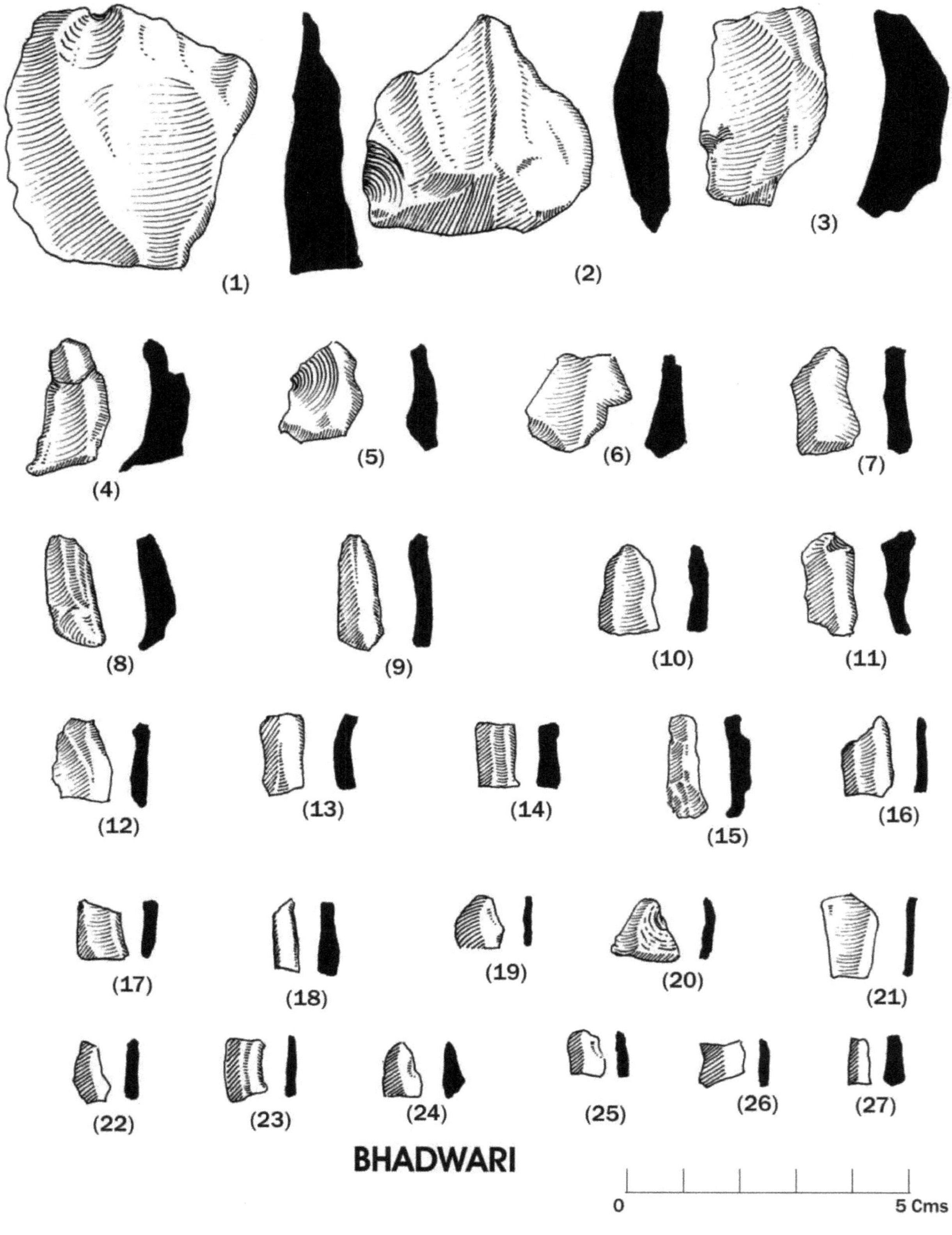

Figure 4.5 – Bhadwari (1-3) chunks; (4-7) flakes; (8-27) blades

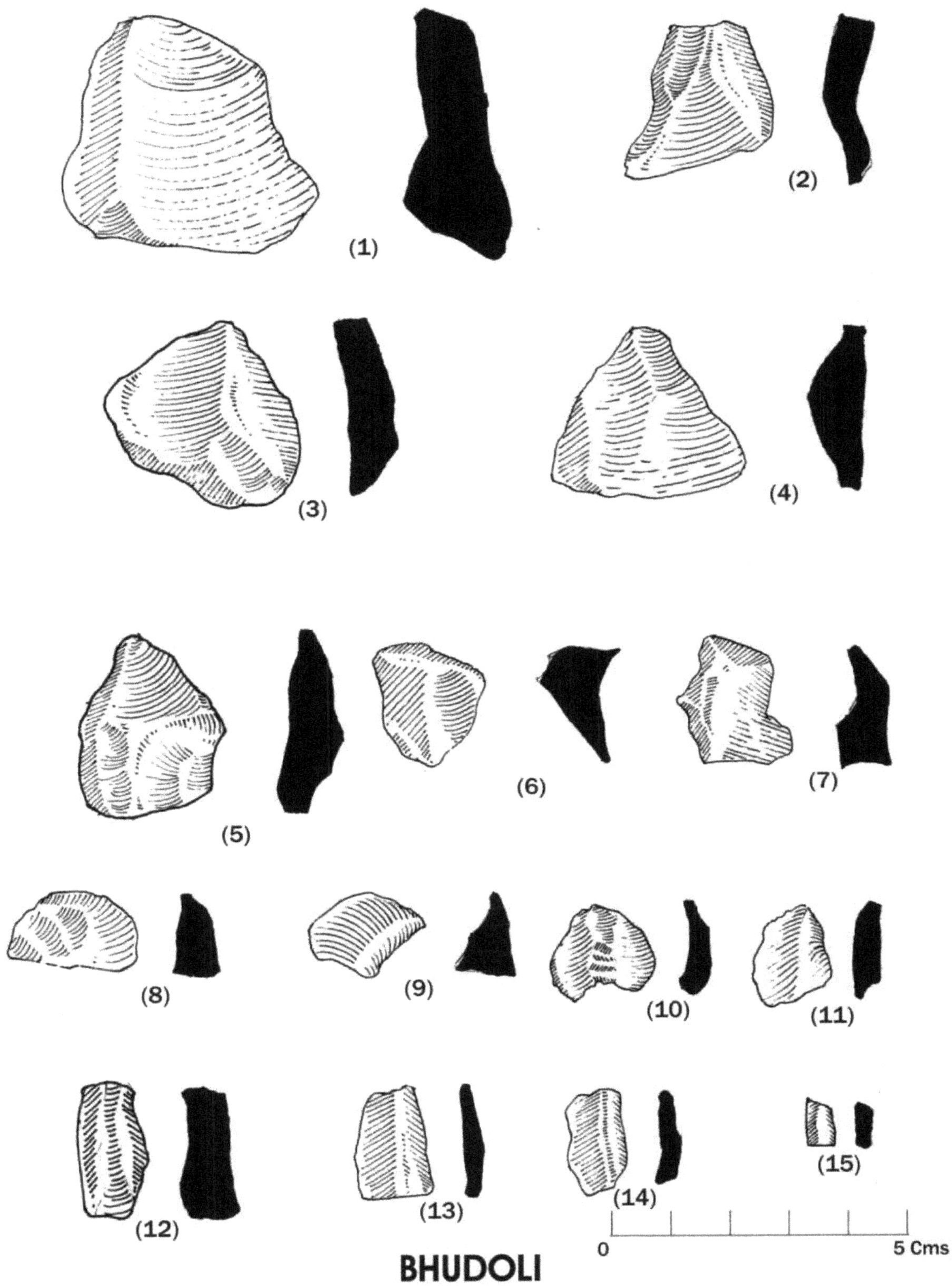

Figure 4. 6– Bhudoli (1-5) chunks; (6-11) flakes; (12-15) blades

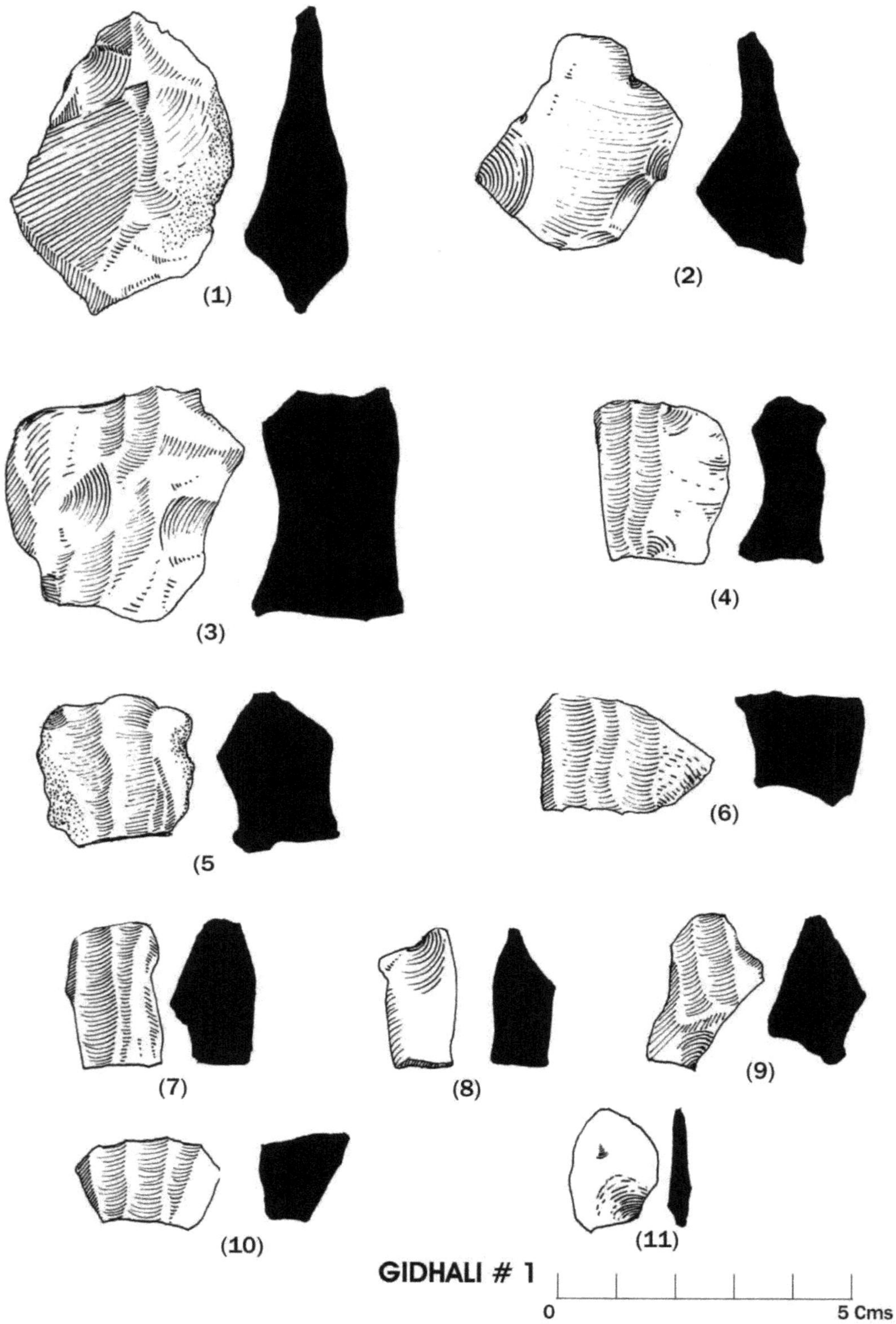

Figure 4.7 – Gidhali No 1 (1,2,9) chunks; (3-7, 10) cores; (8) burin; (11) flake

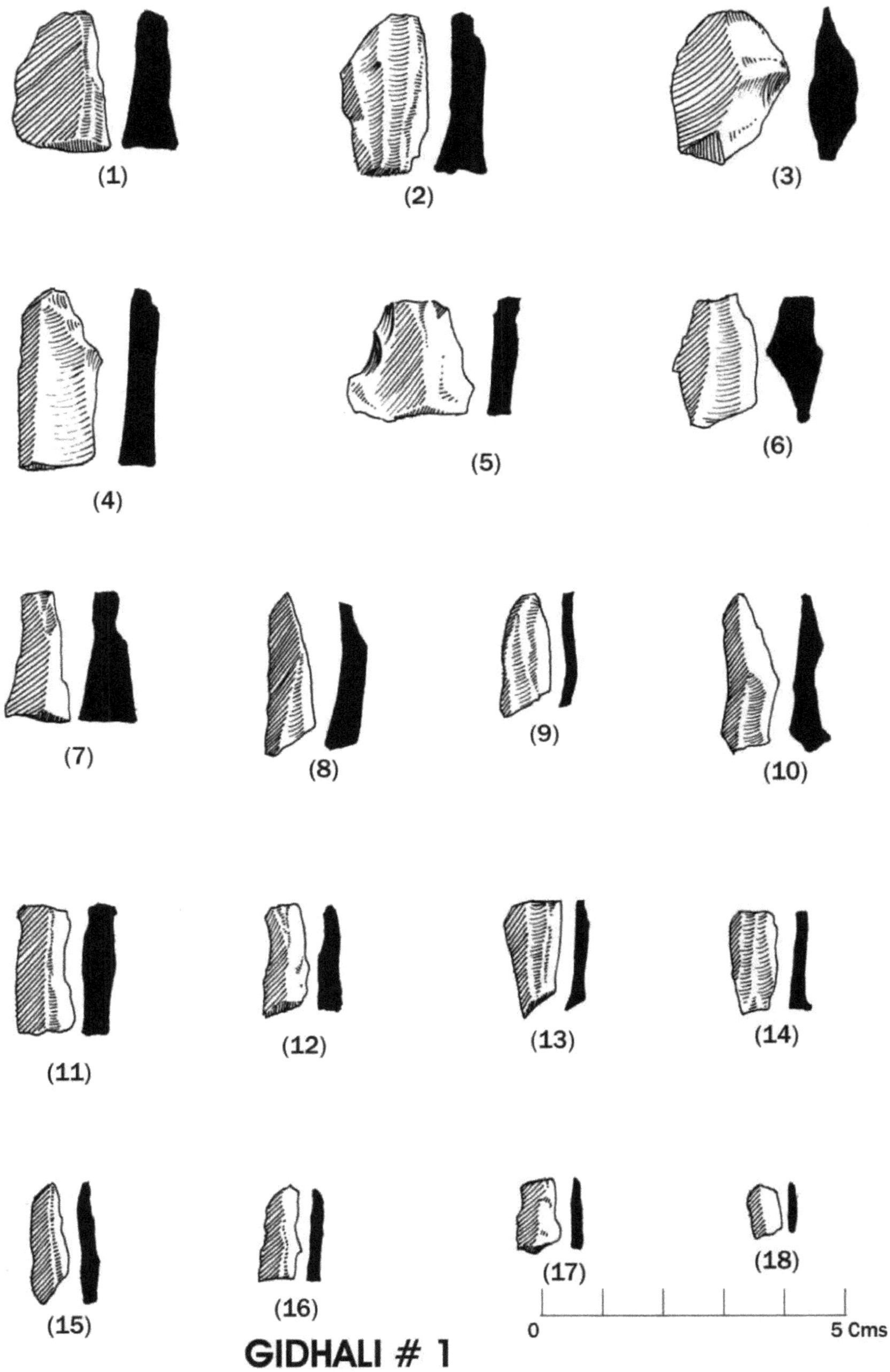

Figure 4.8 – Gidhali No 1 (1,5 - 18) blades; (2) core; (3-4) flakes

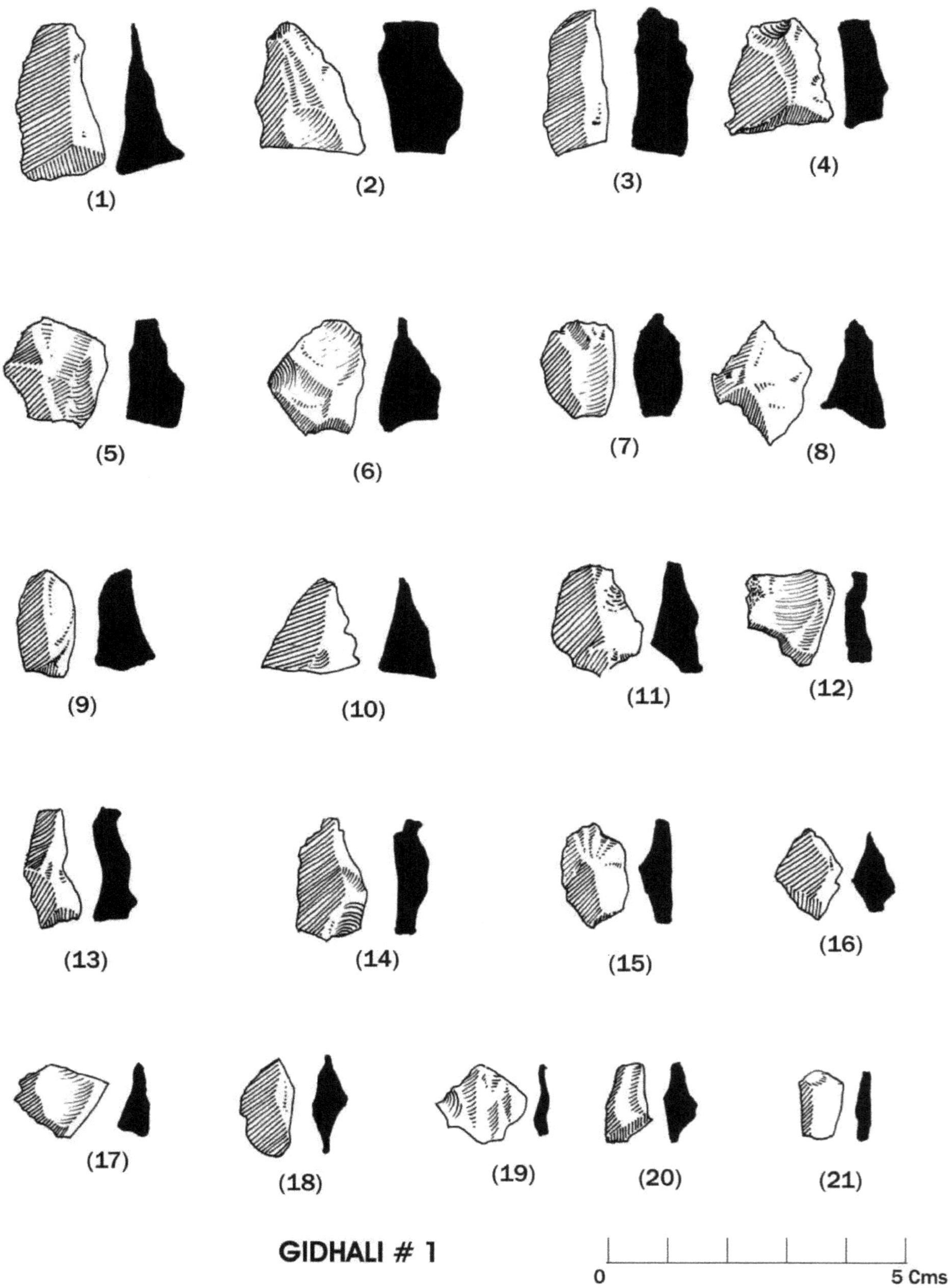

Figure 4.9 – Gidhali No 1 (1-21) flakes

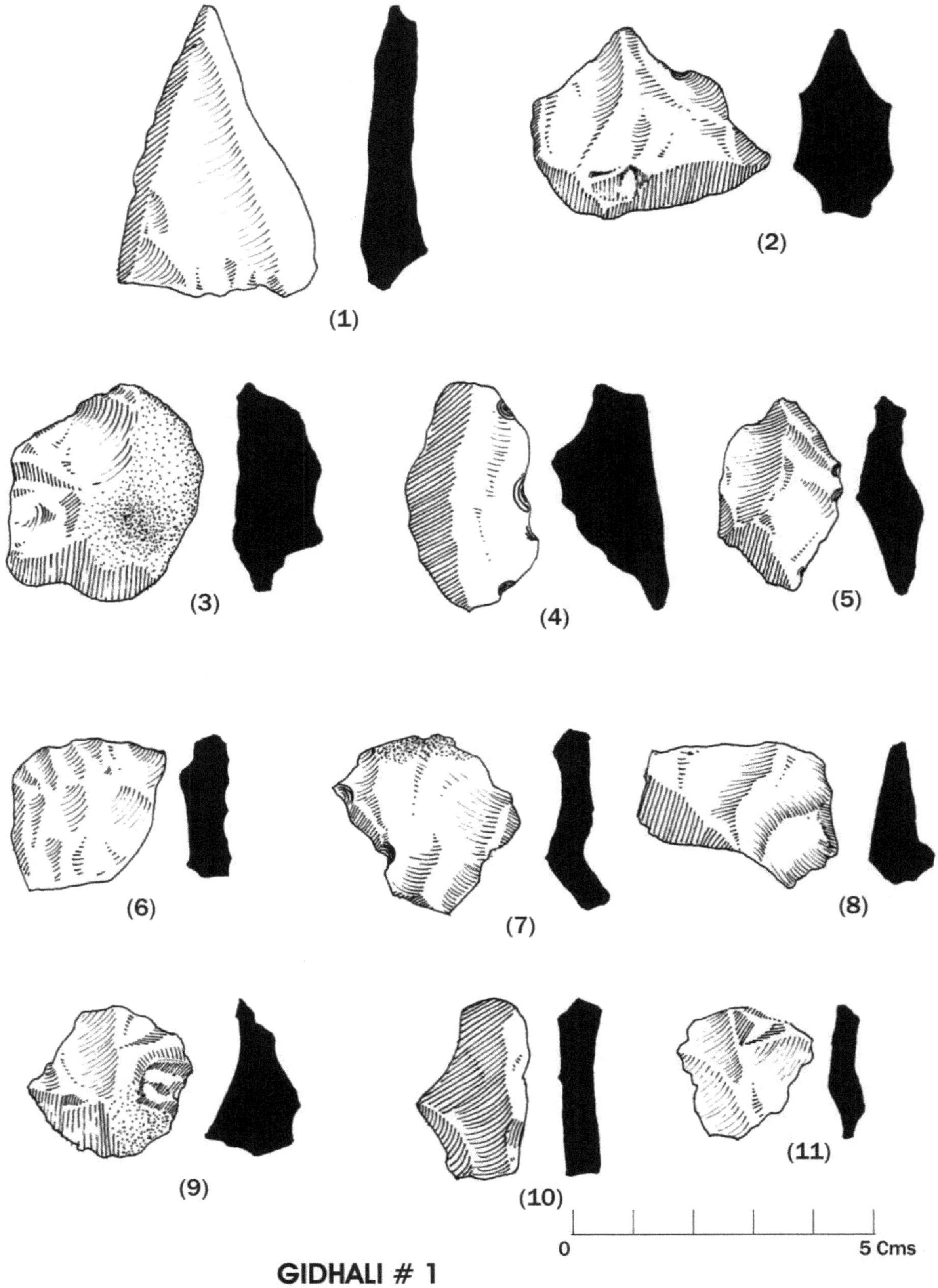

Figure 4.10 – Gidhali No 1 (1 – 11) chunks

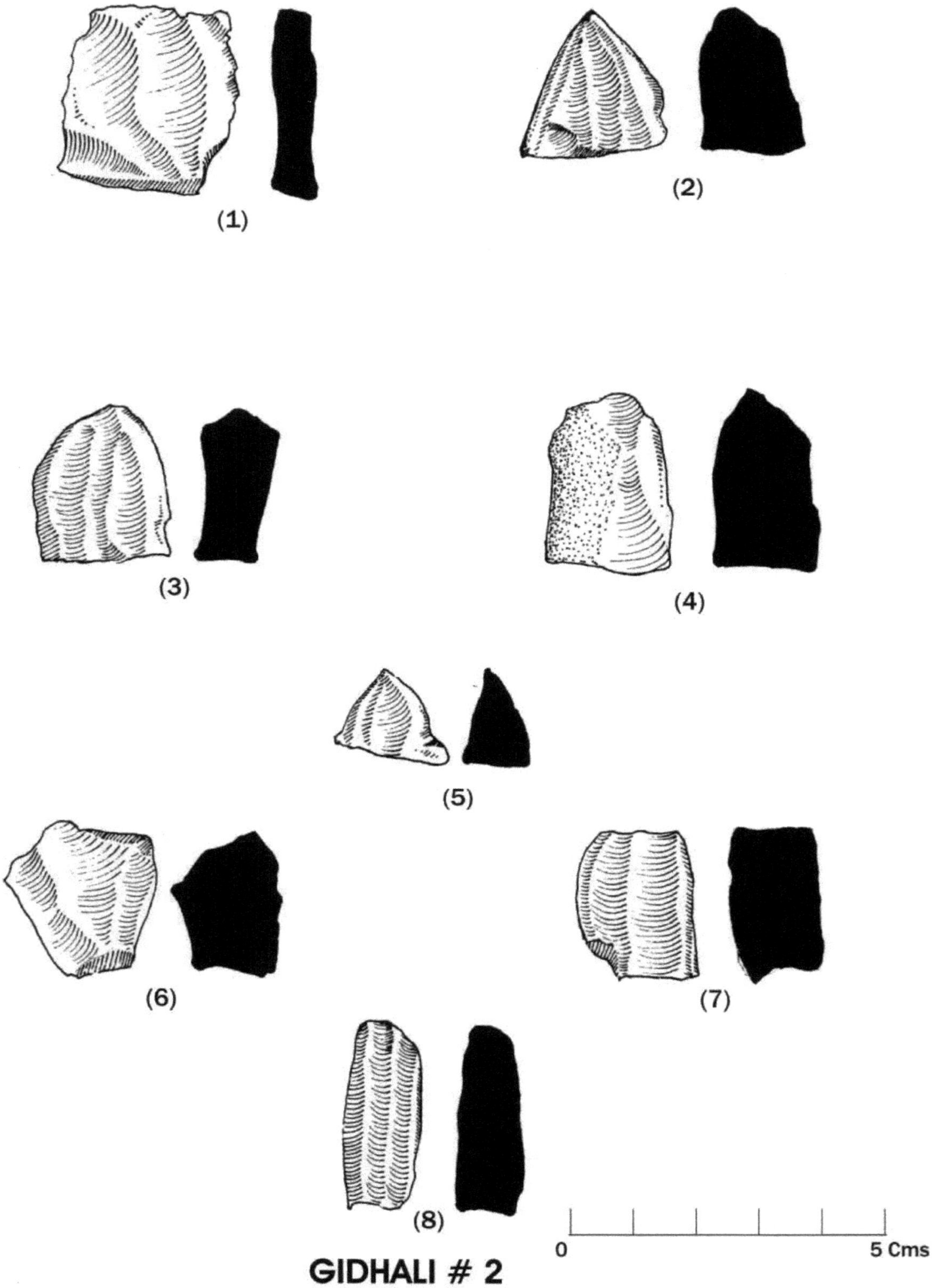

Figure 4.11 – Gidhali No 2 (1 – 8) cores

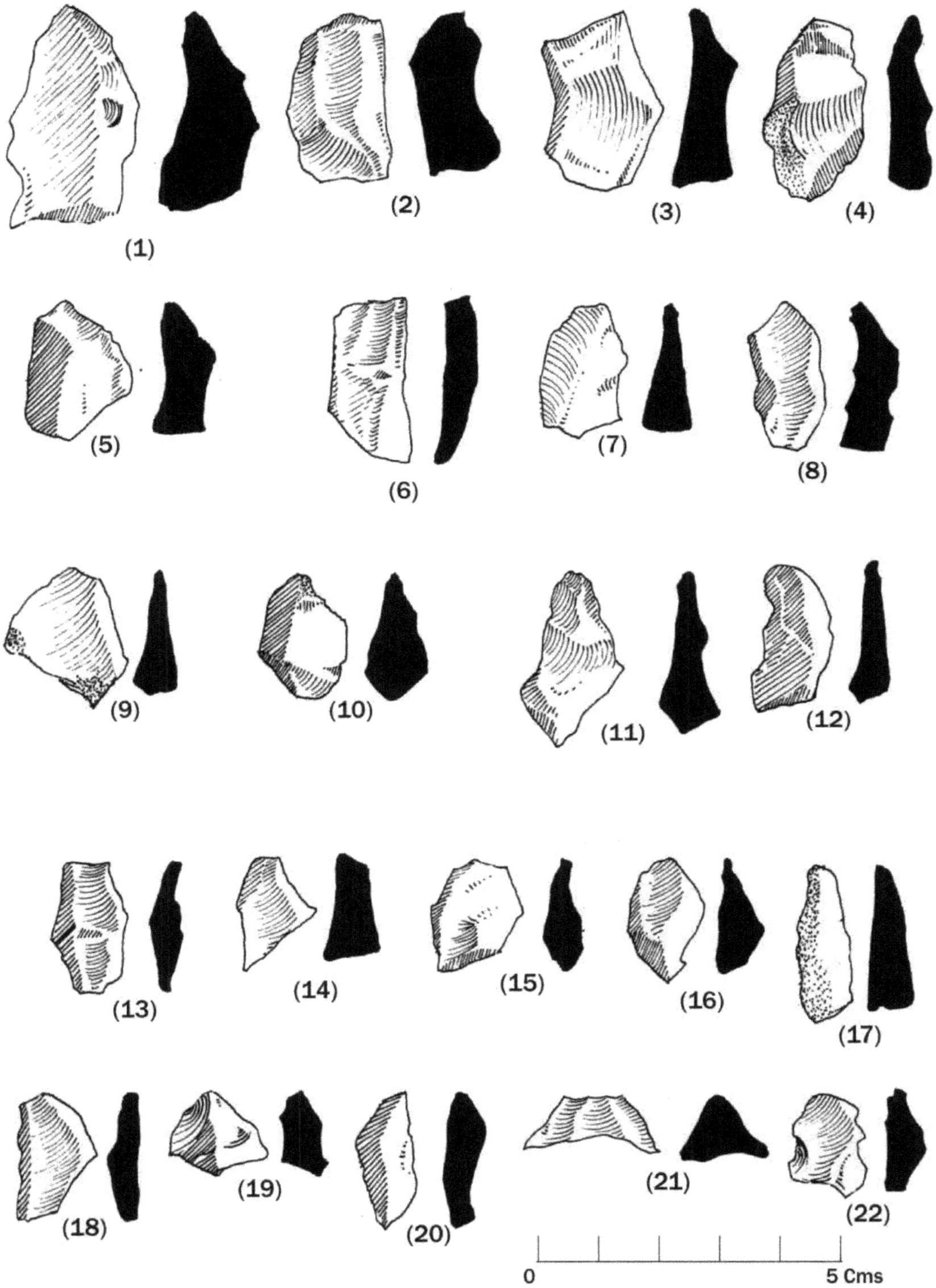

Figure 4.12 – Gidhali No 2 (1 – 22) flakes

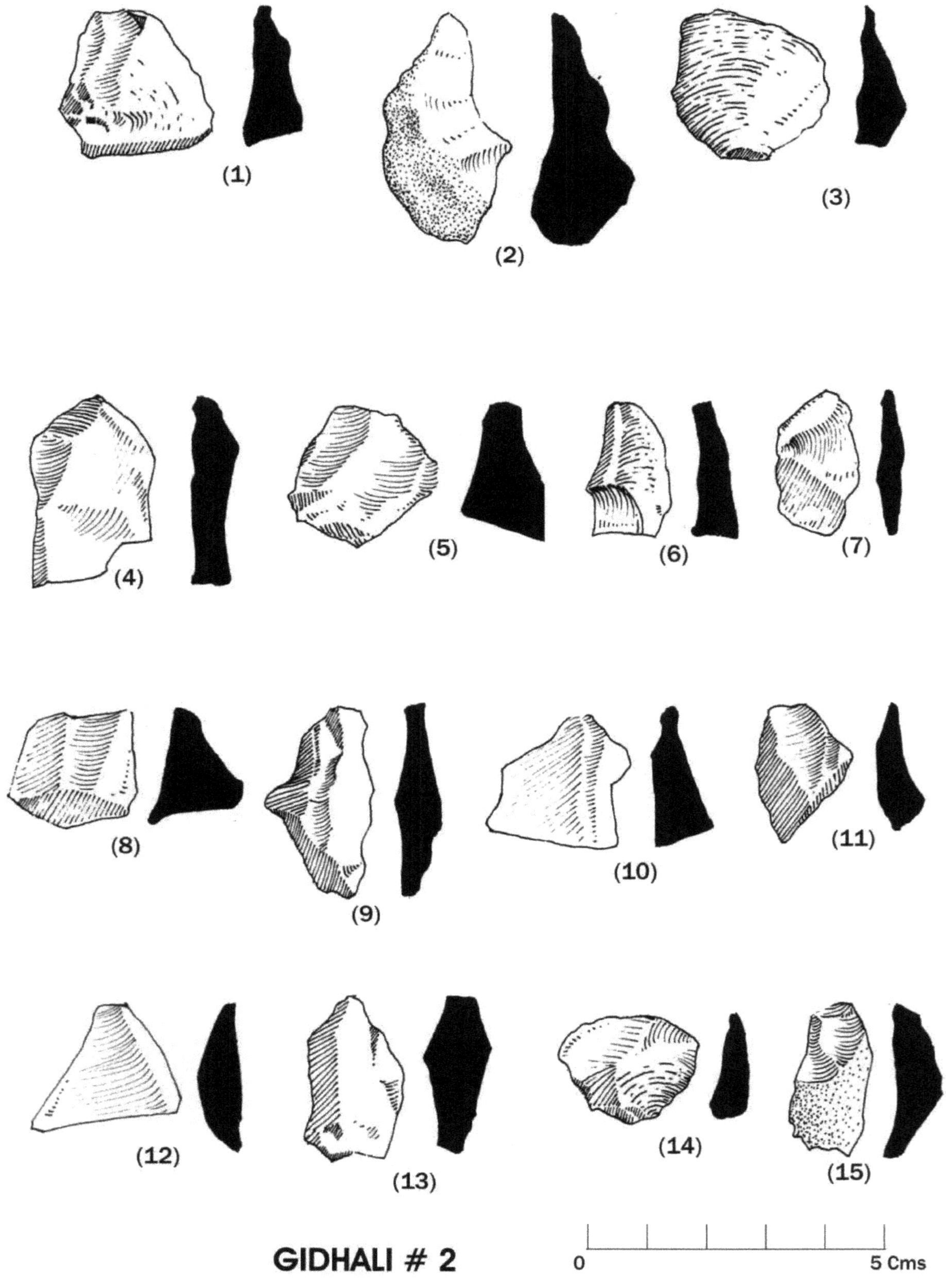

Figure 4.13 – Gidhali No 2 (1 – 15) flakes

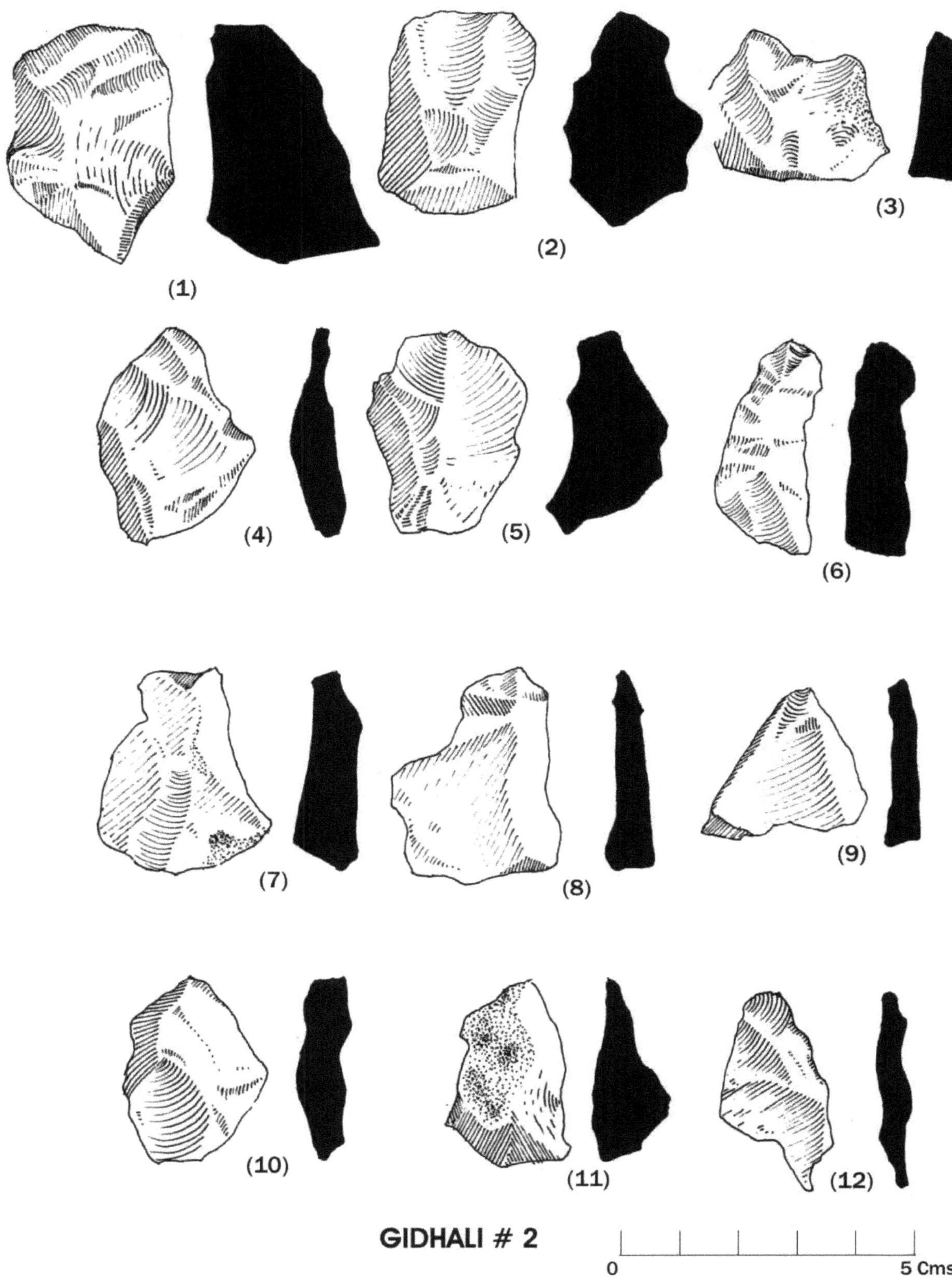

Figure 4.14 – Gidhali No 2 (1 – 12) flakes

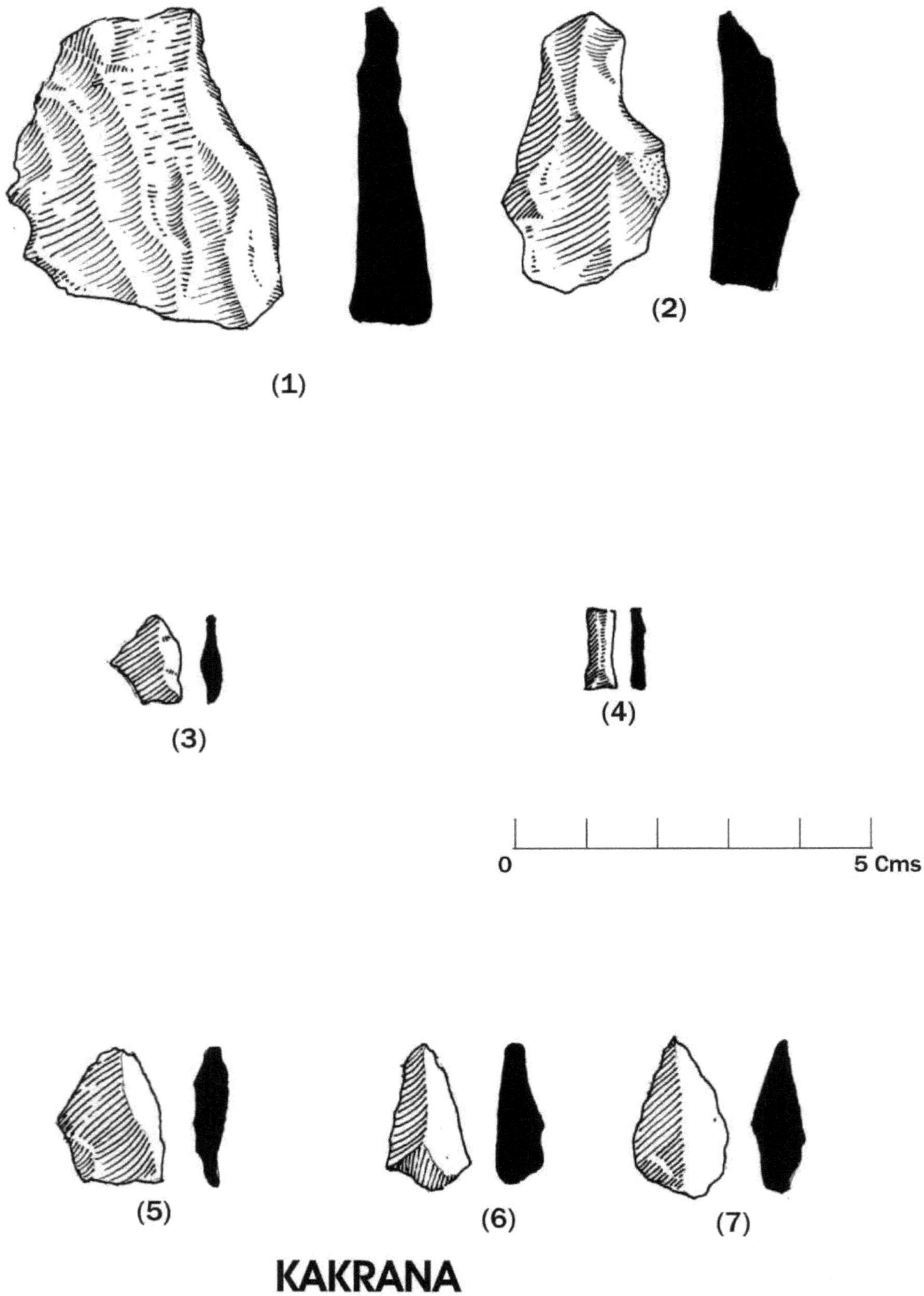

Figure 4.15 – Kakrana (1 – 2) chunks; (3, 5 – 7) flakes; (4) blade

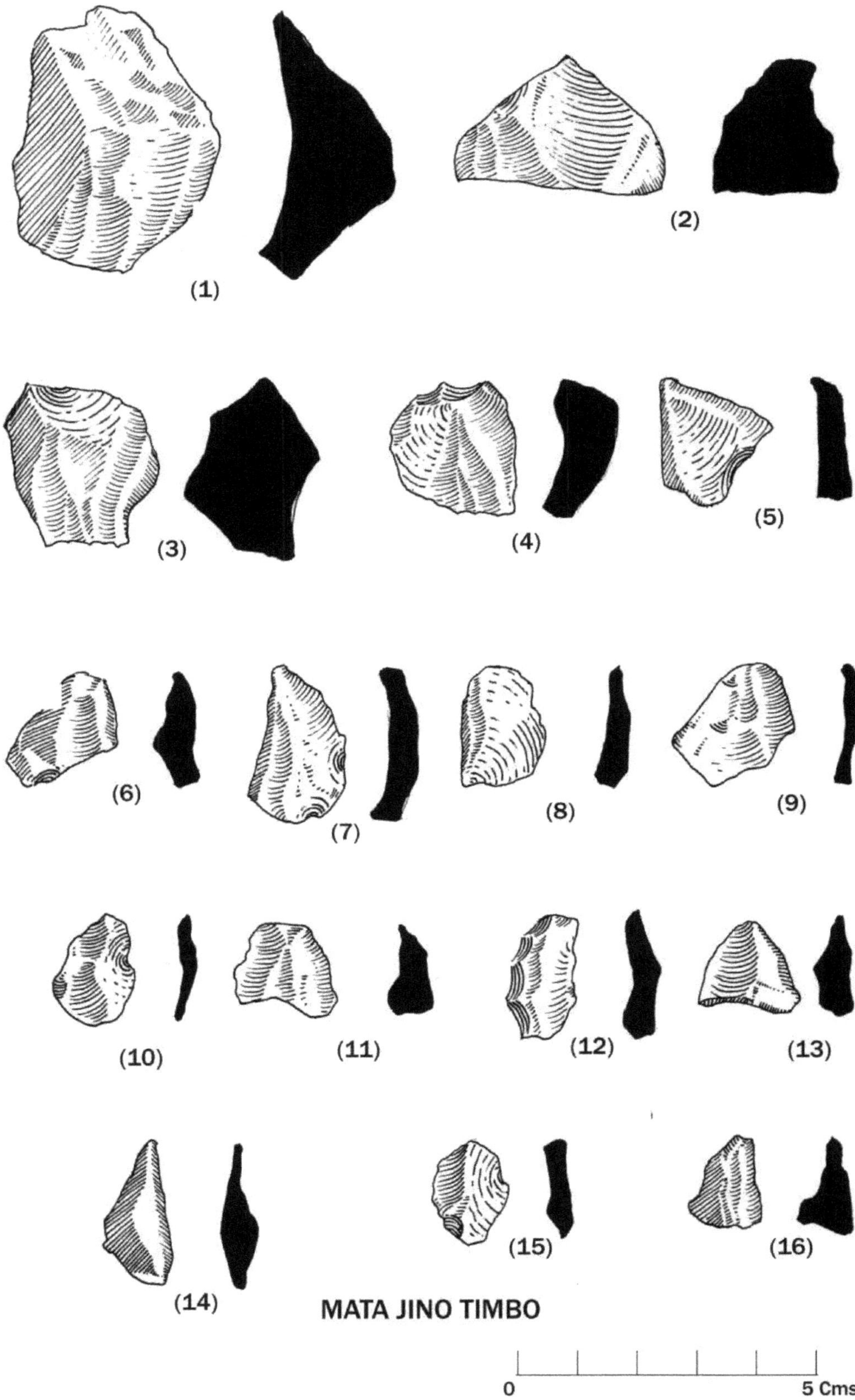

Figure 4.16 – Matajino Timbo (1 – 3) cores; (4 – 16) flakes

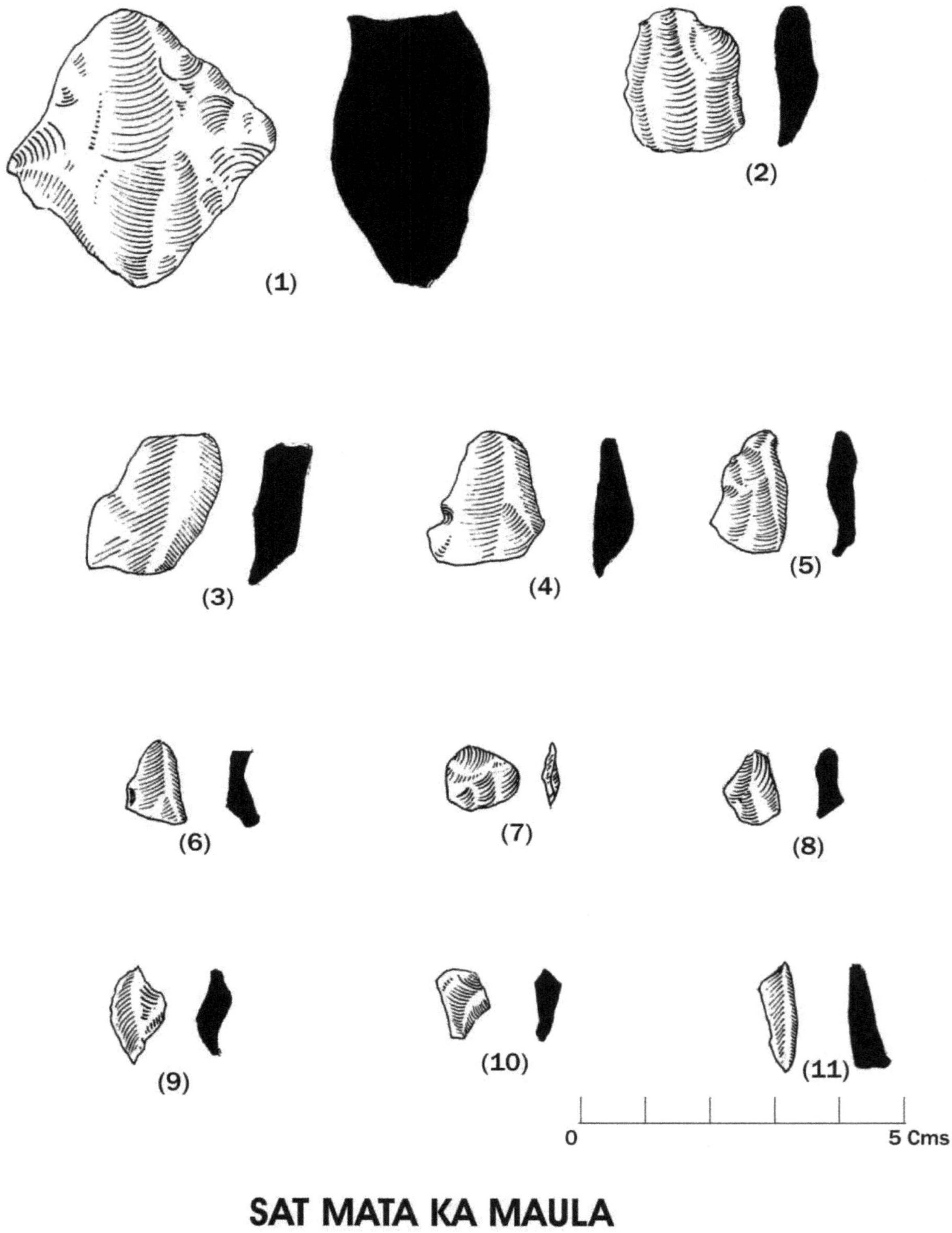

Figure 4.17 – Sat Mata ka Maula (1 – 2) cores; (3 – 5, 11) blades; (6 – 10) flakes

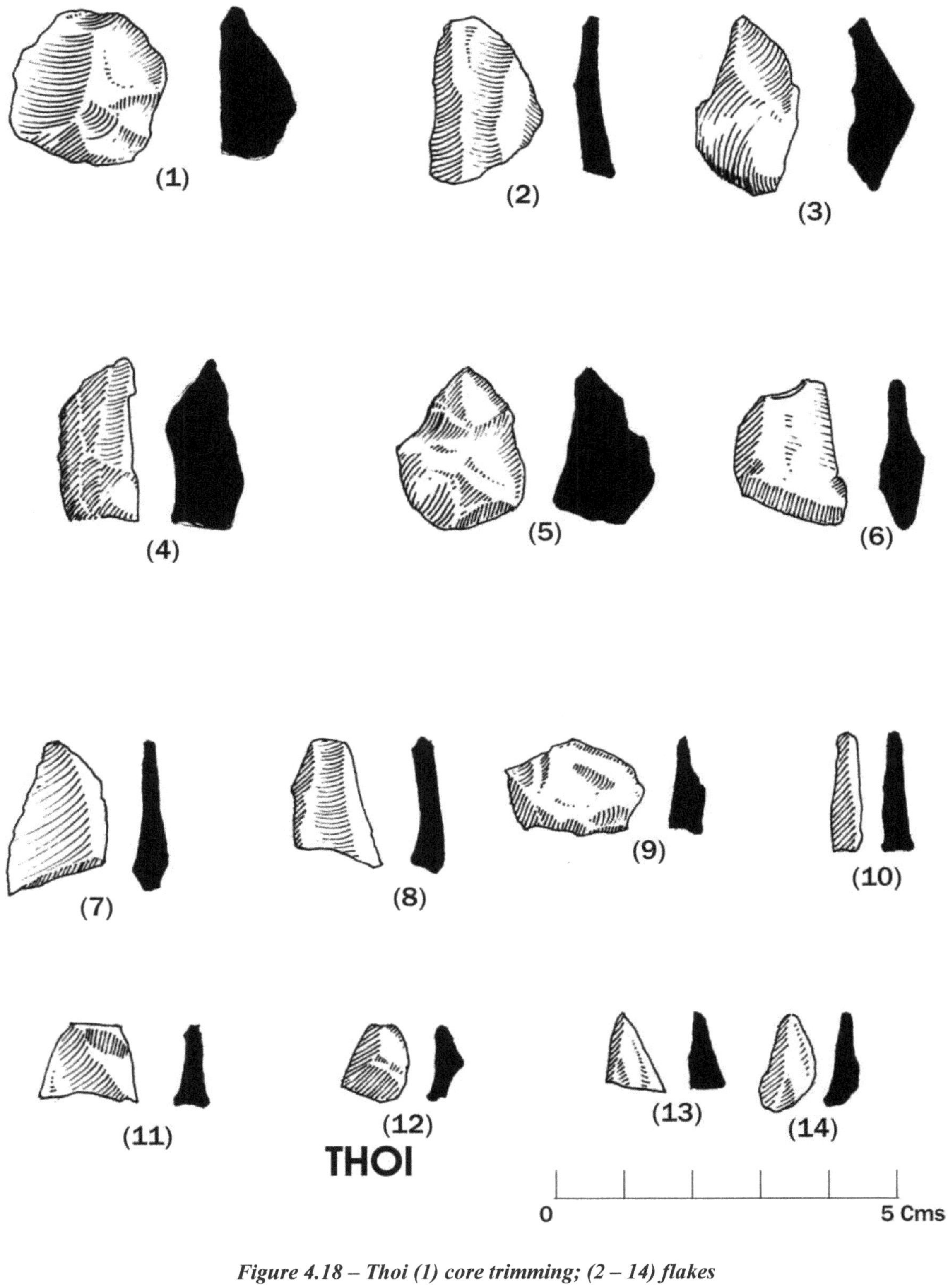

Figure 4.18 – Thoi (1) core trimming; (2 – 14) flakes

MBD or THT = MBD. If the IH = or > one-half the MBD, the form is a Pot.

Bowls: Vessels that have an IH < one-half of MBD are categorized as a Bowl (or if extremely squat and open, a Dish).

Dishes: Vessels that have extremely squat whose IH = or < one-sixth of MBD.

Basins: Vessel having thick, deep and squat dishes.

Handi's shows very dark gray soothing mark on the internal surface with carinated shoulder, corrugated neck and rustication on the body.

Rim Forms: "Rim" refers to the physical configuration of the extreme uppermost edge of a vessel. A vessel can have any possible rim form, ranging from simple unmodified extensions of the body wall to complex modifications built up on the edge of the vessel.

1. GANESHWAR (GNR) (Figure - 4.19 to 4.37)

In the present study, an attempt has been made to examine the explored pottery with the aim to present corpus of Ganeshwar ceramic types.

At Ganeshwar 252 sherds were examine. Out of which 129 sherds were diagnostic and 123 sherds were undiagnostic. Diagnostic sherds show six variances of vessel form: **Pots** 29, **Bowls** 56, **Basins** 34, **Jars** 5, **Dishes** 3 and **Handi's** 2.

Vessel Forms

Pots: Pots shows variance in their body forms having bulbous body, small neck, broad neck, corrugated neck, slanting neck, splayed grooves on neck, flange on neck, horizontal grooves on shoulder and carinated shoulder. The surface enhancement on the body shows incised notches on the shoulder, incised triangular notches on shoulder with black color band above it, incised wavy lines on the shoulder, thumb impression in internal surface on the shoulder, black color band on lip of the rim, burnishing on neck on the external surface and graffiti marks on the internal surface.

There are four variances in rim form, which are further divided into seven sub-forms: (1) Simple everted, slightly curved rounded rim. (2) Simple external projecting, extremely everted tapered rim. (3) External projecting, short, narrow rounded rim. (3a) External projecting, short, narrow triangular rim. (3b) External projecting, short, narrow short quadrangular rim. (3c) External projecting, long, wide beaked rim. (4) Bilateral projecting, equal projection rim.

Bowls: Bowls shows variance in their body forms having convex-sided, carination on the shoulder, blunt carination on the shoulder and corrugation on the neck. The surface enhancement on the body shows; rustication on the body, incised marks on the shoulder, black color band on lip of the rim, burnishing on both the surface and graffiti marks on the internal surface.

There are six variances in rim forms, which are further divided into 13 sub-forms: (1a) Simple everted, slightly curved rounded rim, (1b) Simple external projecting, extremely everted rounded rim, (2a) Simple internal projecting, extremely inverted rounded rim, (2b) Internal projecting, short, narrow triangular rim, (3) Simple perpendicular rounded rim, (4a) External projecting, short, narrow rounded rim, (4b) External projecting, short, narrow triangular rim, (4c) External projecting, short, narrow short beaked rim, (4d) External projecting, long, wide rounded rim, (4e) External projecting, long, wide triangular rim, (5) Channel rim, (6a) Bilateral projecting, equal projection rim (horizontal oriented with rounded top), and (6b) Bilateral projecting, equal projection rim (horizontal oriented with concave depression on top)

Basins: Basins show variance in their body form having carination on shoulder, blunt carination on shoulder, corrugation on neck, sharp corrugation on neck and thick sturdy convex-sided body. The surface enhancement on the body shows incised notches on the shoulder, rustication on the body, black color band on lip of the rim and burnishing on the neck.

There are six variances in rim forms, which are further divided into 10 sub-forms: (1) Simple everted, straight rounded rim. (2) Simple inverted, rounded rim. (3) Simple external projecting extremely everted rounded rim. (4) Simple perpendicular, rounded rim. (5a) External projecting, short, narrow rounded rim. (5b) External projecting, short, narrow triangular rim. (5c) External projecting, short, narrow short beaked rim. (5d) External projecting, short, narrow quadrangular rim. (5e) External projecting, long, wide triangular rim. (5f) External projecting, long, wide triangular rim. (6) Internal projecting, short, narrow, triangular rim.

Jars: The surface enhancement on the body of jars shows; graffiti marks on the internal surface, black color band on lip of the rim and burnishing on both the surface.

There are four variances in rim forms: (1) Simple external projecting, extremely everted rounded rim. (2) External projecting, short, narrow rounded rim. (3) External projecting, short, narrow triangular rim. (4) External projecting, long, wide rounded rim.

Dishes: Dishes having three variances in rim forms: (1) Simple external projecting, extremely everted tapered rim. (2) Simple external projecting, extremely everted rounded rim. (3) External projecting, short, narrow triangular rim.

Handi's shows very dark gray soothing mark on the internal surface with carinated shoulder, corrugated neck and rustication on the body.

There are two variances in rim forms: (1) Internal projecting, short, narrow triangular rim. (2) Bilateral projecting, equal projection rim.

Undiagnostic Sherds

The surface enhancement on undiagnostic sherds shows rustication on the body, burnishing on the surface, black color sooth mark on internal surface and graffiti marks on the internal surface. Incised patterns having notches, triangular notches, nail-headed notches, thumb-headed notches, diamond shape notches, wavy lines, zigzag lines, horizontal lines, oblique lines and hatch lines. The painting shows black color band on lip of the rim, black color band over incised patterns and black color wavy lines over the red surface (Reserve Slip technique).

Bowls Type A Variety 1: convex-sided bowls having one variety and one rim variant:

(a) simple internal projecting, extremely inverted rounded rim

Sample No. 70, 89 and 195.

Fabric	Medium sandy with abundant mica and white inclusion.
Surface enhancement	Sample No. 70 is unslip on both the surfaces. Sample No. 89 having red slip with black color band on lip of the rim and burnishing on both the surfaces. Sample No. 195 having red slip and burnishing on both the surfaces.
Size range	Bowl measures 22 to 26 cms in rim diameters.

Bowls Type B Variety 1: carinated bowls having three varieties, first variety with carinated shoulder, corrugated neck and incised notches and five rim variants:

(a) simple everted, slightly curved rounded rim.
(b) external projecting, short, narrow triangular rim.
(c) external projecting, short, narrow short beaked rim.
(d) external projecting, long, wide triangular rim
(e) internal projecting, short, narrow triangular rim.

Sample No. 13, 14, 15, 20, 24, 25, 161, 165, 173, 206, 207 and 243

Fabric	Medium sandy with abundant mica and white inclusion.
Surface enhancement	Sample No. 13 having red slip on both the surfaces with black color band on lip of the rim. Sample No. 14, 15 and 20 having red slip on both the surfaces with black color band on lip of the rim and burnishing on both the surfaces. Sample No. 24 having reddish yellow slip on both the surfaces. Sample No. 25 having reddish brown slip on both the surfaces with black color band on lip of the rim. Sample No. 173 having red slip on both the surfaces with rustication on exterior surface of the body. Sample No. 161, 165 and 243 having red slip on both the surfaces with rustication on both the surfaces. Sample No. 206 and 207 having red slip on both the surfaces with black color band on lip of the rim and rustication on exterior surface of the body.
Size range	Bowl measures 11 to 30 cms in rim diameters.

Bowls Type B Variety 2: carinated bowls having three varieties, second variety with sharp carinated shoulder and four rim variants:

a) simple perpendicular, straight rounded rim
b) simple external projecting, extremely everted rounded rim
c) external projecting, short, narrow rounded rim
d) external projecting, short, narrow short beaked rim

Sample No. 31, 33, 50, 58 and 196, 203 and 204

Fabric	Medium sandy with abundant mica and white inclusion.
Surface enhancement	Sample No. 31 and 203 having red slip on both the surfaces with black color band on lip of the rim. Sample No. 33 and 204 having red slip on both the surfaces with black color band on lip of the rim and rustication on exterior surface of the body. Sample No. 50, 58 and 196 having red slip on both the surfaces.
Size range	Bowl measures 16 to 36 cm in rim diameter.

Bowls Type B Variety 3: carinated bowls having three varieties, third variety with blunt carinated shoulder and corrugated neck and seven rim variants:

a) simple external projecting, extremely everted rounded rim
b) external projecting, short, narrow rounded rim
c) external projecting, short, narrow triangular rim
d) external projecting, long, wide rounded rim
e) external projecting, long, wide triangular rim
f) internal projecting, short, narrow triangular rim
g) bilateral projecting, equal projection rim

Sample No. 4, 10, 26, 30, 34, 39, 41, 42, 43, 44, 46, 52, 53, 60, 76, 92, 93, 95, 96, 160, 162,166, 172 and 234

Fabric	Fine to Medium to Coarse sandy with abundant mica, white inclusion and organic matter.
Surface enhancement	Sample No. 4 and 172 having red slip on both the surfaces with graffiti marks on the interior surface. Sample No. 34, 44, 46, 160 and 162 having red slip on both the surfaces. Sample 42 and 46 having red slip on the exterior surface and reddish yellow slip on the interior surface with black color band on lip of the rim. Sample No. 43 having red slip on both the surfaces with black color band on

	lip of the rim. Sample No. 53 having light red slip on the interior surface. Sample No. 76, 92, 95 and 96 unslipped on both the surfaces. Sample No. 93 having very dusky red slip on both the surfaces.
Size range	Bowl measures 17 to 32 cms in rim diameter.

Pots Type A Variety 1: small size pots with rim diameter ranges between 11 to 16 cms having three varieties, first variety with short neck, incised notches and bulbous body and four rim variants:

a) simple everted, slightly curved rounded rim
b) simple external projecting, extremely everted tapered rim
c) simple external projecting, extremely everted rounded rim
d) external projecting, short, narrow triangular rim

Sample No. 17, 21, 87, 88, 109, 158, 167 and 248

Fabric	Fine to Medium sandy with abundant mica and white inclusion.
Surface enhancement	Sample No. 17 having light red slip on exterior surface. Sample No. 21 having red slip and burnishing on the exterior surface. Sample No. 167 having yellowish red slip on both the surfaces with black color band on lip of the rim and burnishing on the exterior surface. Sample No. 87 having red slip on the interior surface with black color band on lip of the rim and burnishing on the exterior surface. Sample No. 158 having red slip on exterior surface with black color band on lip of the rim and burnishing on the exterior surface. Sample No. 88 having red slip on both the surfaces with black color band on lip of the rim and burnishing on the exterior surface. Sample No. 248 and 109 having red slip on exterior surface.
Size range	Pot measures 11 to 16 cms in rim diameter.

Pots Type A Variety 2: small size pots with rim diameter ranges between 11 to 16 cms having three varieties, second variety with long neck and bulbous body and three rim variants:

(a) simple external projecting, extremely everted rounded rim
(b) external projecting, short, narrow rounded rim
(c) external projecting, short, narrow triangular rim

Sample No. 51, 201 and 247

Fabric	Medium sandy with abundant mica and white inclusion.
Surface enhancement	Sample No. 51 unslipped on both the surfaces. Sample No. 201having red slip on both the surfaces with thumb impression on the interior surface. Sample No. 247 having red slip and reddish yellow slip on the exterior surface with graffiti marks on the interior surface.
Size range	Pot measures 11 to 14 cms in rim diameter

Pots Type A Variety 3: small size pots with rim diameter ranges between 11 to 16 cms having three varieties, third variety with corrugated neck and bulbous body and one rim variant:

(a) simple everted, straight rounded rim

Sample No. 250

Fabric	Medium sandy with abundant mica and white inclusion.
Surface enhancement	Sample No. 250 having red slip on the exterior surface.
Size range	Pot measures 13 cm in rim diameter

Pots Type B Variety 1: medium size pots with rim diameter ranges between 16 to 27 cms having four varieties, first variety with short neck and bulbous body and three rim variantsn:

(a) simple everted, straight rounded rim
(b) simple external projecting, extremely everted rounded rim
(c) external projecting, short, narrow triangular rim

Sample No. 40, 86, 236, 244, 246

Fabric	Fine to Medium sandy with abundant mica and white inclusion.
Surface enhancement	Sample No. 40, 236 and 244 having red slip on the exterior surface. Sample No. 246 and 247 having red slip and reddish yellow slip on the exterior surface with graffiti marks on the interior surface. Sample No. 86 having red slip on both the surfaces.
Size range	Pot measures 16 to 22 cms in rim diameter

Pots Type B Variety 2: medium size pots with rim diameter ranges between 16 to 27 cms having four varieties, second variety with short neck, incised notches and bulbous body and three rim variants:

(a) simple external projecting, extremely everted tapered rim
(b) simple external projecting, extremely everted rounded rim
(c) external projecting, short, narrow rounded rim

Sample No. 6, 9, 99, 157, 198 and 199

Fabric	Medium sandy with abundant mica, white inclusion and organic matter.

Surface enhancement	Sample No. 201and 157 having red slip on both the surfaces with thumb impression on the interior surface. Sample No. 9 having light red slip on the exterior surface. Sample No. 198 having red slip on the exterior surface. Sample No. 6 and 199 having light red slip and burnishing on the exterior surface.
Size range	Pot measures 18 to 26 cms in rim diameters.

Pots Type B Variety 3: medium size pots with rim diameter ranges between 16 to 27 cms having four varieties, third variety with long neck and bulbous body and one rim variant:
(a) simple everted, straight rounded rim
Sample No. 202

Fabric	Medium sandy, white inclusion and organic matter.
Surface enhancement	Sample No. 202 having red slip on the exterior surface.
Size range	Pot measures 17 cm in rim diameter.

Pots Type B Variety 4: medium size pots with rim diameter ranges between 16 to 27 cms having four varieties, fourth variety with carinated, corrugated, flanged and slanting neck and bulbous body and five rim variants:
(a) simple external projecting, extremely everted tapered rim
(b) simple external projecting, extremely everted rounded rim
(c) external projecting, short, narrow triangular rim
(d) external projecting, long, wide beaked rim
(e) bilateral projecting, equal projection rim
Sample No. 36, 37, 83, 101, 103 and 233

Fabric	Fine to Medium to Coarse sandy with abundant mica, white inclusion and organic matter.
Surface enhancement	Sample No. 36 having red slip on the exterior surface with black color band on lip of the rim. Sample No. 37, 101 and 103 unslipped on both the surfaces. Sample No. 83 and 233 having red slip on both the surfaces.
Size range	Pot measures 12 to 27 cms in rim diameter.

Jars, with long neck and bulbous body having four rim variants:
(a) simple external projecting, extremely everted rounded rim
(b) external projecting, short, narrow rounded rim
(c) external projecting, short, narrow triangular rim
(d) external projecting, long, wide rounded rim
Sample No. 1, 3, 49, 51, 84 and 85

Fabric	Medium sandy with abundant mica and white inclusion.
Surface enhancement	Sample No. 1 having red slip on the exterior surface with graffiti marks on the interior surface. Sample No. 3 having red slip on exterior surface with black color band on lip of the rim and graffiti marks on interior surface. Sample No. 84 having light red slip and burnishing on both the surfaces. Sample No. 49 having red slip on both the surfaces with black color band on lip of the rim. Sample 85 having red slip on both the surfaces. Sample No. 51 unslipped on both the surfaces.
Size range	Pot measures 11 to 14 cms in rim diameter

Basins Type A Variety 1: carinated basins having one variety carinated shoulder, corrugated neck and incised notches and nine rim variants:
(a) simple perpendicular, straight rounded rim
(b) simple external projecting, extremely everted rounded rim
(c) external projecting, short, narrow rounded rim
(d) external projecting, short, narrow triangular rim
(e) external projecting, short, narrow short beaked rim
(f) external projecting, short, narrow short quadrangular rim
(g) external projecting, long, wide rounded rim
(h) external projecting, long, wide triangular rim
(i) internal projecting, short, narrow triangular rim
Sample No. 5, 7, 8, 10, 11, 12, 16, 18, 23, 28, 29, 55, 82, 159, 163, 197, 200, 205 and 232

Fabric	Medium to Coarse sandy with abundant mica and white inclusion.
Surface enhancement	Sample No. 7 having red slip on both the surfaces with black color band on lip of the rim, burnishing and rustication on exterior surface of the body. Sample No. 8 and 16 having reddish yellow slip with black color band on lip of the rim and burnishing on the neck. Sample No. 10 having red slip on both the surfaces. Sample No. 11 having red slip on interior surface up to the neck with black color band on lip of the rim, burnishing on interior surface and rustication on the exterior surface of the body. Sample No. 12 having light red slip on both the surfaces with burnishing on the neck and rustication on exterior surface of the body. Sample No. 18 having light red slip.

Surface enhancement	Sample No. 23 having light yellowish brown slip on both the surfaces with black color band on lip of the rim and burnishing on the neck. Sample No. 28 having red slip on both the surfaces with rustication on exterior surface of the body. Sample No. 29 having red slip on both the surfaces. Sample No. 55 having red slip on both the surfaces with black color band on lip of the rim. Sample No. 82 having red slip on both the surfaces. Sample No. 159 having red slip on both the surfaces with black color band on lip of the rim, burnishing on both the surface and rustication on exterior surface of the body. Sample No. 163 having red slip on both the surfaces with burnishing on both the surface and rustication on exterior surface on the body. Sample No. 197 having red slip on both the surfaces. Sample No. 200 and 205 having red slip on both the surfaces. Sample No. 18 and 200 having black color band on lip of the rim. Sample No. 28 and 232 having red slip on both the surfaces with rustication on exterior surface of the body.
Size range	Basin measures 24 to 38 cms in rim diameter.

Basins Type B Variety 1: thick sturdy basins having three rim variants:
(a) simple everted, straight rounded rim
(b) simple inverted, rounded rim
(c) external projecting, short, narrow triangular rim
Sample No. 107, 108 and 241

Fabric	Medium sandy with abundant mica and white inclusion.
Surface enhancement	Sample No. 107 and 108 unslipped on both the surfaces. Sample No. 241 having light red slip on the exterior surface.
Size range	Basin measures 28 to 32 cms in rim diameter.

Handi's having one variety and two rim variants:
(a) internal projecting, short, narrow triangular rim
(b) bilateral projecting, equal projection rim
Sample No. 75 and 234

Fabric	Medium sandy with abundant mica and white inclusion.
Surface enhancement	Sample No. 75 and 241 unslipped on both the surfaces.
Size range	Handi's measures 25 cm in rim diameter.

Dishes having one variety and three rim variants:
(a) simple external projecting, extremely everted tapered rim
(b) simple external projecting, extremely everted rounded rim
(c) external projecting, short, narrow triangular rim
Sample No. 59, 74 and 240

Fabric	Fine to Medium sandy with abundant mica and white inclusion.
Surface enhancement	Sample No. 59 having red slip on both the surfaces. Sample No. 74 having reddish yellow slip on the interior surface. Sample No. 240 having red slip on both the surfaces.
Size range	Dish measures 17 to 30 cms in rim diameter.

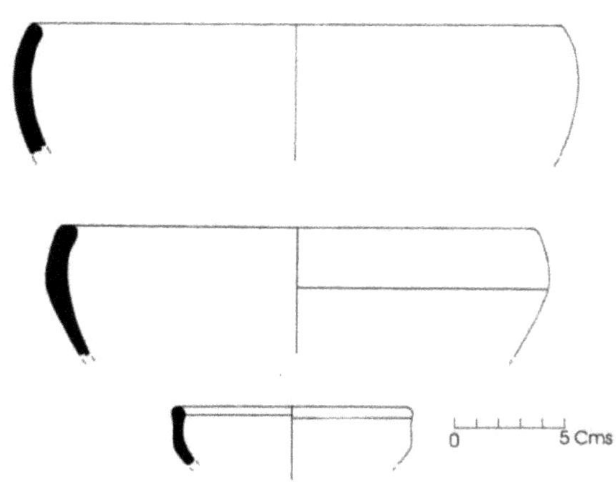

Figure 4.19 – Ganeshwar Bowl Type A (1)

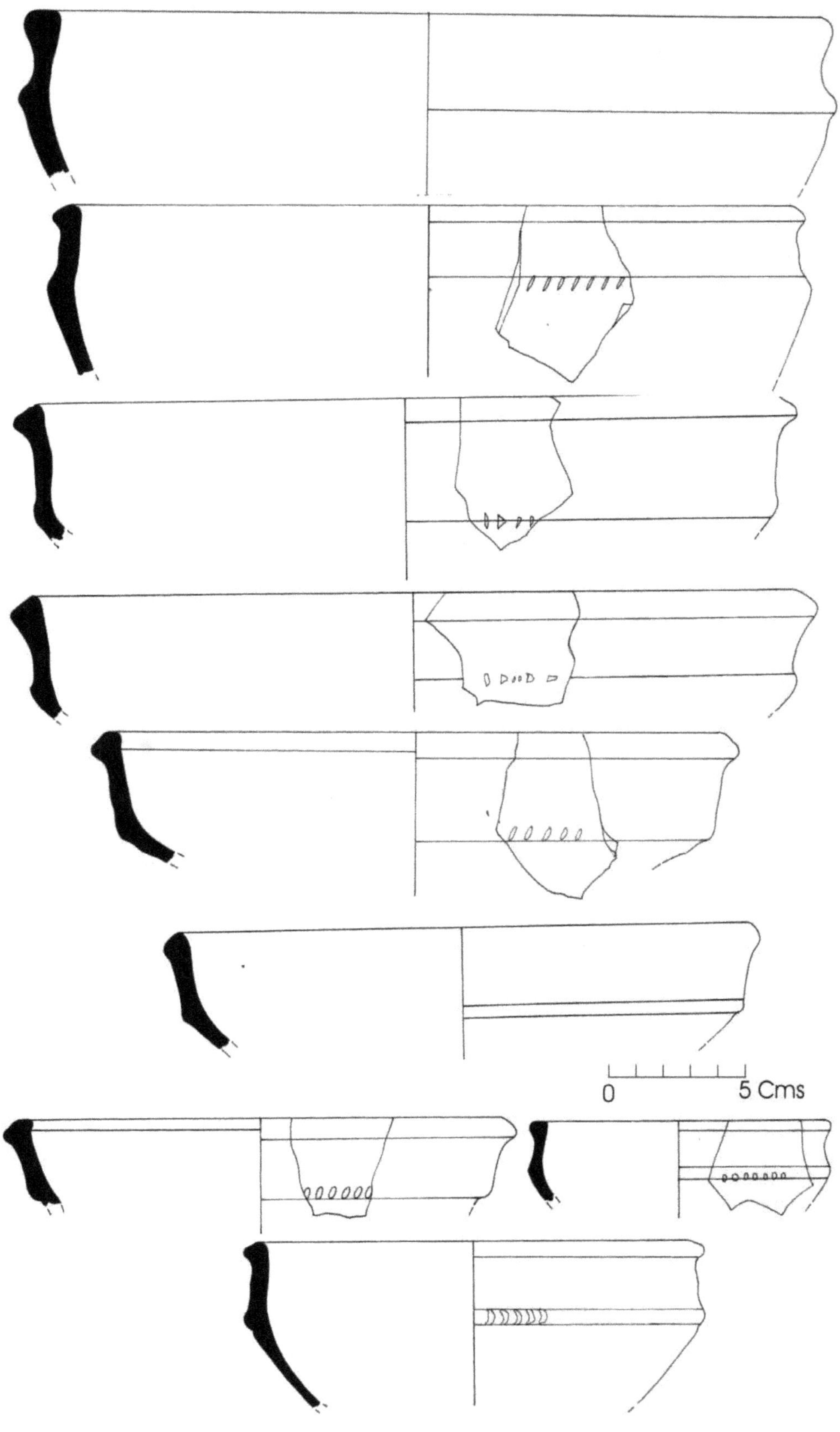

Figure 4.20 – Ganeshwar Bowl Type B (1)

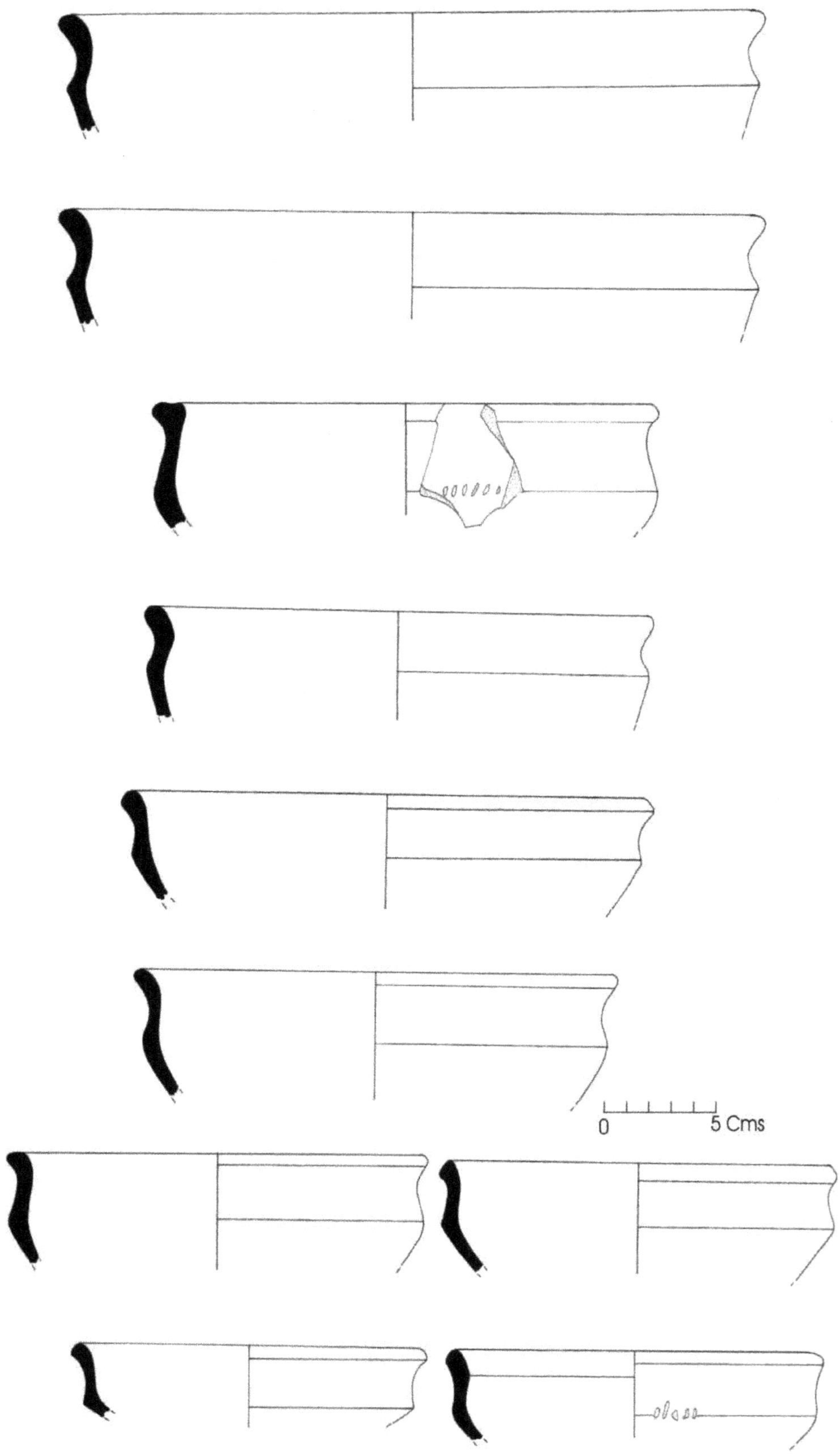

Figure 4.21 – Ganeshwar Bowl Type B (2)

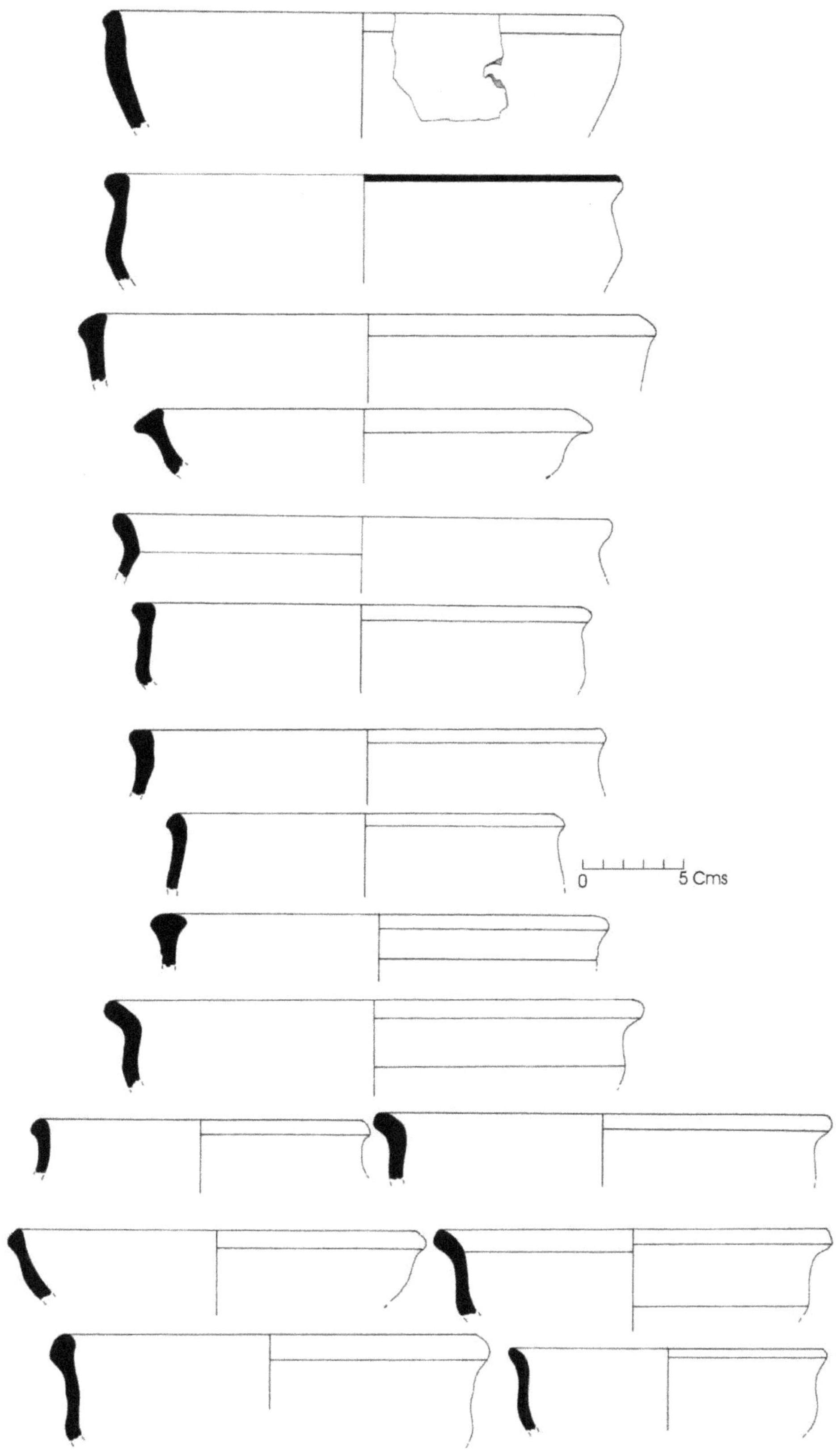

Figure 4.22 – Ganeshwar Bowl Type B (3)

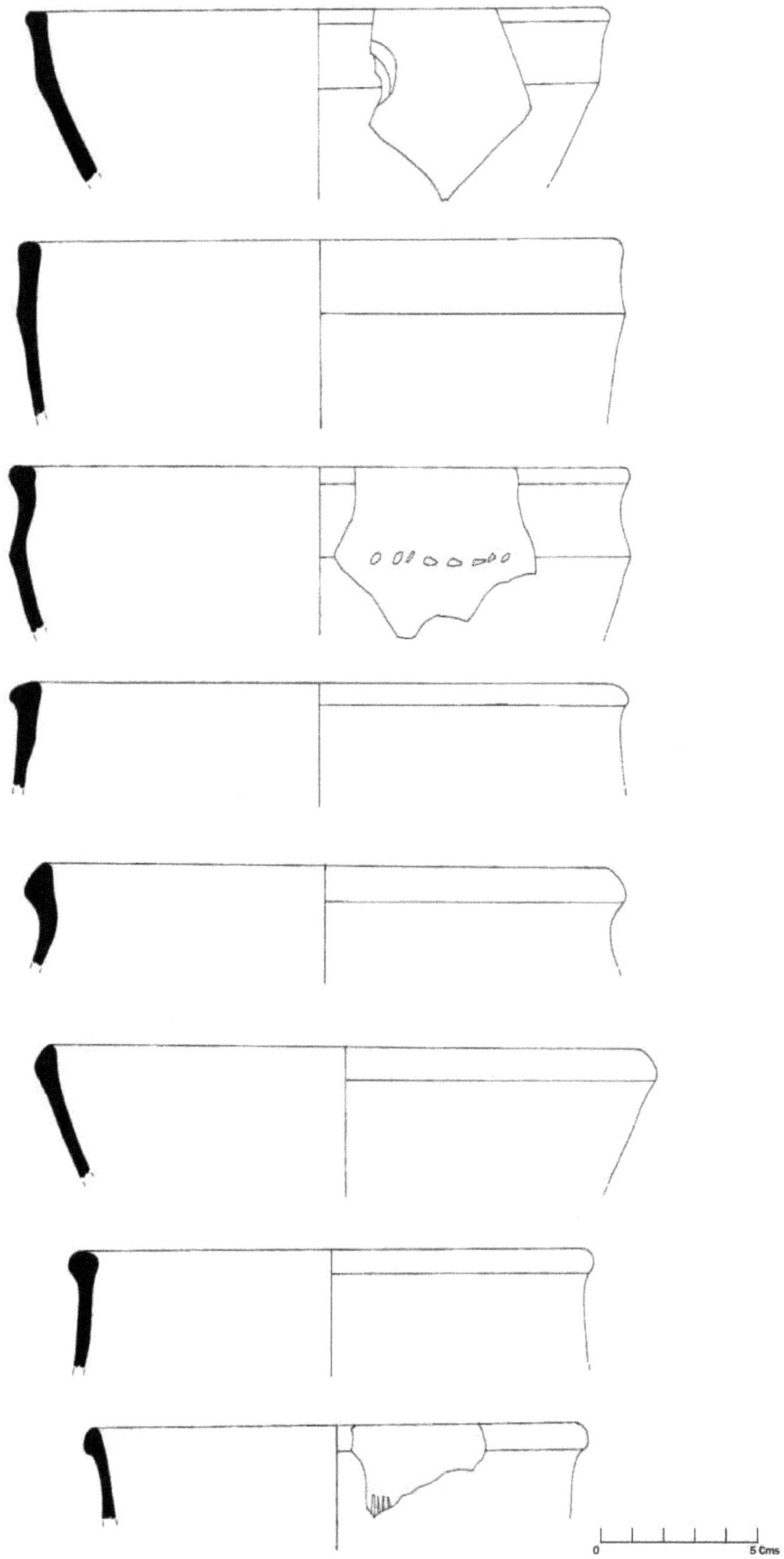

Figure 4.23 – Ganeshwar Bowl Type B (3)

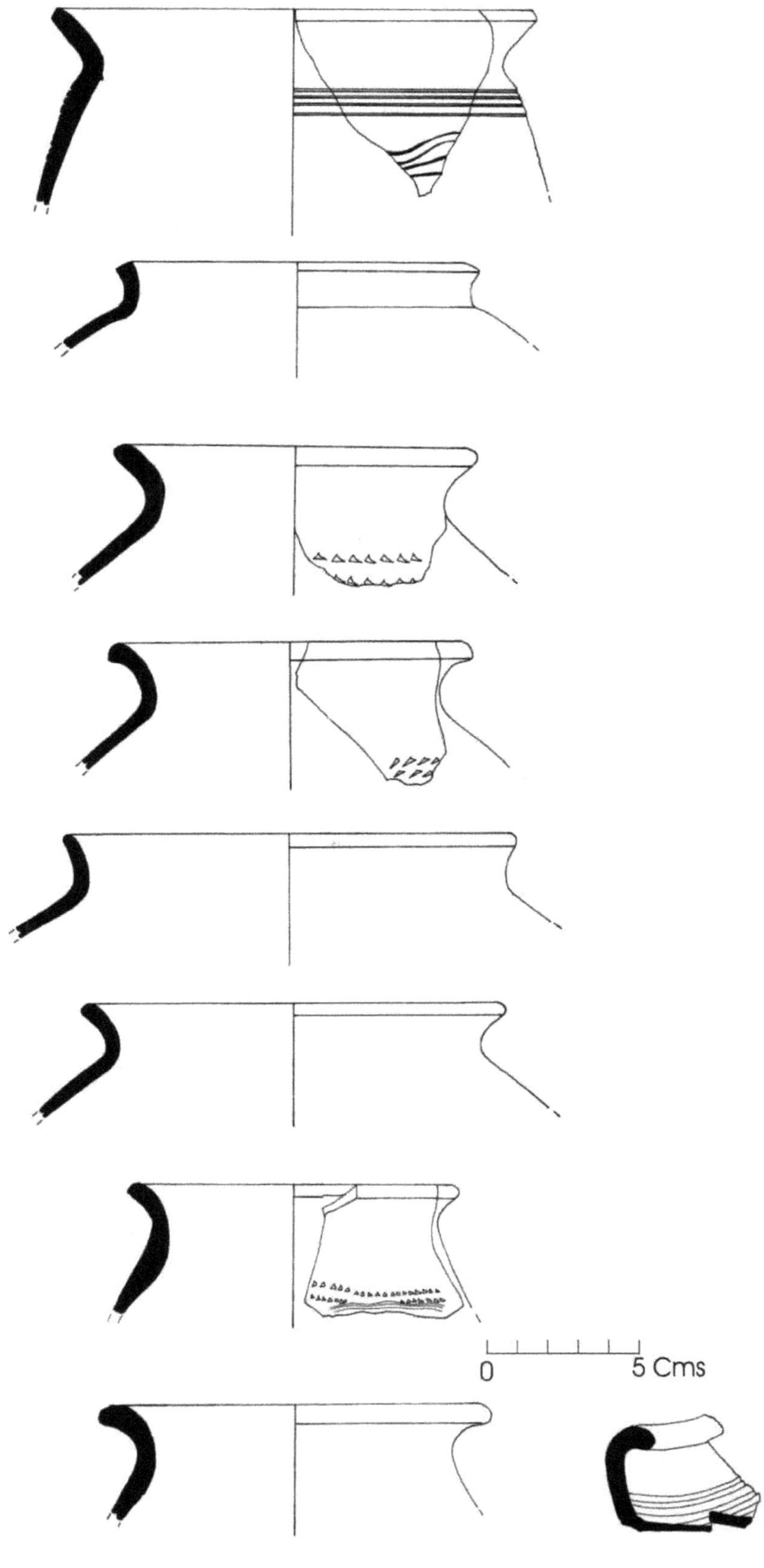

Figure 4.24 – Ganeshwar Pot Type A (1)

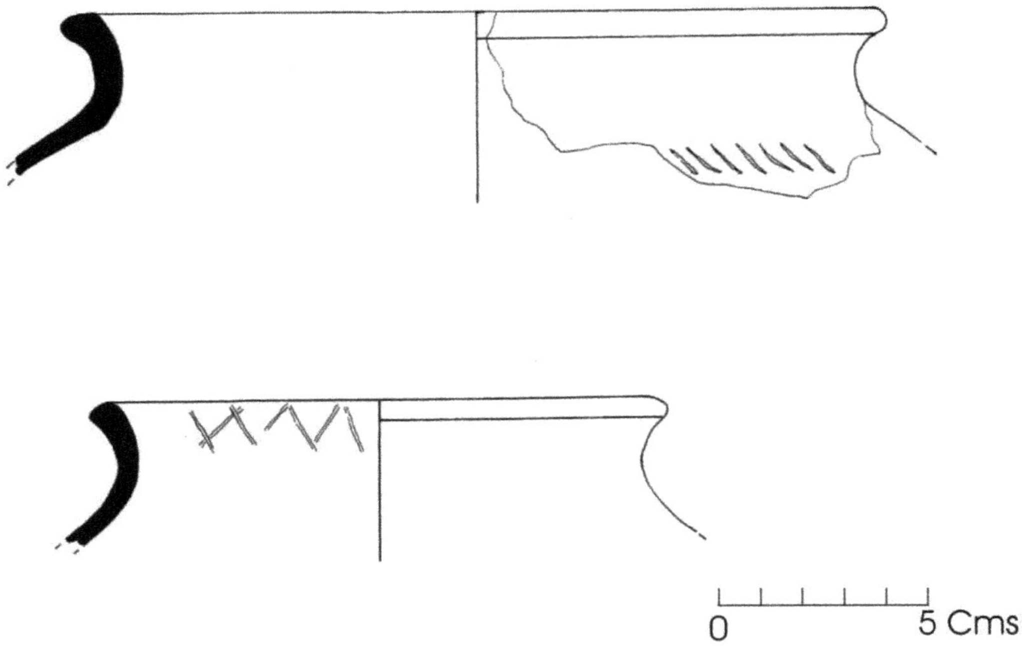

Figure 4.25 – Ganeshwar Pot Type A (2)

Figure 4.26 – Ganeshwar Pot Type A (3)

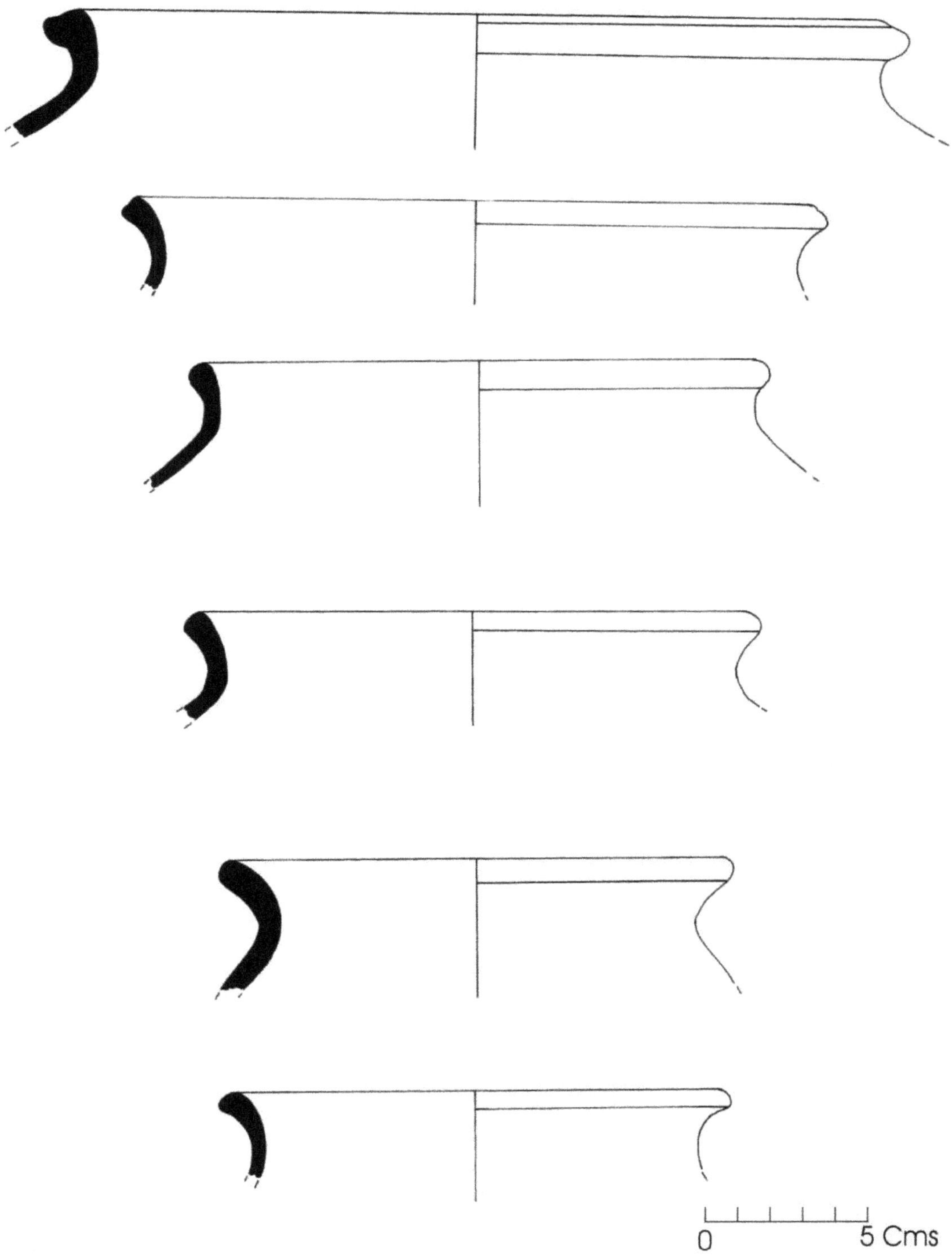

Figure 4.27 – Ganeshwar Pot Type B (1)

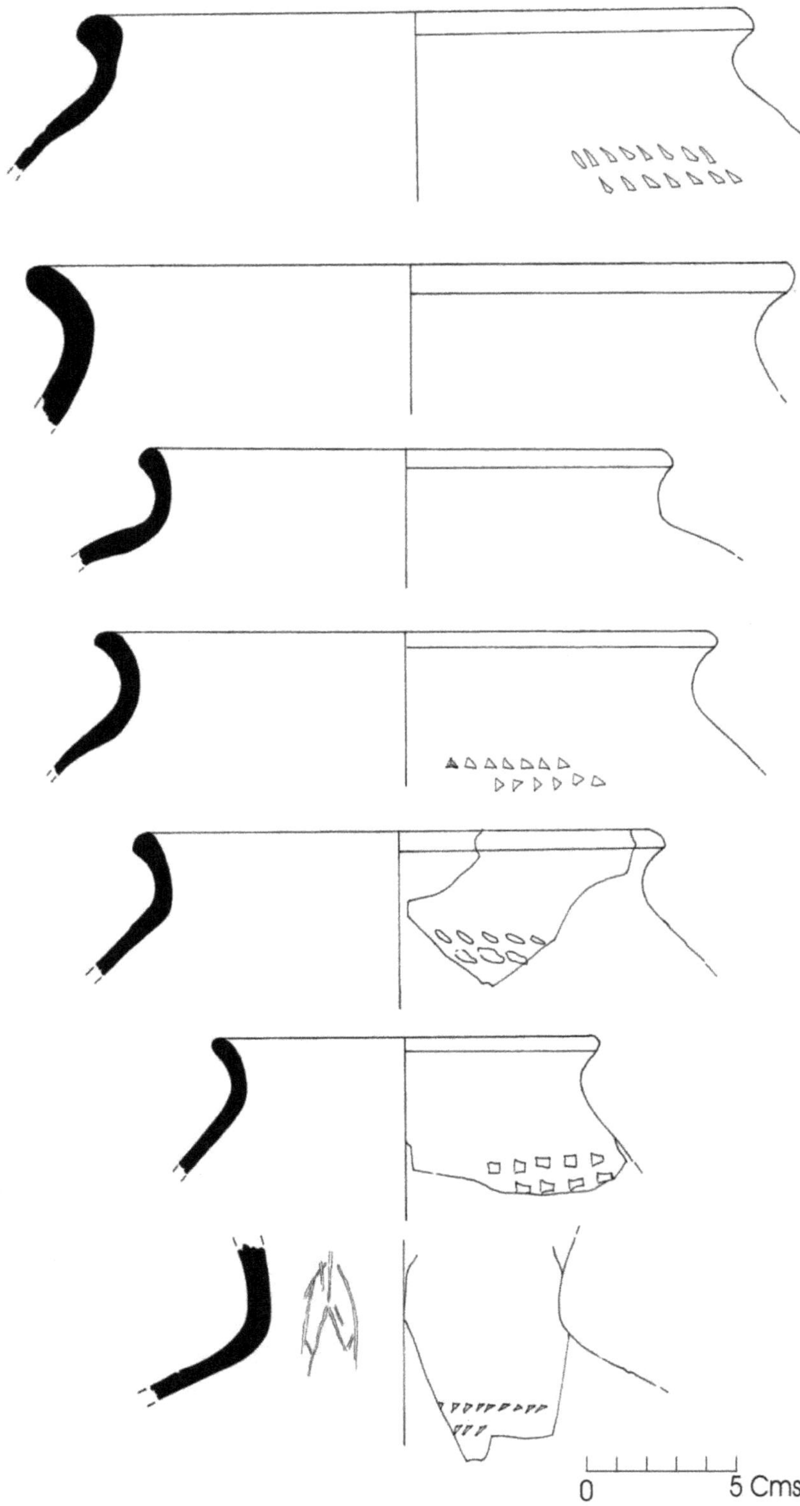

Figure 4.28 – Ganeshwar Pot Type B (2)

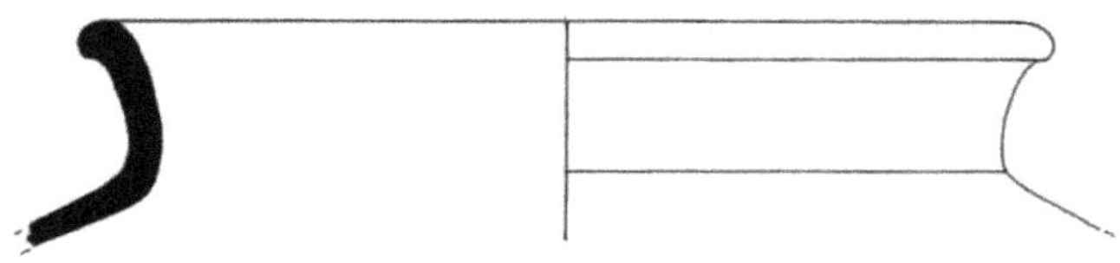

Figure 4.29 – Ganeshwar Pot Type B (3)

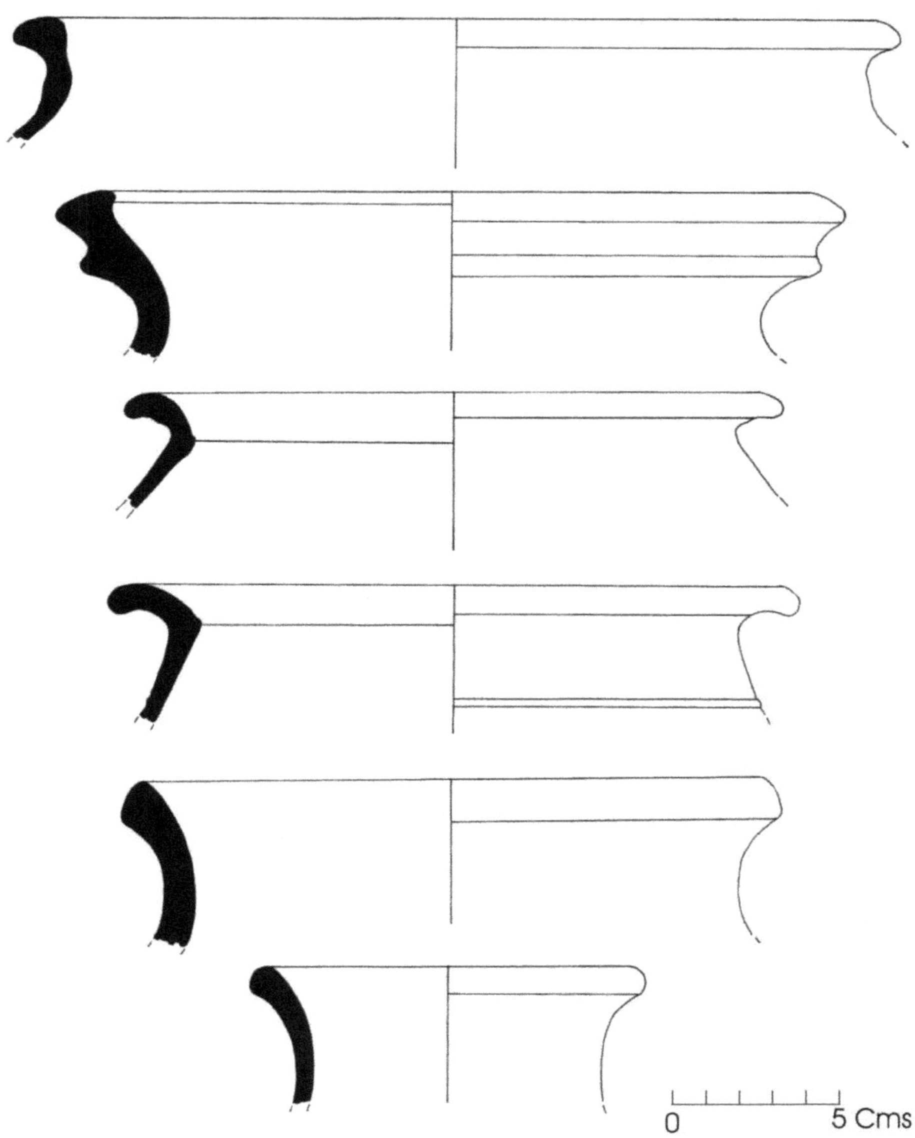

Figure 4.30 – Ganeshwar Pot Type B (4)

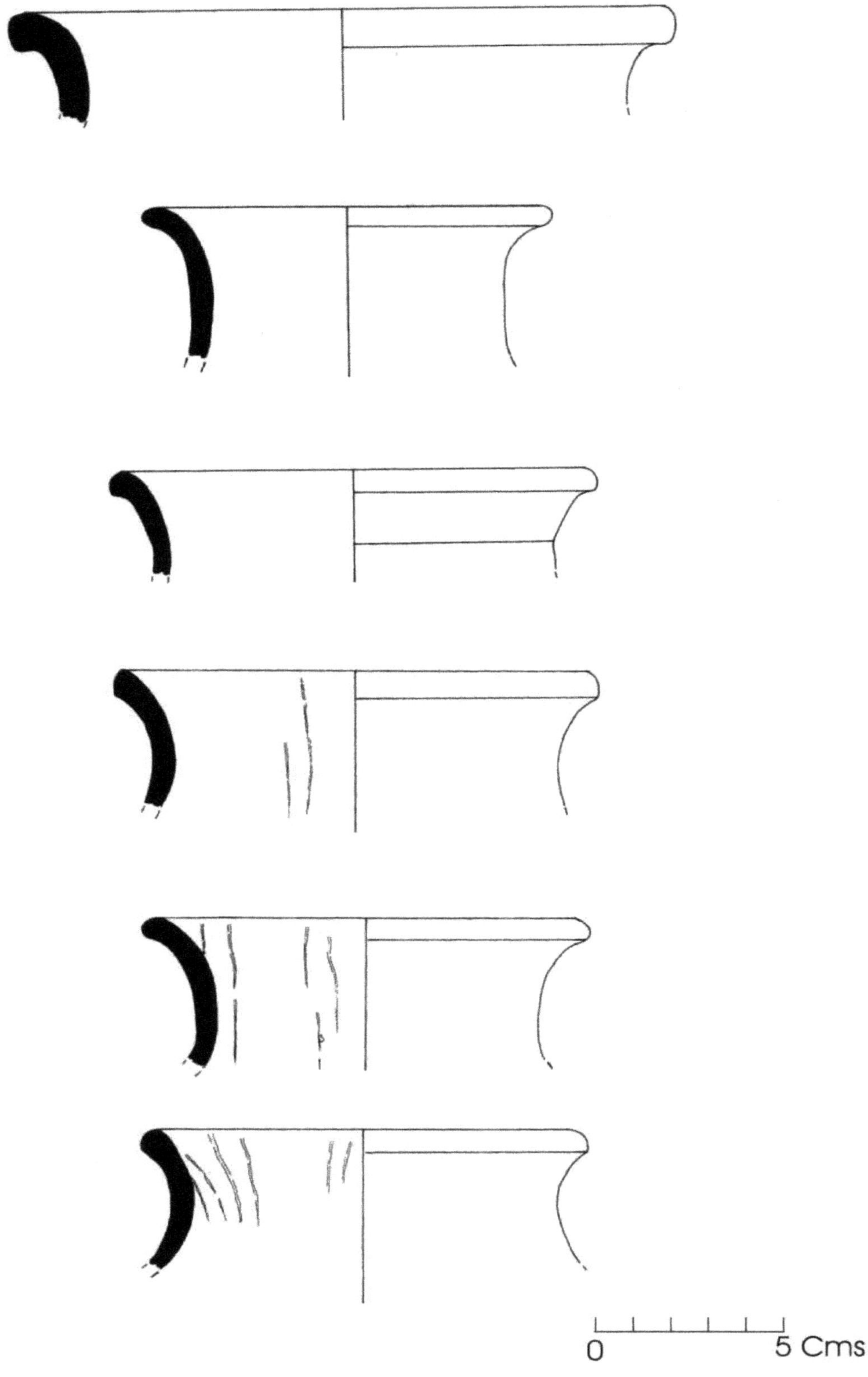

Figure 4.31 – Ganeshwar Jars

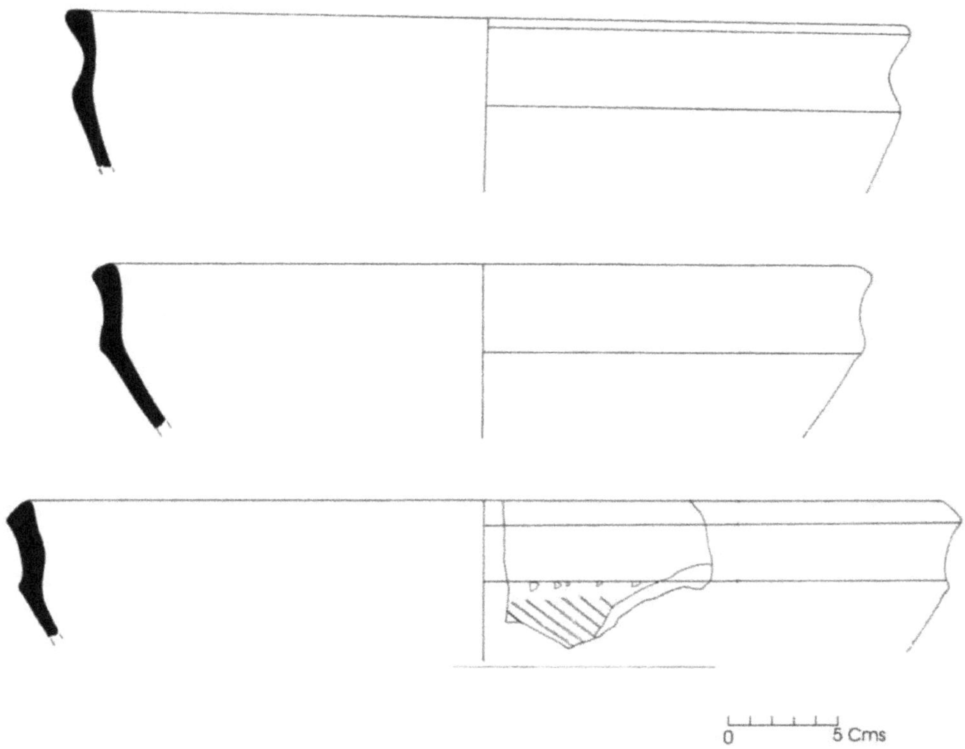

Figure 4.32 – Ganeshwar Basin Type A (1)

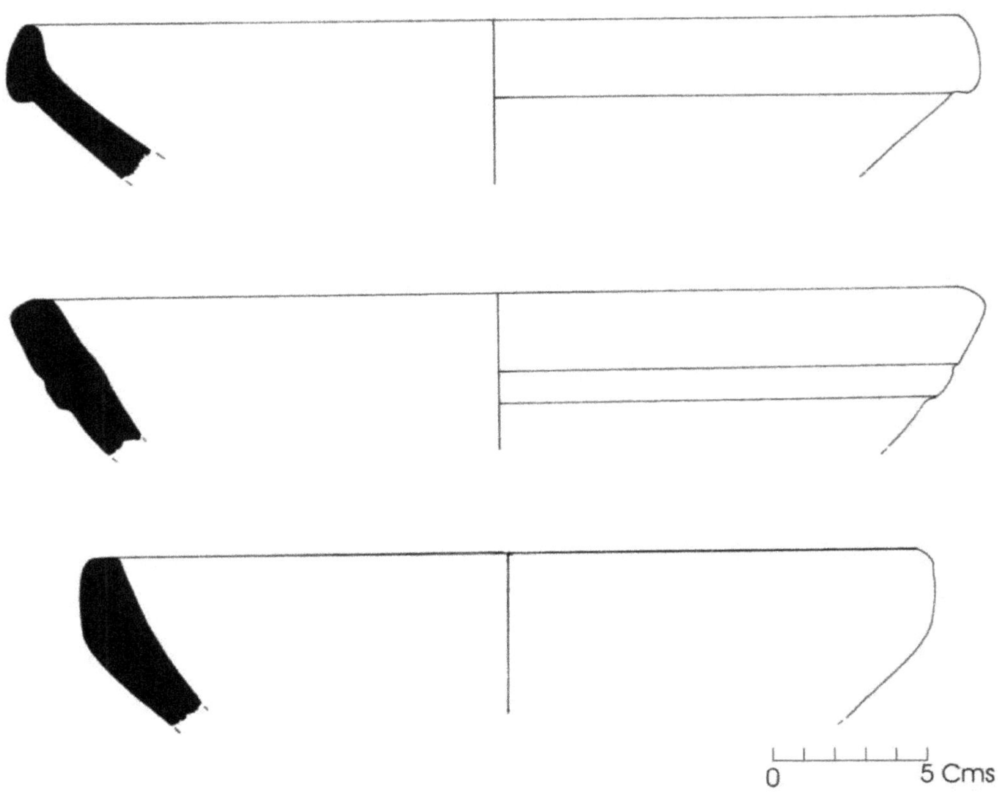

Figure 4.33 – Ganeshwar Basin Type B (1)

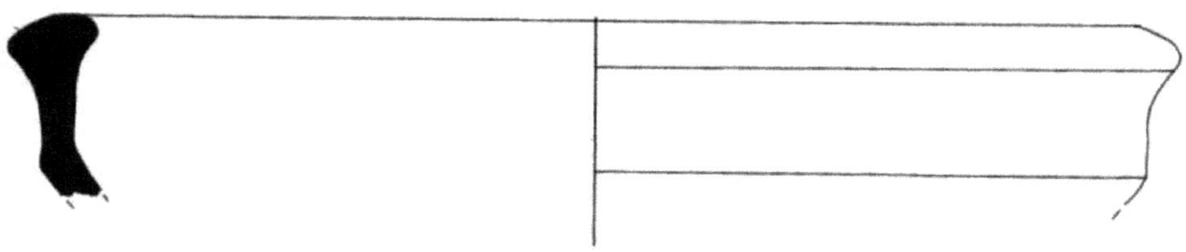

Figure 4.34 – Ganeshwar Handi

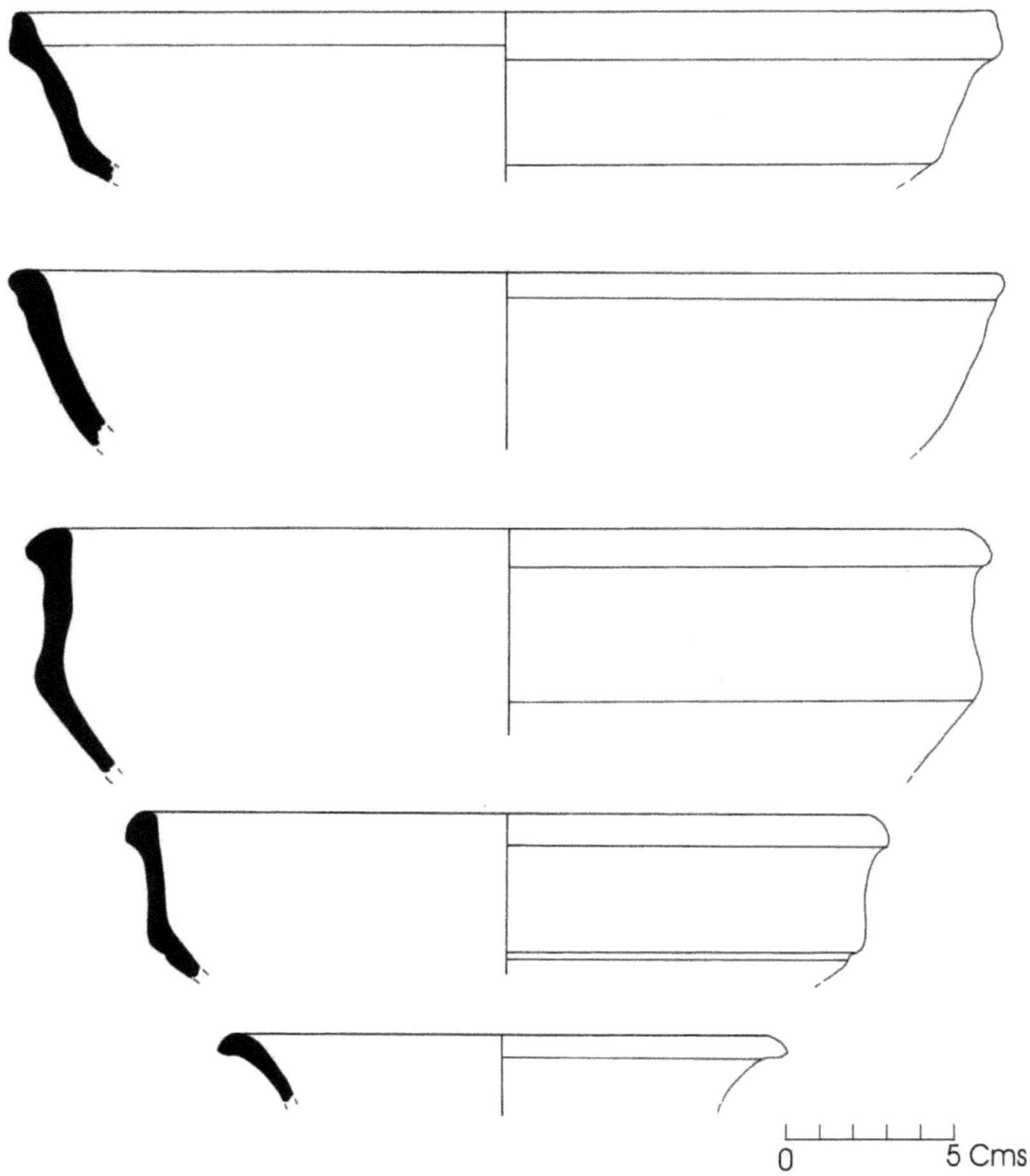

Figure 4.35 – Ganeshwar Dishes

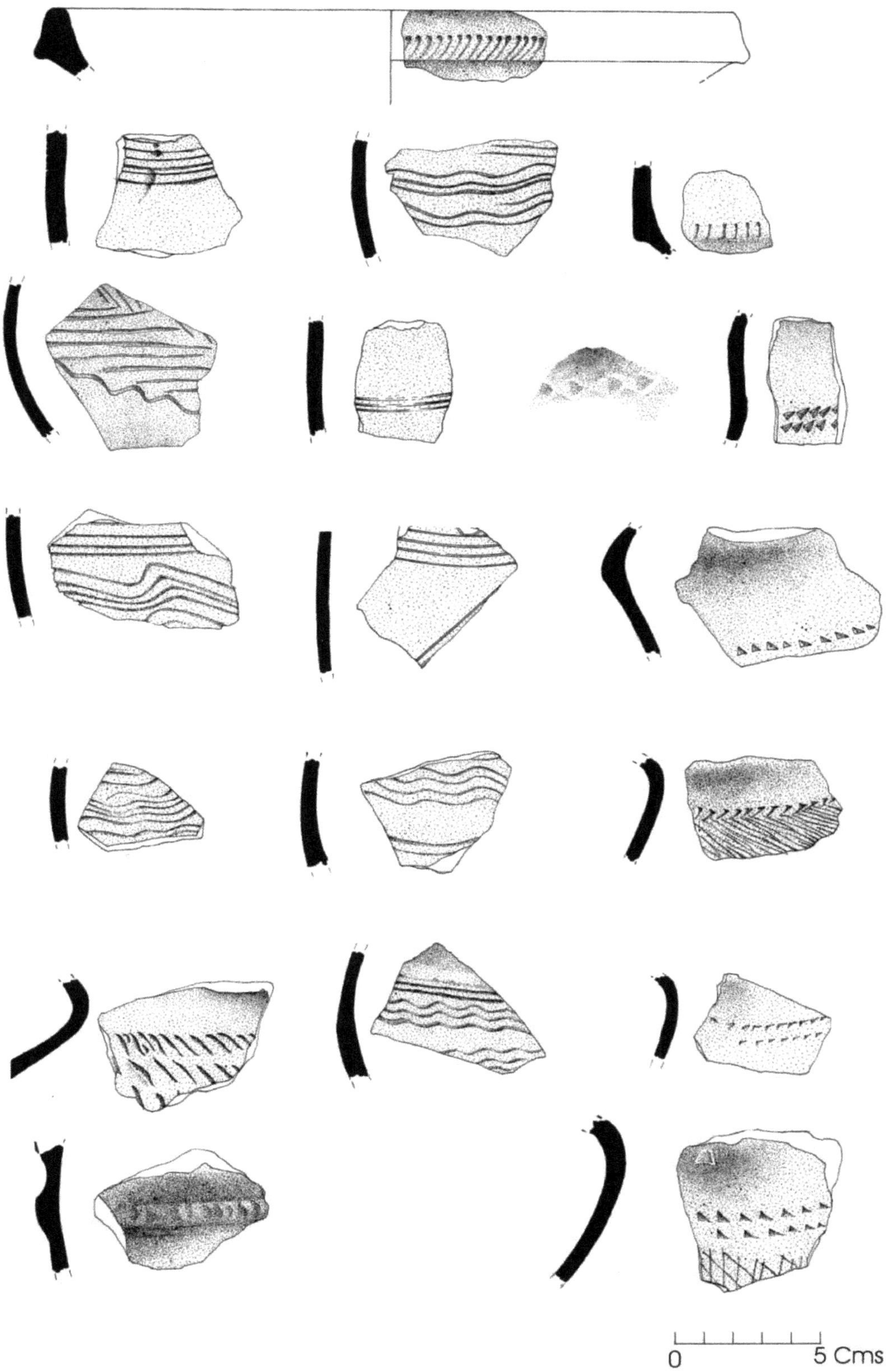

Figure 4.36 – Ganeshwar Incised Designs

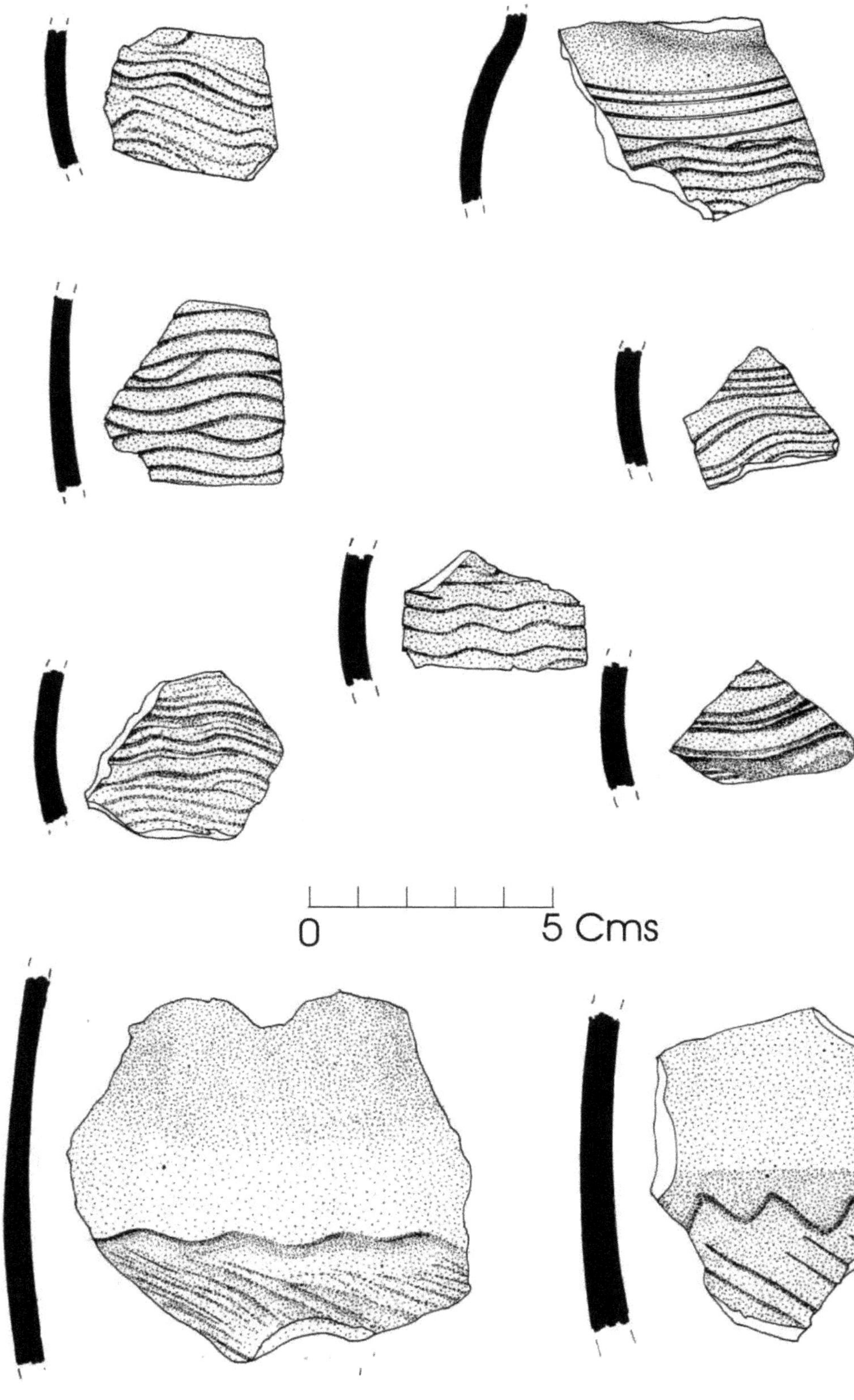

Figure 4.37 – Ganeshwar Incised Designs

2. BAVADI KI JHAMIN (BKJ) (Figure - 4.38)

Bowls Type A Variety 1: convex-sided bowls having one variety and one rim variant:
(a) simple inverted, curved tapered rim.
Sample No. 17

Fabric	Medium sandy with Tiny Mica and White inclusion.
Surface enhancement	Sample No. 17 with external ledge and corrugation on the Neck having reddish yellow slip on both the surfaces.
Size range	Bowl measures 22 cm in rim diameter.

Bowls Type B Variety 3: carinated bowls having one variety blunt carinated shoulder and three rim variants:
(a) external projecting, short, narrow triangular rim
(b) internal projecting, short, narrow triangular rim
(c) bilateral projecting, equal projection rim
Sample No. 18, 20 and 24

Fabric	Medium sandy with Tiny Mica and White inclusion.
Surface enhancement	Sample No. 18 having red slip on both the surfaces. Sample No. 20 having light red slip on both the surfaces. Sample No. 24 having red slip on both the surfaces with black color band on the interior surface.
Size range	Bowls measures 17 to 20 cms in rim diameter.

Bowls Type C Variety 1: straight-sided bowls having one variety and one rim variant:
(a) internal projecting, short, narrow triangular rim
Sample No. 19, 21, 22 and 23

Fabric	Medium sandy with Tiny Mica and White inclusion.
Surface enhancement	Sample No. 19 having reddish brown slip on both the surfaces. Sample No. 21 and 22 unslip on both the surfaces. Sample No. 23 having light red slip on both the surfaces.
Size range	Bowls measures 14 to 18 cms in rim diameter.

Pots Type A Variety 1: small size pots with rim diameter ranges between 12 to 16 cms having one variety with short neck and four rim variants:
a) external projecting, short, narrow, short quadrangular rim
b) external projecting, short, narrow rounded rim
c) external projecting, short, narrow tapered rim
d) external projecting, long, wide triangular rim
Sample No. 1, 2, 5, 8, 9, 11, 12, 13 and 14

Fabric	Medium to Coarse sandy with Tiny Mica and White inclusion.
Surface enhancement	Sample No. 1, 5, 9, 11 and 12 having red slip on both the surfaces. Sample No. 1 with black color band on Neck on the exterior surface. Sample No. 2 and 13 having light red slip on both the surfaces. Sample No. 8 and 14 unslip on both the surfaces.
Size range	Pots measures 12 to 16 cms in rim diameters.

Pots Type B Variety 1: medium size pots with rim diameter ranges between 18 to 20 cms having one variety with short neck and three rim variants:
a) external projecting, short, narrow rounded rim
b) external projecting, short, narrow triangular rim
c) external projecting, short, narrow tapered rim
Sample No. 3, 4, 6, 7 and 10

Fabric	Medium sandy with Tiny Mica and White inclusion.
Surface enhancement	Sample No. 3, 4, 6, 7 and 10 having red slip on both the surfaces.
Size range	Pots measures 18 to 20 cms in rim diameters.

Basins Type A Variety 2: carinated basins having two varieties, first variety with blunt carinated shoulder and one rim variant:
(a) internal projecting, short, narrow triangular rim
Sample No. 15 and 16

Fabric	Medium sandy with Tiny Mica and White inclusion.
Surface enhancement	Sample No. 15 having light red slip on both the surfaces. Sample No. 16 having red slip on both the surfaces.
Size range	Basins measures 25 to 26 cms in rim diameter.

Basins Type A Variety 3: carinated basins having two varieties, second variety with angular shoulder and one rim variant:
(a) internal projecting, short, narrow triangular rim
Sample No. 25

Fabric	Medium sandy with Mica and White inclusion.
Surface enhancement	Sample No. 25 having red slip on both the surfaces.
Size range	Basin measures 46 cm in rim diameter.

Dishes having one variety angular shoulder and two rim variants:
(a) simple everted slightly curved rounded rim
(b) simple everted straight rounded rim
Sample No. 26 and 27

Fabric	Medium sandy with Tiny Mica and White inclusion.
Surface enhancement	Sample No. 26 and 27 slip worn-out from both the surfaces.
Size range	Dishes measures 17 to 18 cms in rim diameter.

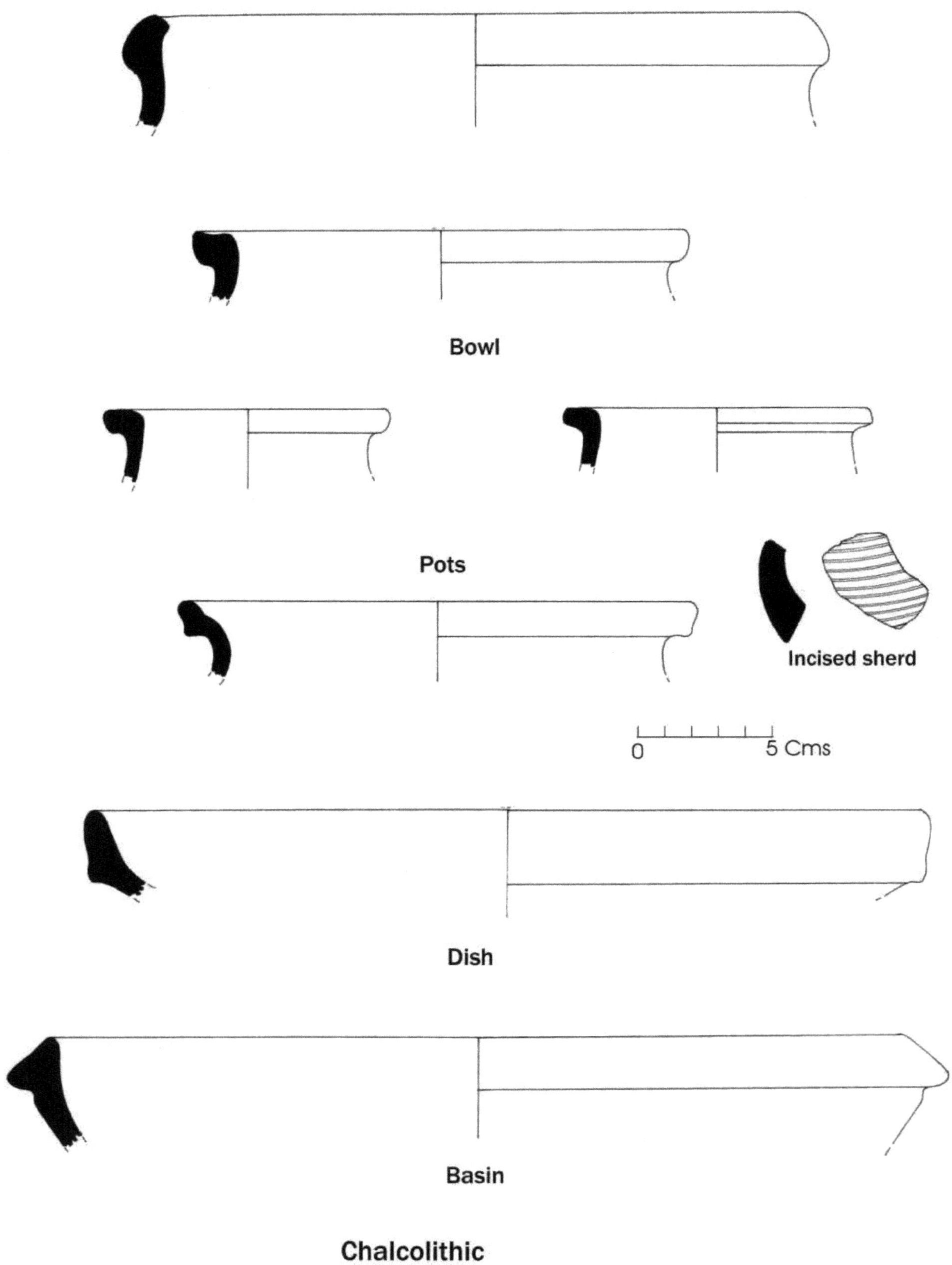

Figure 4.38 – Bavadi ki Jhamin

Undiagnostic Sherd: having vertical incised lines on the exterior surface
Sample No. 28

Fabric	Fine sandy.
Surface enhancement	Sample No. 28 having very dark gray slip on the exterior surface and unslip on the interior surface.

3. BEED KI DHANI (BKD) (Figure - 4.39)

Bowls Type C Variety 1: straight-sided bowls having one variety and one rim variant:
(a) simple everted, slightly curved rounded rim
Sample No. 6

Fabric	Medium sandy with Mica and White inclusion.
Surface enhancement	Sample No. 6 having Red Slip on both the surfaces
Size range	Bowl measures 15 cm in rim diameter.

Pots Type A Variety 1: small size pots with rim diameter ranges between 13 to 14 cms having two varieties, first variety with short neck and two rim variants:
(a) simple external projecting, extremely everted rounded rim
(b) external projecting, short, narrow triangular rim
Sample No. 4 and 5

Fabric	Medium to Coarse sandy with Tiny Mica and White inclusion
Surface enhancement	Sample No. 4 and 5 having Red Slip on both the surfaces
Size range	Pots measures 13 to 14 cms in rim diameters.

Pots Type A Variety 2: small size pots with rim diameter ranges between 13 to 14 cms having two varieties, second variety with long neck and one rim variant:
(a) bilateral projecting, equal projection rim
Sample No. 3

Fabric	Medium sandy with Tiny Mica, White inclusion and Organic matter
Surface enhancement	Sample No. 3 Slip Worn-Out from both the surfaces
Size range	Pot measures 14 cm in rim diameters.

Pots Type B Variety 1: medium size pots with rim diameter ranges between 18 to 20 cms having two varieties, first variety with short neck and one rim variant:
(a) external projecting, short, narrow, short quadrangular rim
Sample No. 2

Fabric	Medium sandy with Tiny Mica and White inclusion.
Surface enhancement	Sample No. 2 having red slip on both the surfaces.
Size range	Pot measures 18 cm in rim diameters.

Pots Type B Variety 3: medium size pots with rim diameter ranges between 18 to 20 cms having two varieties, second variety with long and corrugated neck having one rim variant:
(a) external projecting, long, wide quadrangular rim
Sample No. 1

Fabric	Medium sandy with Tiny Mica, White inclusion and Organic matter
Surface enhancement	Sample No. 1 having red slip on both the surfaces.
Size range	Pot measures 20 cm in rim diameters.

4. BHADWARI (BWR) (Figure - 4.40)

Bowls Type B Variety 3: carinated bowls having two varieties, first variety with blunt carinated shoulder and one rim variant:
(a) internal projecting, short, narrow triangular rim
Sample No. 8

Fabric	Medium sandy with White inclusion.
Surface enhancement	Sample No. 8 having Gray Slip on both the surfaces
Size range	Bowl measures 22 cm in rim diameter.

Bowls Type B Variety 4: carinated bowls having two varieties, second variety with angular shoulder and four rim variants:
(a) simple perpendicular, straight tapered rim
(b) external projecting, short, narrow rounded rim
(c) external projecting, long, wide triangular rim
(d) bilateral projecting, equal projection rim
Sample No. 4, 6, 7 and 9

Fabric	Medium sandy with Mica and White inclusion
Surface enhancement	Sample No. 4 and 6 having Red Slip on the both the surfaces Sample No. 7 having Gray Slip on both the surfaces Sample No. 9 unslip on both the surfaces
Size range	Bowls measures 8 to 19 cms in rim diameter.

Bowls Type C Variety 1: straight-sided bowl having one variety and one rim variant:
(a) external projecting, long, wide triangular rim
Sample No. 5

Fabric	Medium sandy with Mica and White inclusion.
Surface enhancement	Sample No. 5 having Red Slip on both the surfaces with incised horizontal lines on lip of the rim
Size range	Bowl measures 29 cm in rim diameter.

Pots Type A Variety 1: small size pots with rim diameter ranges between 15 to 21 cms having two varieties, first variety with short and ribbed neck and bulbous body and two rim variants:
(a) external projecting, short, narrow triangular rim
(b) bilateral projecting, equal projection rim

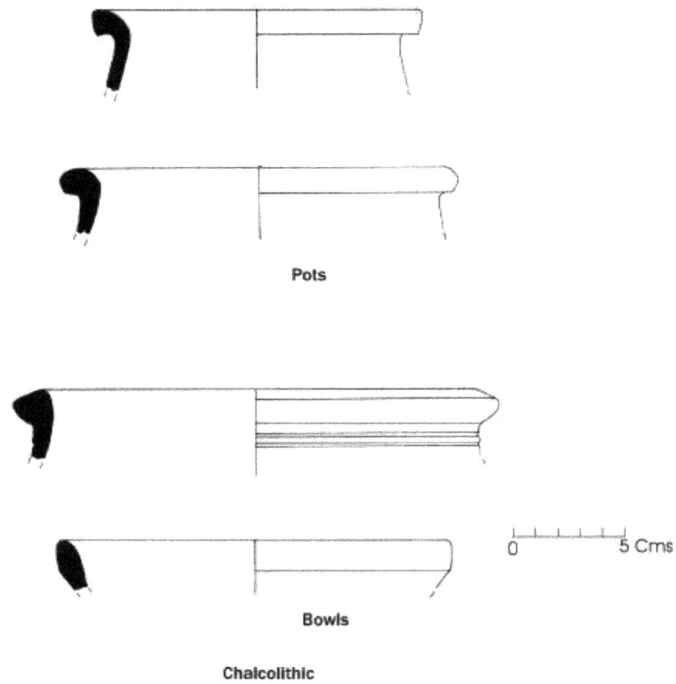

Figure 4.39 – Beed ki Dhani

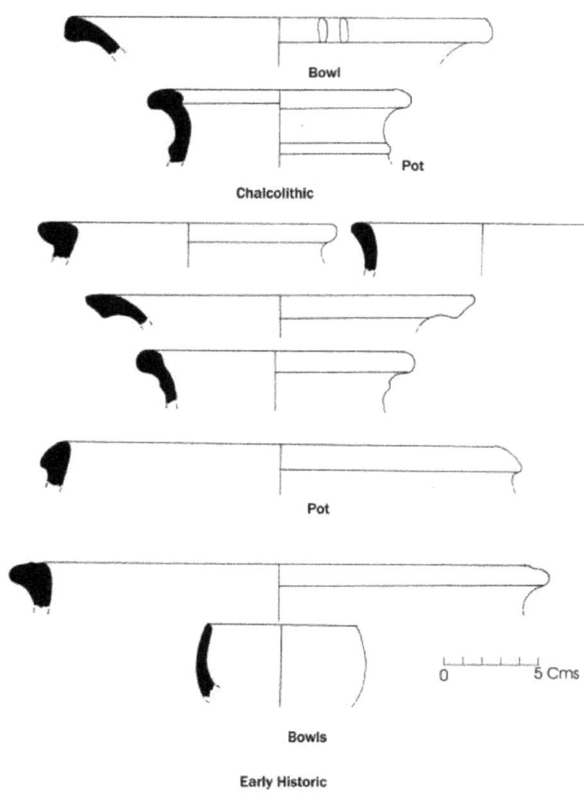

Figure 4.40 – Bhadwari

Sample No 2 and 3.

Fabric	Medium sandy with Mica, White inclusion and Organic matter
Surface enhancement	Sample No. 2 and 3 having Red Slip on both the surfaces
Size range	Pots measures 15 to 21 cms in rim diameters.

Pots Type A Variety 1: small size pots with rim diameter ranges between 15 to 21 cms having two varieties, first variety with short and ribbed neck and bulbous body and two rim variants:
(a) external projecting, short, narrow triangular rim
(b) bilateral projecting, equal projection rim
Sample No. 2 and 3

Fabric	Medium sandy with Mica, White inclusion and Organic matter
Surface enhancement	Sample No. 2 and 3 having Red Slip on both the surfaces
Size range	Pots measures 15 to 21 cms in rim diameters.

Pots Type A Variety 4: small size pots with rim diameter ranges between 15 to 21 cms having two varieties, second variety with carinated shoulder and one rim variant:
(a) external projecting, short, narrow, short beaked rim
Sample No. 1

Fabric	Medium sandy with Mica and White inclusion
Surface enhancement	Sample No. 1 having red Slip on both the surfaces
Size range	Pot measures 14 cm in rim diameters.

5. BHAGEGAWALA TILA (BWT) (Figure - 4.41)

Bowls Type A Variety 1: convex-sided bowls having one variety and three rim variants:
(a) simple perpendicular, straight rounded rim
(b) internal projecting, short, narrow triangular rim
(c) internal projecting, long, wide beaked rim
Sample No. 16, 17 and 19

Fabric	Medium sandy with Mica
Surface enhancement	Sample No. 16 and 19 having Red Slip on both the surfaces Sample No. 17 having Light red Slip on both the surfaces
Size range	Bowls measures 15 to 35 cms in rim diameter.

Bowls Type C Variety 1: straight-sided bowls having one variety and one rim variant:
(a) simple external projecting, extremely everted rounded rim
Sample No. 18

Fabric	Medium sandy with White inclusion.
Surface enhancement	Sample No. 18 having Red Slip on both the surfaces
Size range	Bowl measures 16 cm in rim diameter.

Pots Type A Variety 1: small size pots with rim diameter ranges between 9 to 22 cms having one variety with short neck and bulbous body and six rim variants.
(a) simple external projecting, extremely everted rounded rim
(b) external projecting, short, narrow, rounded rim
(c) external projecting, short, narrow, short beaked rim
(d) external projecting, short, narrow, short quadrangular rim
(e) external projecting, short, narrow tapered rim
(f) external projecting, long, wide quadrangular rim
Sample No. 1, 5, 6, 8, 11, 12, 13 and 15

Fabric	Medium sandy with Tiny Mica, White inclusion and Organic matter
Surface enhancement	Sample No. 1 Unslip on both the surfaces Sample No. 5, 6, 12 and 13 having Red Slip on both the surfaces Sample No. 8 having Red Slip on the exterior surface and Brown Slip on the interior surface Sample No. 11 having Light Red Slip on the interior surface and Red Slip on the exterior surface Sample No. 15 having Gray Slip on both the surfaces
Size range	Pots measures 9 to 22 cms in rim diameters.

Pots Type B Variety 1: medium size pots with rim diameter ranges between 17 to 35 cms having one variety with short and ribbed neck and bulbous body and four rim variants:

a. simple perpendicular, straight rounded rim
b. external projecting, short, narrow, short quadrangular rim
c. over turned small, external projecting tapered rim
d. internal projecting, short, narrow triangular rim

Sample No. 2, 3, 4, 10, 20 and 21

Fabric	Medium sandy with Tiny Mica, White inclusion and Organic matter
Surface enhancement	Sample No. 2, 3 and 20 having Red Slip on both the surfaces Sample No. 4 Unslip on both the surfaces Sample No. 10 having Reddish Yellow Slip on both the surfaces Sample No. 21 having Light Red Slip on both the surfaces
Size range	Pots measures 17 to 24 cms in rim diameters.

Basins Type A Variety 2: carinated basins having two varieties, first variety with blunt carinated shoulder and one rim variant:
(a) bilateral projecting, equal projection rim
Sample No. 26

Fabric	Medium sandy with Tiny Mica and White inclusion.
Surface enhancement	Sample No. 26 having Red Slip on both the surfaces

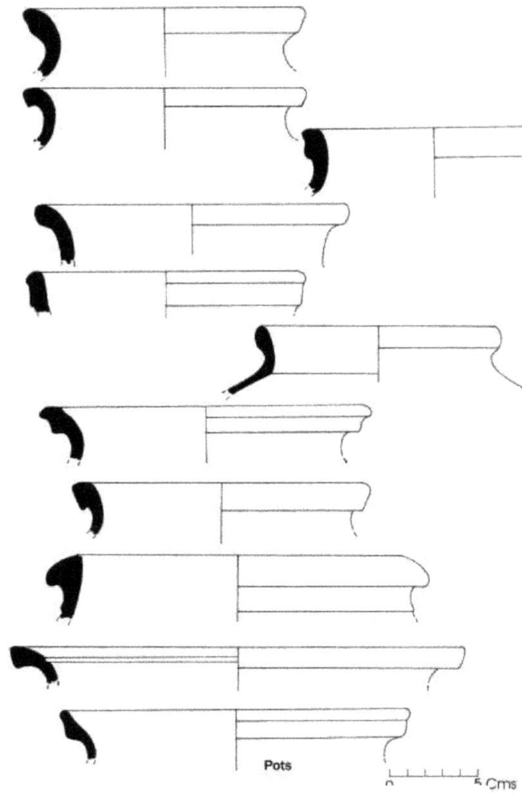

Figure 4.41(a) – Bhagegawala Tila

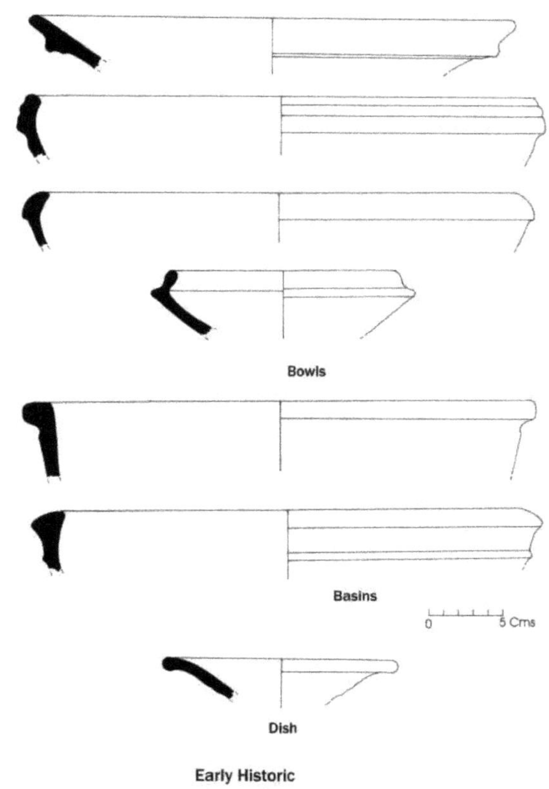

Figure 4.41(b) – Bhagegawala Tila

Fabric	Medium sandy with Tiny Mica and White inclusion.
Surface enhancement	Sample No. 26 having Red Slip on both the surfaces
Size range	Basin measures 35 cms in rim diameter.

Basins Type A Variety 2: carinated basins having two varieties, first variety with blunt carinated shoulder and one rim variant:
(a) bilateral projecting, equal projection rim
Sample No. 26

Fabric	Medium sandy with Tiny Mica and White inclusion.
Surface enhancement	Sample No. 26 having Red Slip on both the surfaces
Size range	Basin measures 35 cms in rim diameter.

Basins Type A Variety 3: carinated basins having two varieties, second variety with angular shoulder and three rim variants:
(a) external projecting, short, narrow, short quadrangular rim
(b) internal projecting, short, narrow triangular rim
(c) bilateral projecting, equal projection rim
Sample No. 22, 23, 27 and 29

Fabric	Medium sandy with Tiny Mica and White inclusion.
Surface enhancement	Sample No. 22, 23 and 29 having Red Slip on both the surfaces Sample No. 27 having Light Red Slip on both the surfaces
Size range	Basins measures 24 to 35 cms in rim diameter.

Dishes having one variety with angular shoulder and one rim variant:
(a) simple everted slightly curved rounded rim
sample no. 28

Fabric	Medium sandy with White inclusion.
Surface enhancement	Sample No. 28 unslip on both the surfaces
Size range	Dish measures 16 cm in rim diameter.

6. BIHARIPUR (Figure – 4.42)

Bowls Type A Variety 1: convex-sided bowls having one variety with corrugated neck and four rim variants:
(a) simple perpendicular, straight tapered rim
(b) simple perpendicular, straight squared rim
(c) simple inverted, curved tapered rim
(d) simple internal projecting, extremely inverted rounded rim
Sample No. 8, 9 10, 13 and 15

Fabric	Medium sandy with Mica and White inclusion.
Surface enhancement	Sample No. 8, 13 and 15 having red slip on both the surfaces. Sample No. 9 and 10 having light red slip on both the surfaces.
Size range	Bowl measures 14 to 32 cms in rim diameter.

Bowls Type B Variety 2: carinated bowls having two varieties, first variety with sharp carinated shoulder and two rim variants:
(a) external projecting, long, wide rounded rim
(b) overturned small, external projecting, triangular rim
Sample No. 6 and 14

Fabric	Medium sandy with Mica and White inclusion.
Surface enhancement	Sample No. 6 and 14 having red slip on both the surfaces.
Size range	Bowl measures 14 to 16 cms in rim diameter.

Bowls Type B Variety 3: carinated bowls having two varieties, second variety with blunt carinated shoulder and two rim variants:
(a) simple inverted, curved tapered rim
(b) simple external projecting, extremely everted rim
Sample No. 11 and 12

Fabric	Medium sandy with Mica and White inclusion.
Surface enhancement	Sample No. 11 having dark gray slip on both the surfaces. Sample No. 12 having red slip on both the surfaces.
Size range	Bowl measures 14 to 20 cms in rim diameter.

Pots Type A Variety 2: small size pots with rim diameter 13 cm having one variety with long neck and bulbous body and two rim variants:
(a) external projecting, short, narrow rounded rim
(b) external projecting, long, wide rounded rim
Sample No. 2 and 4

Fabric	Medium sandy with Mica and White inclusion.
Surface enhancement	Sample No. 2 and 4 having light red slip on both the surfaces
Size range	Pot measures 13 cm in rim diameters.

Pots Type B Variety 1: medium size pots with rim diameter ranges between 16 to 22 cms having three varieties, first variety with short neck and bulbous body and one rim variant:
(a) external projecting, short, narrow, short triangular rim
Sample No. 5 and 7

Fabric	Medium to Coarse sandy with Mica and White inclusion.
Surface enhancement	Sample No. 5 Having red slip on both the surfaces. Sample No. 7 having dark brown slip on both the surfaces. Nail-headed incised notches on neck with black color band on neck on the interior surface.
Size range	Pot measures 16 to 22 cms in rim diameters.

Pots Type B Variety 3: medium size pots with rim diameter ranges between 16 to 22 cms having three

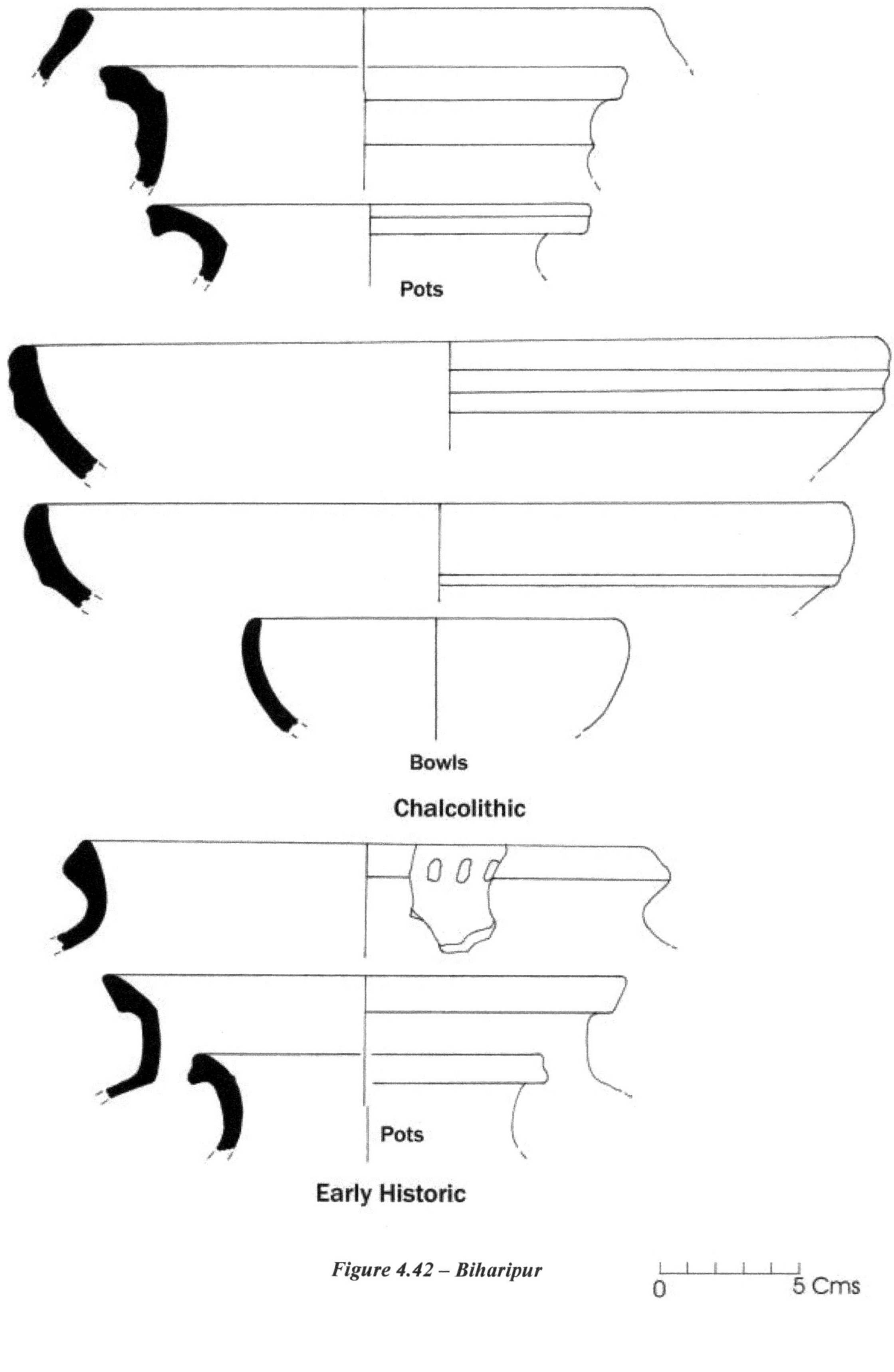

Figure 4.42 – Biharipur

varieties, second varierty with long neck and bulbous body and one rim variant:
(a) external projecting, short, narrow, short triangular rim
Sample No. 1

Fabric	Medium sandy with White inclusion.
Surface enhancement	Sample No. 1 having reddish yellow slip on both the surfaces.
Size range	Pot measures 19 cm in rim diameters.

Pots Type B Variety 4: medium size pots with rim diameter ranges between 16 to 22 cms having three varieties, third variety with carinated shoulder and corrugated neck and one rim variant:
(a) external projecting, short, narrow, short quadrangular rim
Sample No. 3

Fabric	Medium sandy with White inclusion.
Surface enhancement	Sample No. 3 having red slip on both the surfaces.
Size range	Pot measures 19 cm in rim diameters.

Undiagnostic Sherd: carinated sherd with incised horizontal lines on neck and rustication on the body with oblique applied bands of sandy texture
Sample No. 16

Fabric	Medium sandy with Mica, White inclusion and organic matter.
Surface enhancement	Sample No. 16 having red slip on the exterior surface.

Painted Sherds: having black color horizontal bands on the exterior surface
Sample No. 18, 19 and 20

Fabric	Medium sandy with Mica and White inclusion.
Surface enhancement	Sample No. 18 and 20 having red slip on the exterior surface. Sample No. 19 having reddish brown slip on the exterior surface.

7. GADHKA (MAHAWA) (Figure - 4.43)

Bowls Type A Variety 1: convex-sided bowls having one variety and one rim variant:
(a) simple inverted, curved tapered rim
Sample No. 10 and 11

Fabric	Coarse sandy with Tiny Mica and White inclusion.
Surface enhancement	Sample No. 10 and 11 Unslip on both the surfaces
Size range	Bowls measures 8 cm in rim diameter.

Bowls Type B Variety 3: carinated bowls having one variety with blunt carinated shoulder and one rim variant:
(a) bilateral projecting, equal projection rim
Sample No. 5

Fabric	Medium sandy with Mica and White inclusion.
Surface enhancement	Sample No. 5 having Red Slip on both the surfaces
Size range	Bowl measures 27 cm in rim diameter

Bowls Type C Variety 1: straight-sided bowls having one variety straight-sided bowls and one rim variant:
(a) internal projecting, short, narrow rounded rim
Sample No. 8 and 9

Fabric	Medium sandy with Tiny Mica, White inclusion and organic matter
Surface enhancement	Sample No. 8 having Light Red Slip on the interior surface and Unslip on the exterior surface Sample No. 9 having Red Slip on both the surfaces
Size range	Bowls measures 20 cms in rim diameter

Pots Type A Variety 1: small size pots with rim diameter ranges between 13 to 21 cms having two varieties, first variety with short neck and one rim variant:
(a) external channel on the top of the rim
Sample No. 7

Fabric	Medium sandy with Mica and White inclusion.
Surface enhancement	Sample No. 7 having Pale Red Slip on both the surfaces
Size range	Pot measures 21 cm in rim diameters.

Pots Type A Variety 4: small size pots with rim diameter ranges between 13 to 21 cms having two varieties, second variety with carinated shoulder and one rim variant
(a) external projecting, long, wide rounded rim
Sample No. 1

Fabric	Coarse sandy with Mica, White inclusion and Organic matter
Surface enhancement	Sample No. 1 Unslip on both the surfaces
Size range	Pot measures 13 cm in rim diameters.

Pots Type B Variety 1: medium size pots with rim diameter ranges between 18 to 22 cms having one variety short and ribbed neck and bulbous body and four rim variants:
(a) simple external projecting, extremely everted rounded rim
(b) external projecting, short, narrow, short quadrangular rim
(c) external projecting, short, narrow tapered rim
(d) bilateral projecting, equal projection rim
Sample No. 2, 3, 4 and 6

Fabric	Medium sandy with mica, white inclusion and organic matter.
Surface enhancement	Sample No. 2 having light red slip on both the surfaces Sample No. 3 unslip on both the surfaces Sample No. 4 having light red slip on the exterior surface and unslip on the

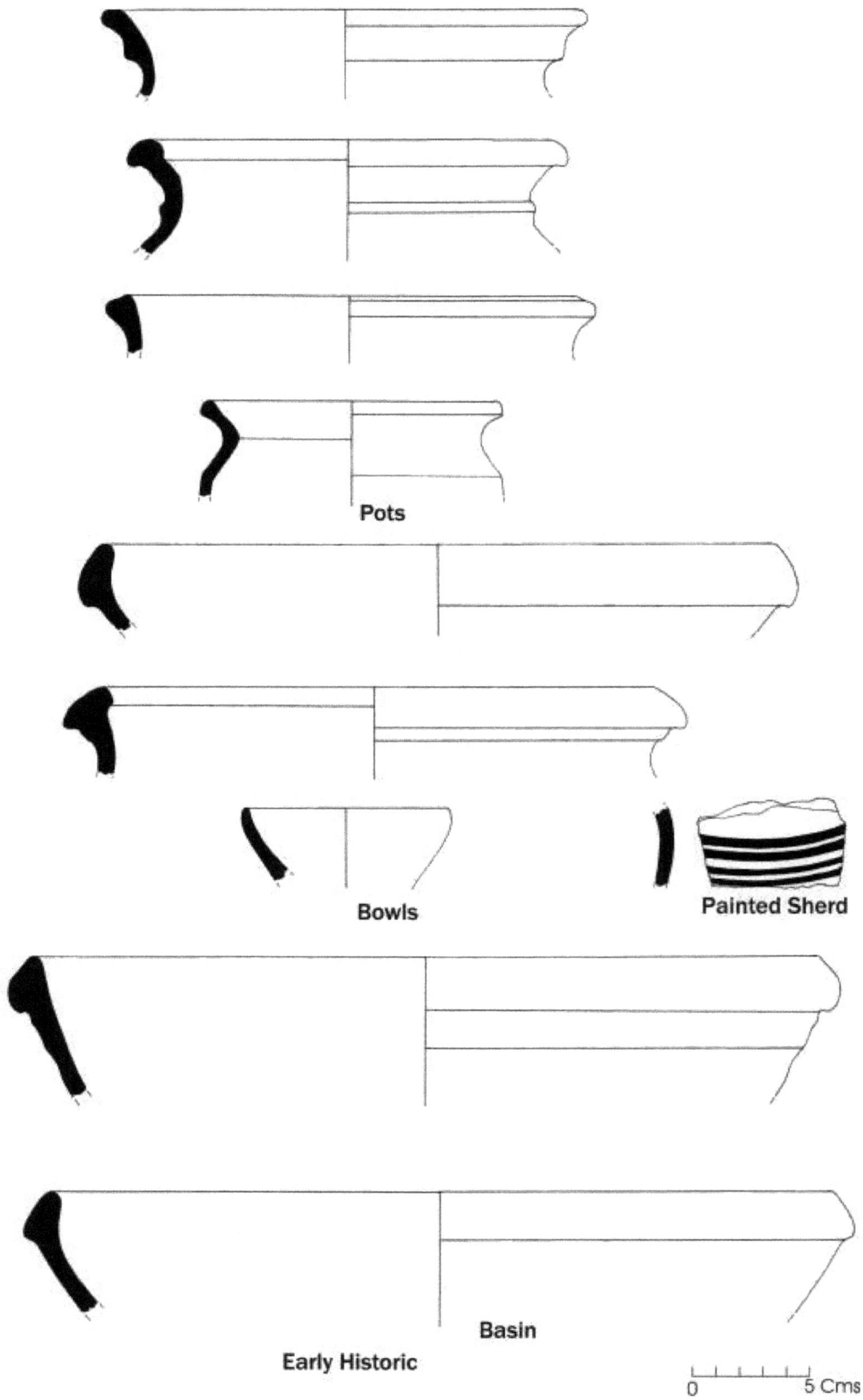

Figure 4.43 – Gadhka

	interior surface
	Sample No. 6 having reddish brown slip on both the surfaces
Size range	Pot measures 18 to 22 cms in rim diameters.

Basins Type A Variety 3: carinated basins having one variety with angular shoulder and three rim variants:
(a) over turned small external projecting rim
(b) internal projecting, short, narrow rounded rim
(c) bilateral projecting, equal projection rim
Sample No. 12, 13, 14 and 15

Fabric	Medium sandy with Tiny Mica and White inclusion.
Surface enhancement	Sample No. 12, 13 and 14 having Light Red Slip on both the surfaces Sample No. 15 having Red Slip on both the surfaces
Size range	Basins measures 20 to 36 cms in rim diameter.

Undiagnostic Painted Sherds:
Sample No. 16, 17 and 18

Fabric	Medium sandy with Mica and White inclusion.
Surface enhancement	Sample No. 16 and 17 having Pale Red Slip on the exterior surface and Unslip on the interior surface with black color band over pale red surface Sample No. 18 having red Slip on the exterior surface and Unslip on the interior surface with black color band over red surface

8. GIDHALI (Figure - 4.44)

Bowls Type B Variety 2: carinated bowls having two varieties, first variety with sharp carinated shoulder and one rim variant:
(a) external projecting, long, wide triangular rim
Sample No. 5

Fabric	Medium sandy with Mica and White inclusion.
Surface enhancement	Sample No. 5 having red slip on both the surfaces.
Size range	Unmeasured

Bowls Type B Variety 3: carinated bowls having two varieties, second variety with blunt carinated shoulder and one rim variant:
(a) simple external projecting, extremely everted rounded rim
Sample No. 3

Fabric	Medium sandy with Mica and White inclusion.
Surface enhancement	Sample No.3 having red slip on the exterior surface.
Size range	Bowl measures 22 cm in rim diameter.

Type C Variety 1: straight-sided bowls having one variety and two rim variants:
(a) external projecting, long, wide triangular rim
(b) bilateral projecting, equal projection rim
Sample No. 4, 6 and 7

Fabric	Medium sandy with Mica and White inclusion.
Surface enhancement	Sample No. 4, 6 and 7 having Red Slip on both the surfaces.
Size range	Bowl measures 24 cm in rim diameter.

Pots Type B Variety 1: medium size pots with rim diameter ranges between 18 to 22 cms having one variety with short neck and bulbous body and two rim variants:
(a) simple external projecting, extremely everted rounded rim
(b) external projecting, short, narrow triangular rim
Sample No. 1 and 2

Fabric	Medium sandy with Mica and White inclusion.
Surface enhancement	Sample No. 1 and 2 unslip on both the surfaces.
Size range	Pot measures 18 to 22 cms in rim diameters.

Undiagnostic Sherds: having horizontal and wavy applied bands of sandy texture on the body
Sample No. 10, 12 and 14

Fabric	Medium sandy with Mica
Surface enhancement	Sample No. 10, Sherd having horizontal lines and rustication on the exterior surface of the body. Sample No. 12, Sherd having rustication on the exterior surface of the body with appliqué bands of sandy texture. Sample No. 14, Sherd having rustication on the exterior surface of the body with horizontal appliqué bands of sandy texture.

Undiagnostic Sherd: having external ledge on the body and vertical applied bands of sandy texture on the body
Sample No. 9

Fabric	Medium sandy with Mica.
Surface enhancement	Sample No. 9 having red slip on both the surfaces.

Sample No. 11

Fabric	Medium sandy with Mica.
Surface enhancement	Sample No. 11, Sherd having incised oblique notches on the exterior surface.

9. GOGAJI WALI DONGRI (Figure - 4.45)

Bowls Type B Variety 1: carinated bowls having one variety with carinated shoulder and one rim variant:
(a) bilateral projecting, equal projection rim
Sample No. 7

Fabric	Medium sandy with White inclusion

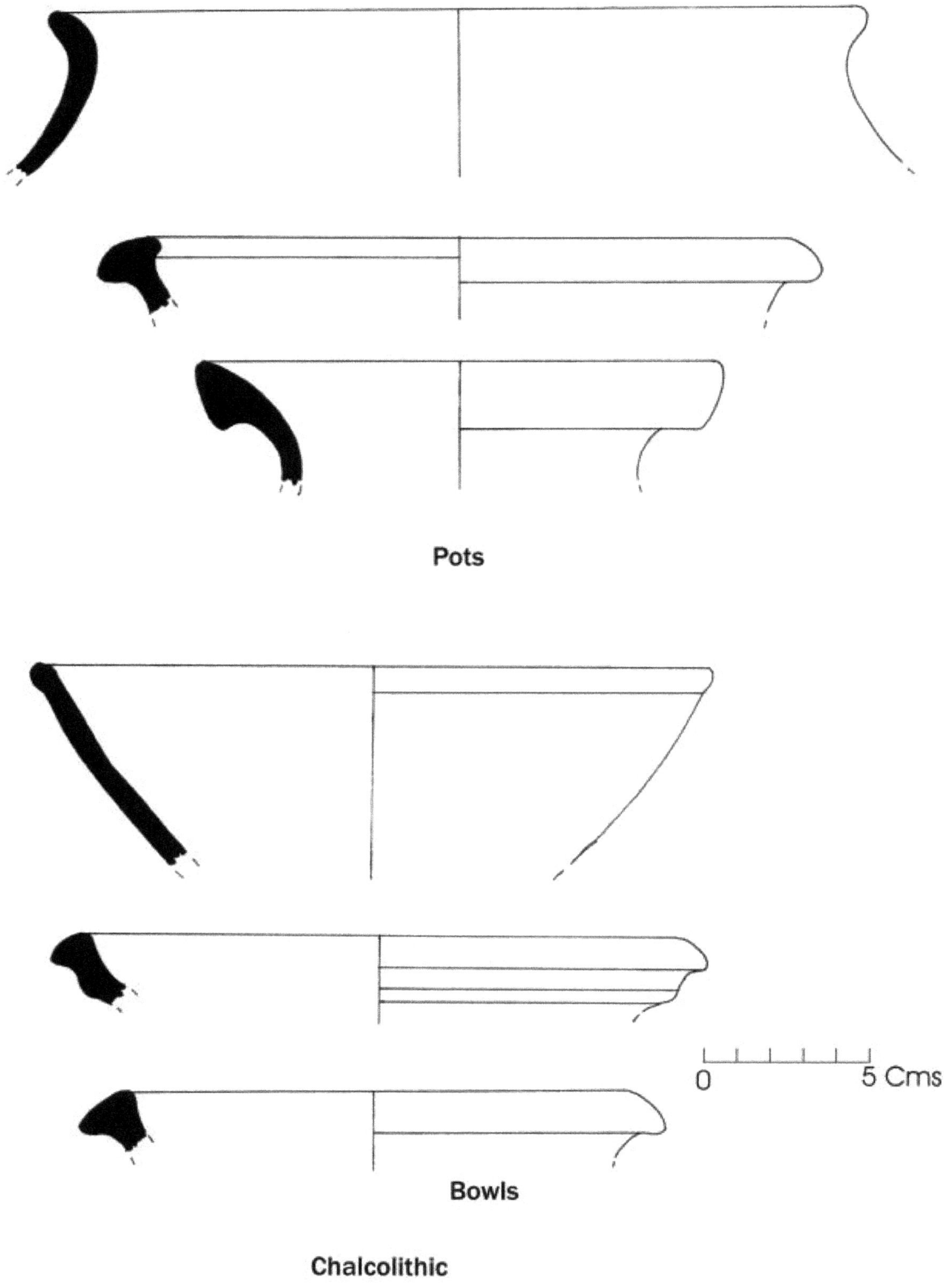

Figure 4.44 – Gidhali

	and Organic matter
Surface enhancement	Sample No. 7 Slip Worn-Out from both the surfaces.
Size range	Bowl measures 25 cm in rim diameter.

Bowls Type C Variety 1: straight-sided bowls having one variety and three rim variants:
(a) external projecting, long, wide rounded rim
(b) internal projecting, short, narrow triangular rim
(c) external ledge angular rim
Sample No. 9, 10 and 11

Fabric	Medium sandy with White inclusion and Organic matter
Surface enhancement	Sample No. 9, 10 and 11 having Red Slip on both the surfaces
Size range	Bowls measures 18 to 23 cms in rim diameter.

Pots Type A Variety 1: small size pots with rim diameter ranges between 12 to 15 cms having one variety and four rim variants
(a) simple everted, slightly curved rim
(b) external projecting, short, narrow rounded rim
(c) external projecting, short, narrow, short quadrangular rim
(d) internal projecting, short, narrow tapered rim
Sample No. 1, 4, 5 and 6

Fabric	Medium sandy with White inclusion and Organic matter
Surface enhancement	Sample No. 1Light Red Slip on both the surfaces with black color band on the neck and on the lip of the rim Sample No. 4 Slip Worn-Out from both the surfaces Sample No. 5 and 6 having Red Slip on both the surfaces. Sample No. 6 with black color band on the neck
Size range	Pots measures 12 to 15 cms in rim diameters.

Pots Type B Variety 1: medium size pots with rim diameter 18 cm having one variety with short and ribbed neck and bulbous body and one rim variant:
(a) bilateral projecting, equal projection rim
Sample No. 2 and 3

Fabric	Medium sandy with White inclusion and Organic matter
Surface enhancement	Sample No. 2 having Red Slip on both the surfaces Sample No. 3 Slip Worn-Out from both the surfaces
Size range	Pot measures 18 to 22 cms in rim diameters.

Basins Type A Variety 1: carinated basins having two varieties, first variety with carinated shoulder and one rim variant:
(a) external projecting, short, narrow quadrangular rim
Sample No. 12

Fabric	Medium sandy with White inclusion
Surface enhancement	Sample No. 12 Unslip on both the surfaces with Rustication on the exterior surface
Size range	Basin measures 32 cms in rim diameter.

Basins Type A Variety 3: carinated basins having two varieties, second variety with angular shoulder having one rim variant:
(a) bilateral projecting, equal projection rim
Sample No. 8

Fabric	Medium sandy with White inclusion
Surface enhancement	Sample No. 8 Unslip on both the surfaces
Size range	Basin measures 34 cms in rim diameter.

Undiagnostic Sherds: having rustication on the body with applied bands of wet design
Sample No. 13

Fabric	Medium sandy with White inclusion
Surface enhancement	Sample No. 13 having unslip on both the surfaces

Undiagnostic Sherds: having rustication on the body with applied horizontal bands of sandy texture
Sample No. 15, 16 and 17

Fabric	Medium sandy with White inclusion
Surface enhancement	Sample No. 15 having Red Slip on both the surfaces Sample No. 16 and 17 Unslip on both the surfaces

Undiagnostic Sherd: having black color bands on the exterior surface
Sample No. 14

Fabric	Medium sandy with Mica.
Surface enhancement	Sample No. 14 having Red Slip on the exterior surface and Unslip on the interior surface

Undiagnostic Sherd: having thumb impression on the interior surface
Sample No. 19

Fabric	Medium sandy with Mica.
Surface enhancement	Sample No. 19 Unslip on both the surfaces

10. GUJJARO KI DHANI (Figure - 4.46)

Bowls Type A Variety 1: convex–sided bowl having one variety with ribbed neck and one rim variant:
(a) simple inverted, curved rounded rim.

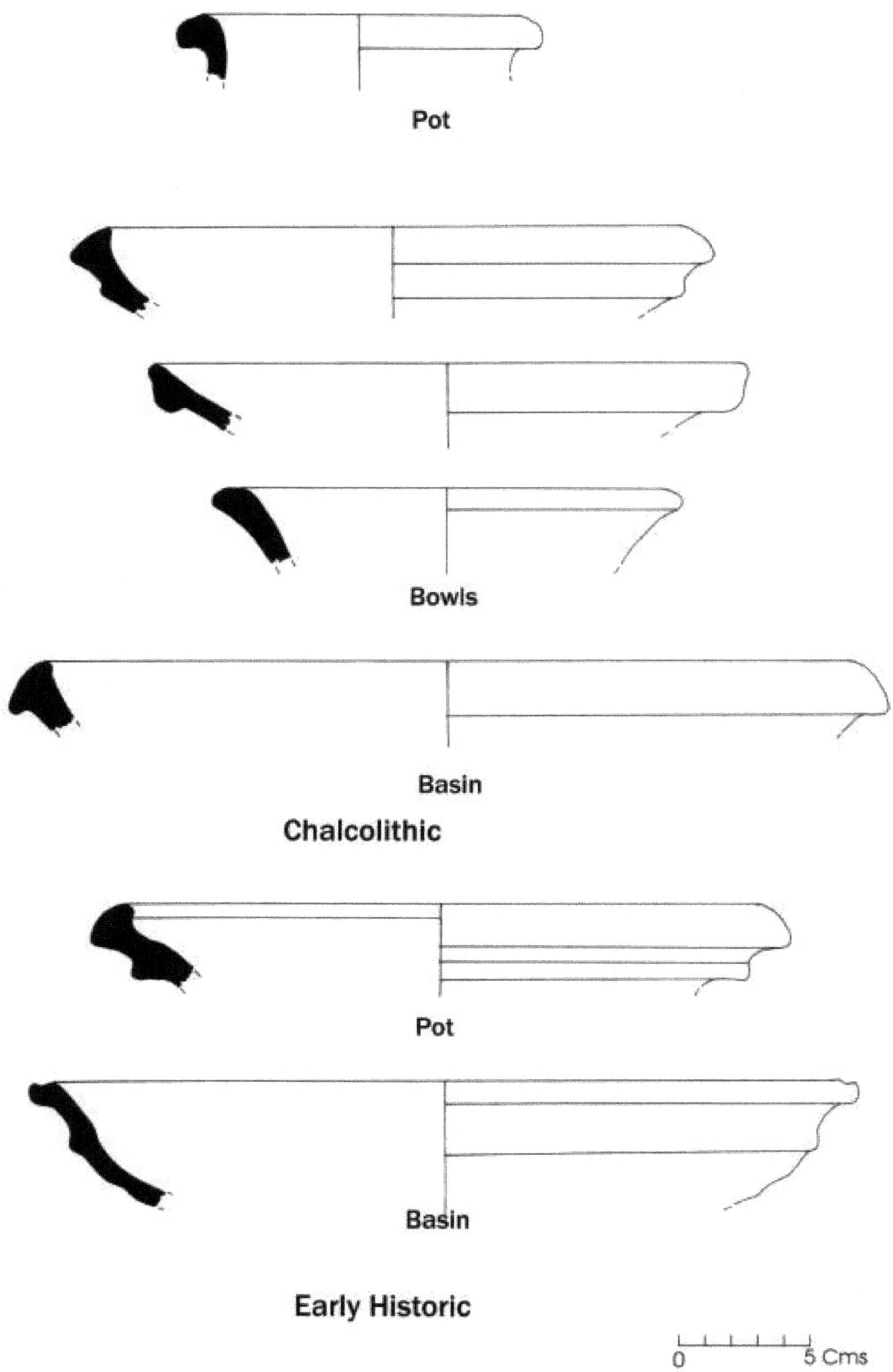

Figure 4.45 – Gogajiwali Dongri

Sample No. 12

Fabric	Medium sandy with White inclusion
Surface enhancement	Sample No. 12 having Red Slip on the exterior surface and Light Red Slip on the interior surface.
Size range	Bowl measures 16 cms in rim diameter.

Bowls Type B Variety 1: carinated bowls having one variety with carinated shoulder and two rim variants:
(a) simple internal projecting, extremely everted rounded rim
(b) internal projecting, short, narrow triangular rim
Sample No. 7, 13 and 14

Fabric	Medium sandy with Mica and White inclusion.
Surface enhancement	Sample No. 7, 13 and 14 having Red Slip on both the surfaces. Sample 14 having external ledge on the neck.
Size range	Bowls measures 15 to 23 cms in rim diameter.

Bowls Type C Variet 1: straight-sided bowls having one variety and two rim variants:
(a) external projecting, long, wide triangular rim
(b) bilateral projecting, equal projection rim
Sample No. 9, 10 and 16

Fabric	Medium sandy with Mica and White inclusion.
Surface enhancement	Sample No. 9 having Light Red Slip on the exterior surface and Unslip on the interior surface. Sample No. 10 having Red Slip on the interior surface and Unslip on the exterior surface. Sample No. 16 having Red Slip on both the surfaces.
Size range	Bowls measures 14 to 27 cms in rim diameter.

Pots Type A Variety 1: small size pots with rim diameter ranges 18 cm having one variety with short neck and one rim variant:
(a) external projecting, short, narrow rounded rim
Sample No. 6

Fabric	Medium sandy with White inclusion and Organic matter.
Surface enhancement	Sample No. 6 having Red Slip on both the surfaces.
Size range	Pot measures 18 cms in rim diameters.

Pots Type B Variety 1: medium size pots with rim diameter ranges between 18 to 28 cms having one variety with short neck and bulbous body and four rim variants:
(a) simple perpendicular, straight rounded rim
(b) external projecting, short, narrow tapered rim
(c) external projecting, long, wide rounded rim
(d) internal projecting, short, narrow triangular rim
Sample No. 1, 2, 3, 4, 5 and 8

Fabric	Fine to Medium sandy with Mica and White inclusion.
Surface enhancement	Sample No. 1 having Red Slip on the exterior surface. Sample No. 2, 3, 5 and 8 having Red Slip on both the surfaces. Sample No. 4 having Light Red Slip on both the surfaces.
Size range	Pots measures 18 to 28 cms in rim diameters.

Bowls Variety 1: bowls with angular shoulder having two rim variants:
(a) simple everted, straight rounded rim
(b) internal projecting, short, narrow triangular rim
Sample No. 11 and 15

Fabric	Medium sandy with Mica and White inclusion.
Surface enhancement	Sample No. 11 and 15 having Red Slip on both the surfaces.
Size range	Bowl measures 21 to 32 cms in rim diameter.

Lamps: having one variety and two rim variants:
(a) simple inverted curved rounded rim
(b) internal projecting, extremely inverted tapered rim
Sample No. 17 and 20

Fabric	Medium sandy with Mica and White inclusion.
Surface enhancement	Sample No. 17 and 20 unslip on both the surfaces.
Size range	Lamp measures 5 to 8 cms in rim diameter.

Undiagnostic Painted Sherds:
Sample No. 21, 22 and 23

Fabric	Fine to Medium sandy with Mica
Surface enhancement	Sample No. 21 and 22, Sherds having black color horizontal bands on the exterior surface. Sample No. 23, Sherd having reserved slip technique having black color band over red surface.

11. JASI KA BAS (GOVINDPURA) (Figure - 4.47)

Bowls Type A Variety 1: convex-sided bowls having two varieties, first variety and one rim variant:
(a) simple externally beveled rim
Sample No. 21, 22, 24, 26, 31, 35, 36, 39, 41 and 49

Fabric	Medium sandy with tiny Mica, White inclusion and Organic matter
Surface enhancement	Sample No. 21, 22, 36, 39, 41 and 49 slip worn-out from both the surfaces Sample No. 35 having Red Slip on the interior surface and unslip on the

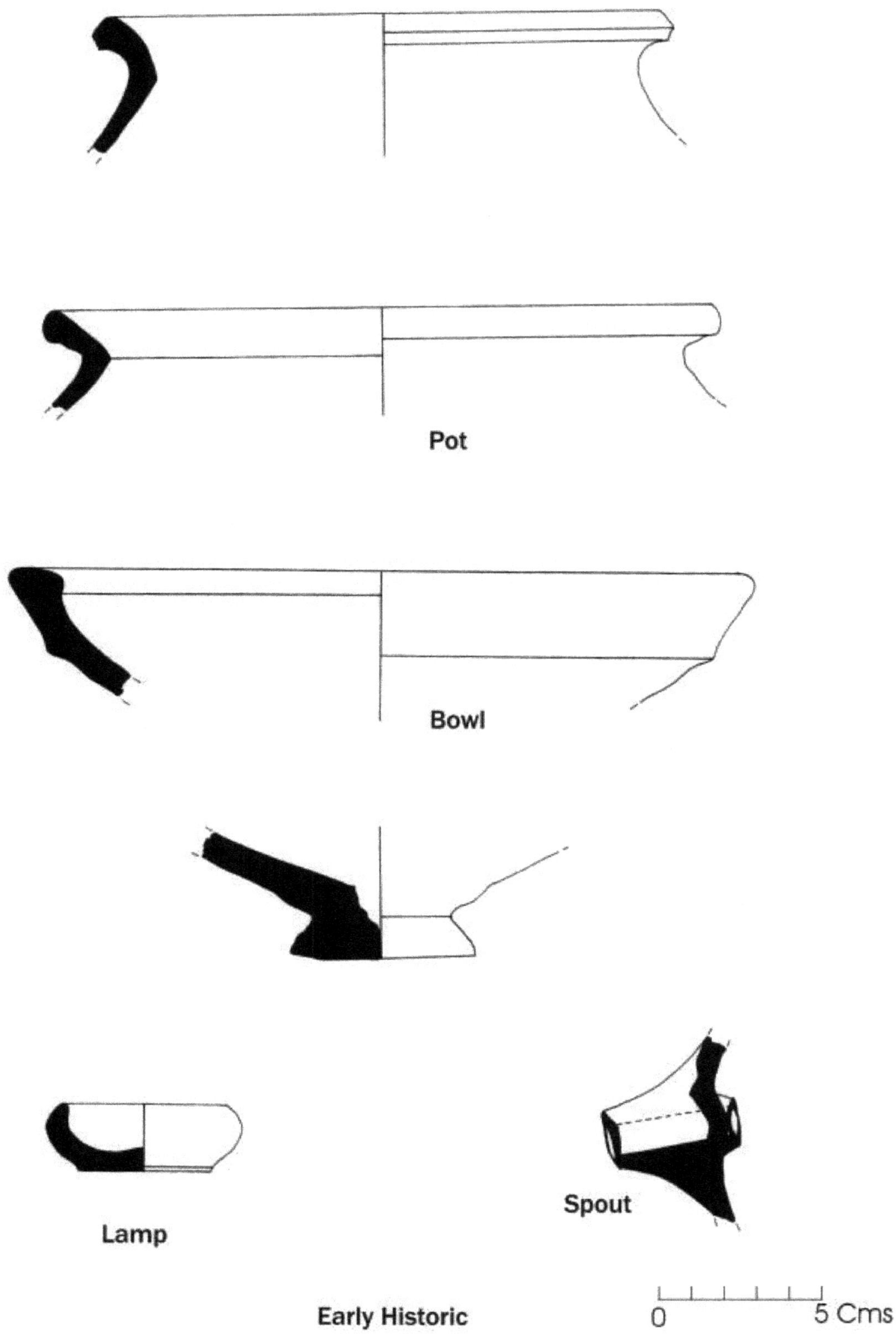

Figure 4.46 – Gujjaro ki Dhani

	exterior surface
	Sample No. 24 having red slip on interior surface and slip worn-out on exterior surface Sample No. 26 having red slip on exterior surface and slip worn-out on interior surface Sample No. 31 having brown slip on both the surfaces.
Size range	Bowls measures 18 to 28 cms in rim diameter.

Bowls Type A Variety 2: convex-sided bowls having two varieties, second variety and three rim variants:
(a) simple everted, straight rounded rim
(b) simple inverted, curved rounded rim
(c) simple internal projecting, extremely inverted rounded rim
Sample No. 27, 30, 32, 34, and 23

Fabric	Fine to Medium sandy with tiny mica and White inclusion
Surface enhancement	Sample No. 30 having Brown Slip on both the surfaces Sample No. 27, 32, 34 and 36 slip worn-out from both the surfaces
Size range	Bowls measures 14 to 26 cms in rim diameter.

Dishes: having one variety and two rim variants:
(a) simple everted, straight rounded rim
(b) external projecting, long, wide rounded rim
Sample No. 13, 25 and 28

Fabric	Medium sandy with tiny Mica, White inclusion and organic matter
Surface enhancement	Sample No. 13, 25 and 28 Slip worn-out from both the surfaces
Size range	Dishes measures 22 to 40 cms in rim diameter.

Dish on Stand: have base diameter 34cms
Sample No. 16

Fabric	Coarse sandy with Tiny Mica and White Inclusion
Surface enhancement	Sample No. 16 slip worn-out from both the surface
Size range	Dish-on-stand measures 34 cms in base diameter.

Pots Type A Variety 1: small size pots with rim diameter ranges between 11 to 16 cms have one variety with angular shoulder and one rim variant:
(a) simple perpendicular, straight rounded rim
(b)simple external projecting, extremely everted rounded rim
Sample No. 7, 8 and 9

Fabric	Medium to Coarse sandy with Tiny Mica, White inclusion and Organic matter
Surface enhancement	Sample No. 7 and 9 having red slip on exterior surface and slip worn-out in interior surface. Sample No. 8 slip worn-out on both the surfaces
Size range	Pots measures 11 to 16 cms in rim diameter.

Pots Type B Variety 1: medium size pots with rim diameter ranges between 19 to 29 cms having one variety and four rim variants:
(a) simple external projecting, extremely everted rounded rim
(b) simple perpendicular rounded rim
(c) external projecting, narrow short beaked rim
(d) external projecting and wide rounded rim
Sample No. 1, 2, 3, 4, 5, 6, 10 and 12

Fabric	Medium to Coarse sandy with Tiny Mica, White inclusion and Organic matter
Surface enhancement	Sample No. 1 and 2 having red slip on exterior surface Sample No. 10 having yellowish red slip on exterior surface. Sample No. 3,4,5,6 and 12 slip worn-out on both the surfaces
Size range	Pots measures 19 to 29 cms in rim diameter.

12. JODHPURA (Figure - 4.48)

Bowls Type B Variety 3: carinated bowls having one variety with blunt carinated shoulder and two rim variants:
(a) external projecting, short, narrow, tapered rim
(b) bilateral projecting, equal projection, large rim
Sample No. 10, 11 and 13

Fabric	Medium sandy with mica and white inclusion.
Surface enhancement	Sample No. 10, 11 and 13 having red slip on both the surfaces
Size range	Bowls measures 16 to 28 cms in rim diameter.

Bowls Type C Variety 1: straight-sided bowls having one variety and two rim variants:
(a) bilateral projecting, equal projection small rim
(b) simple perpendicular, straight rounded rim
Sample No. 12, 30 and 31

Fabric	Fine to Medium sandy with mica and white inclusion.
Surface enhancement	Sample No. 12 having Red Slip on both the surfaces. Sample No. 30 having Red Slip on the exterior surface and Light Red Slip on the interior surface. Sample No. 31 having Red Slip on the interior surface and Reddish Yellow Slip on the exterior surface.
Size range	Bowls measures 20 to 22 cms in rim diameter.

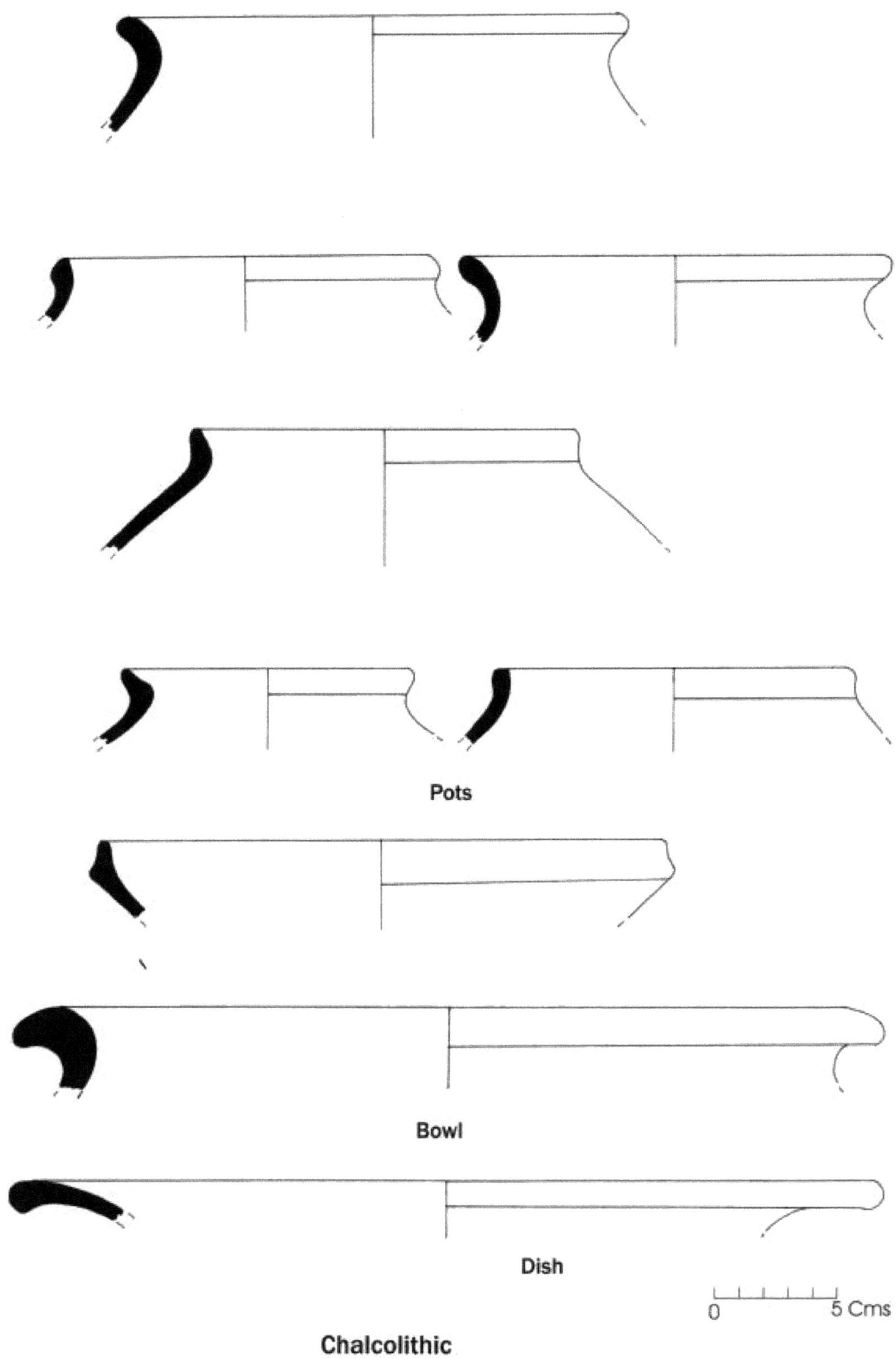

Figure 4.47 – Jasi

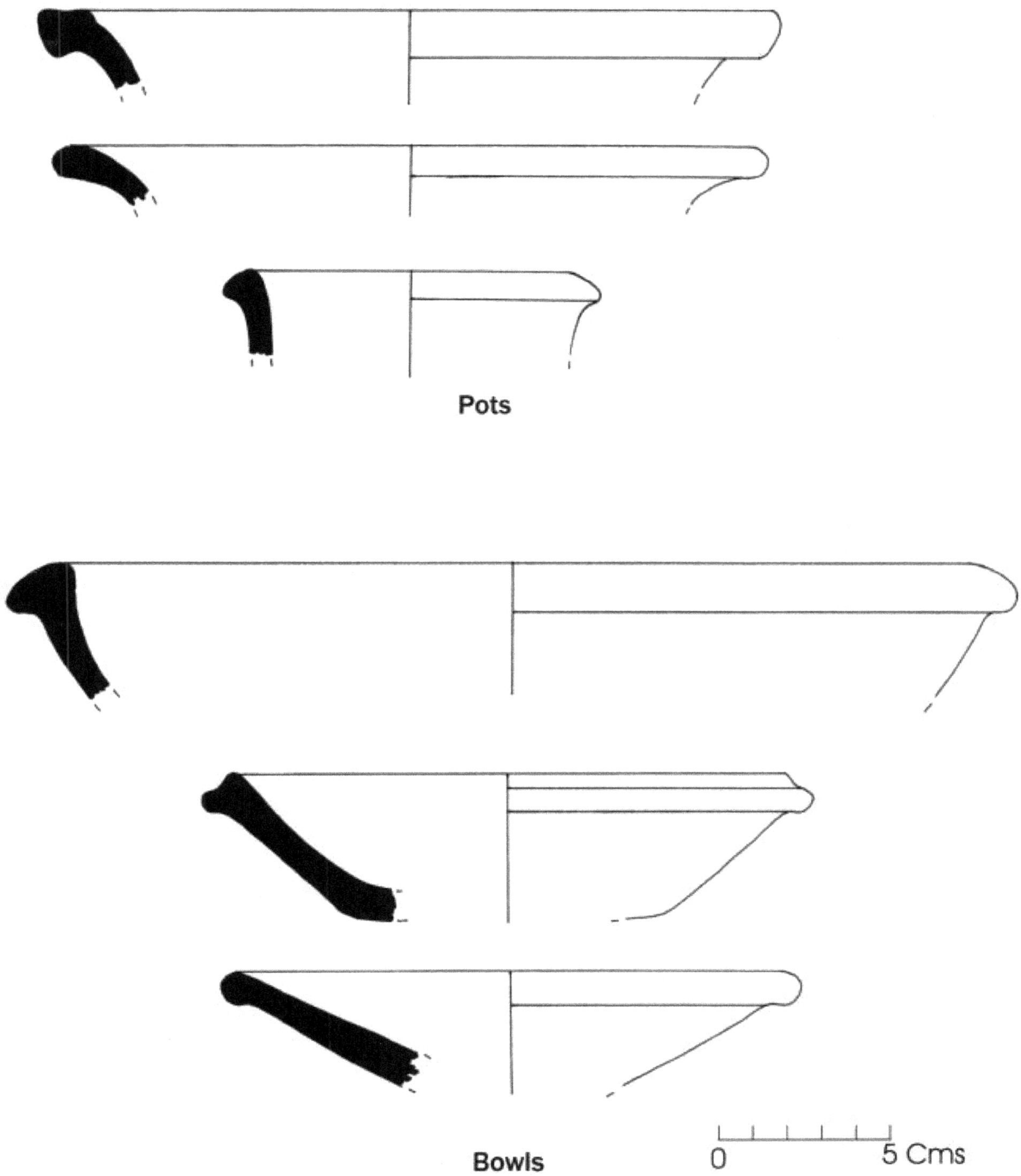

Figure 4.48 – Jodhpura

Pots Type A Variety 1: small size pots with rim diameter ranges between 11 to 16 cms having two varieties, first variety with short neck and bulbous body and three rim variants:
(a) simple perpendicular, straight tapered rim
(b) external projecting, short, narrow rounded rim
(c) external projecting, long, wide rounded rim
Sample No. 4, 7 and 28

Fabric	Medium sandy with mica and white inclusion.
Surface enhancement	Sample No. 4 slip worn-out from both the surfaces. Sample No. 7 and 28 having light red slip on both the surfaces.
Size range	Pot measures 16 cm in rim diameters.

Pots Type A Variety 3: small size pots with rim diameter ranges between 11 to 16 cms having two varieties, second variety with short and corrugated neck and two rim variants:
(a) external projecting, short, narrow short triangular rim
(b) external projecting, short, narrow short beaked rim
Sample No. 5 and 6

Fabric	Medium sandy with mica and white inclusion.
Surface enhancement	Sample No. 5 unslip on both the surfaces with black color band on neck on the exterior surface. Sample No. 6 having light red slip on both the surfaces.
Size range	Pot measures 11 to 12 cms in rim diameters.

Pots Type B Variety 1: medium size pots with rim diameter ranges between 18 to 22 cms having one variety with short neck and bulbous body and three rim variants:
(a) simple external projecting, extremely everted rim
(b) external projecting, short, narrow triangular rim
(c) external projecting, short, narrow, short triangular rim
Sample No. 1, 2, 3, 8 and 9

Fabric	Medium sandy with mica and white inclusion.
Surface enhancement	Sample No. 1 having red slip on the exterior surface and light red slip on the interior surface. Sample No. 2, 8 and 9 having red slip on both the surfaces. Sample No. 3having yellowish red slip on both the surfaces.
Size range	Pot measures 18 to 26 cms in rim diameters.

Basins Type A Variety 1: basins, with carinated shoulder having one variety with incised horizontal grooves on the exterior surface and one rim variant:
(a) internal projecting, short, narrow, triangular rim
Sample No. 14, 15 and 16

Fabric	Medium sandy with mica and white inclusion.
Surface enhancement	Sample No. 14 having red slip on the interior surface and light red slip on the exterior surface. Sample No.15 and 16 having red slip on both the surfaces.
Size range	Basin measures 36 cms in rim diameter.

Dish Type A Variety 1: dish having one variety with carinated neck and one rim variant:
(a) simple everted, straight rounded rim
Sample No. 21

Fabric	Coarse sandy with white inclusion.
Surface enhancement	Sample No. 21 having red slip on both the surfaces
Size range	Dish measures 33 cm n rim diameter

Dishes Type B Variety 1: dishes having one variety and two rim variants:
(a) simple external projecting, extremely everted rim
(b) overturned small external projecting rim
Sample No. 17, 18, 19 and 20

Fabric	Medium sandy with mica and white inclusion
Surface enhancement	Sample No. 17 having light red slip on the interior surface and red slip on the exterior surface. Sample No. 18 having red slip on both the surfaces. Sample No. 19 reddish yellow slip on both the surfaces. Sample No. 20 slip worn-out from both the surfaces.
Size range	Dish measures 20 to 22 cm n rim diameter

Undiagnostic Painted Sherds:
Sample No. 22, 24 and 25

Fabric	Medium sandy with mica and white inclusion
Surface enhancement	Sample No. 22 and 24 having red slip on the exterior surface with black color band over it. Sample No. 25 having yellowish red slip on the exterior surface and red slip on the interior surface with Reserved Slip Technique having red slip over yellowish red slip

Undiagnostic Incised Sherds:
Sample No. 23, 26 and 27

Fabric	Medium sandy with Mica and White inclusion
Surface enhancement	Sample No. 23 Red Slip and Incised Horizontal Grooves on the exterior surface. Sample 26 and 27 having Red Slip and Incised Leave Pattern on the exterior surface.

13. JOGI BADH (BAGOTH) JODHPURA (Figure - 4.49)

Bowls Type A Variety 1: convex-sided bowls having one variety and two rim variants:
(a) simple internal projecting, extremely inverted rim
(b) internal projecting, long, wide beaked rim
Sample No. 13, 14 and 15

Fabric	Medium to Coarse sandy with Mica, White inclusion and Organic matter
Surface enhancement	Sample No. 13 and 14 having Red Slip on the interior surface and Unslip on the exterior surface Sample No. 15 having Red Slip on the exterior surface and Unslip on the interior surface
Size range	Bowls measures 16 to 26 cms in rim diameter.

Bowls Type B Variety 1: carinated bowls having one variety with carinated shoulder and three rim variants:
(a) simple perpendicular, straight rounded rim
(b) external projecting, short, narrow tapered rim
(c) external projecting, long, wide tapered rim
Sample No. 5, 10 and 12

Fabric	Medium sandy with Mica, White inclusion and Organic matter
Surface enhancement	Sample No. 5 having Reddish Yellow Slip on both the surfaces with horizontal incised lines on the neck Sample No. 10 and 12 Unslip on both the surfaces
Size range	Bowls measures 21 to 32 cms in rim diameter.

Bowls Variety 3: bowl with blunt carinated shoulder having one rim variant:
(a) bilateral projecting, equal projection rim
Sample No. 7

Fabric	Medium sandy with Mica, White inclusion and Organic matter
Surface enhancement	Sample No. 7 Slip Worn-Out on both the surfaces
Size range	Bowl measures 15 cm in rim diameter.

Bowls Type C Variety 1: straight-sided bowls having one variety and one rim variant:
(a) external projecting, short, narrow triangular rim
Sample No. 16

Fabric	Medium sandy with Mica, White inclusion and organic matter
Surface enhancement	Sample No. 16 having Red Slip on both the surfaces
Size range	Bowl measures 24 cm in rim diameter.

Pots Type A Variety 1: small size pots with rim diameter ranges between 11 to 15 cms having two varieties, first variety with short neck and bulbous body and one rim variant:
(a) external ledge angular rim
Sample No. 1

Fabric	Medium sandy with Tiny Mica
Surface enhancement	Sample No. 1 having Light Red Slip on the exterior surface and Unslip on the interior surface
Size range	Pot measures 11 cm in rim diameters.

Pots Type A Variety 3: small size pots with rim diameter ranges between 11 to 15 cms having two varieties, second variety with corrugated neck and one rim variant
(a) internal projecting, short, narrow triangular rim
Sample No. 6

Fabric	Medium sandy with Tiny Mica and White inclusion
Surface enhancement	Sample No. 6 Unslip on both the surfaces
Size range	Pot measures 21 cm in rim diameters.

Pots Type B Variety 1: medium size pots with rim diameter ranges between 19 to 34 cms having two varieties, first variety with short neck and bulbous body and one rim variant:
(a) internal projecting, short, narrow triangular rim
Sample No. 2 and 8

Fabric	Medium sandy with Mica, White inclusion and Organic matter
Surface enhancement	Sample No. 2 having Red Slip on the exterior surface and Unslip on the interior surface Sample No. 8 having Reddish Yellow Slip on the exterior surface and unslip on the interior surface
Size range	Pots measures 20 cm in rim diameters.

Pots Type B Variety 3: medium size pots with rim diameter ranges between 19 to 34 cms having two varieties, second variety with corrugated and ribbed neck and one rim variant:
(a) bilateral projecting, equal projection rim
Sample No. 4

Fabric	Medium sandy with White inclusion and Organic matter
Surface enhancement	Sample No. 4 having Light Reddish Brown Slip on both the surfaces
Size range	Pot measures 27 cm in rim diameters.

Basins Type A Variety 3: carinated basins having one variety with angular shoulder and one rim variant:
(a) internal projecting, short, narrow triangular rim
Sample No. 11, 17 and 18

Fabric	Medium sandy with Mica, White inclusion and Organic matter
Surface enhancement	Sample No. 11, 17 and 18 Unslip on both the surfaces Sample No. 11 with wavy appliqué on the rim
Size range	Basins measures 24 to 34 cms in rim diameter.

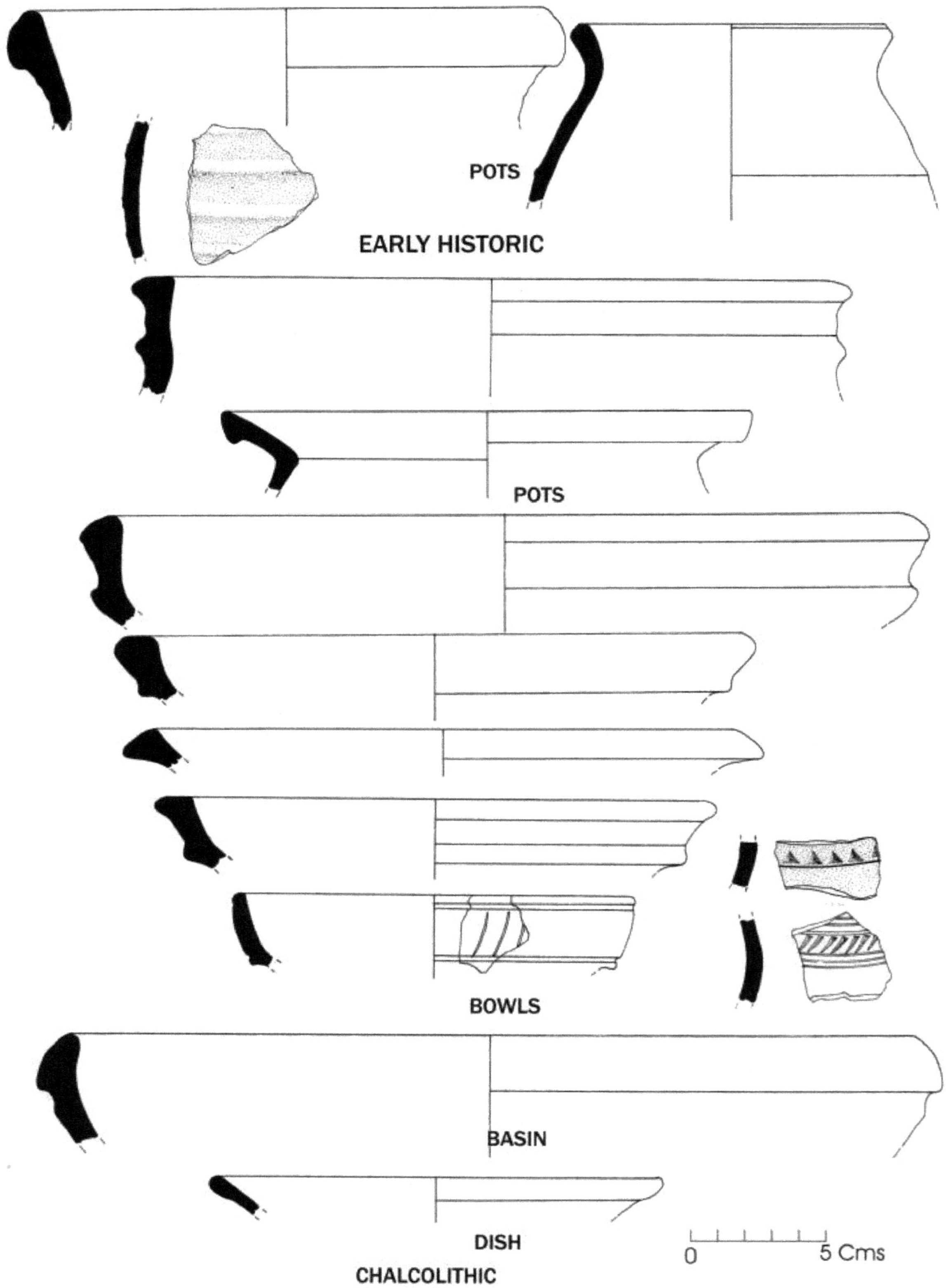

Figure 4.49 – Jogi Badh

Dish Variety 1: dish with angular shoulder having one rim variant:
(a) simple everted straight rounded rim
Sample No. 19

Fabric	Medium sandy with White inclusion.
Surface enhancement	Sample No. 19 Unslip on both the surfaces
Size range	Dish measures 18 cm in rim diameter.

Undiagnostic Sherds: Sherds having rustication on the body with appliqué bands of wet design
Sample No. 20

Fabric	Medium sandy with Mica and White inclusion
Surface enhancement	Sample No. 20 Unslip on both the surfaces

Undiagnostic Sherd: having rustication on the body with horizontal applied bands of sandy texture
Sample No. 21

Fabric	Medium sandy with Mica
Surface enhancement	Sample No. 21 having Light Red Slip on the exterior surface and Unslip on the interior surface

Undiagnostic Sherd: having rows of horizontal lines between which have incised triangular notches
Sample No. 22

Fabric	Medium sandy with Mica and White inclusion
Surface enhancement	Sample No. 22 having Red Slip on the exterior surface and Unslip on the interior

Undiagnostic Sherd: having series of incised triangular notches
Sample No. 23

Fabric	Medium sandy
Surface enhancement	Sample No. 23 having Red Slip on the exterior surface and Unslip on the interior surface

Undiagnostic Sherd: having incised horizontal lines which has been cut by incised vertical lines
Sample No. 24

Fabric	Medium sandy with White inclusion
Surface enhancement	Sample No. 24 Unslip on both the surfaces

Undiagnostic Sherd: having horizontal incised lines
Sample No. 25

Fabric	Medium sandy with White inclusion
Surface enhancement	Sample No. 25 Unslip on both the surfaces

Undiagnostic Sherd: having incised triangular notches
Sample No. 26

Fabric	Medium sandy
Surface enhancement	Sample No. 26 Unslip on both the surfaces

Undiagnostic Sherd: having rows of square stamp design on the interior surface
Sample No. 27

Fabric	Medium sandy
Surface enhancement	Sample No. 27

14. KAKADHIYA (THIKRIYA) (Figure - 4.50)

Bowls Type B Variety 1: carinated bowls having one variety with carinated shoulder unmeasured rim variant:
Sample No. 10

Fabric	Medium sandy with Mica, White inclusion and Organic matter
Surface enhancement	Sample No. 10 having Light Red Slip on both the surfaces
Size range	Bowl with unmeasured rim diameter

Bowls Variety 3: bowls, with blunt carinated shoulder and two rim variants:
(a) external projecting, long, wide beaked rim
(b) internal projecting, short, narrow triangular rim
Sample No. 8 and 20

Fabric	Medium sandy with Mica, White inclusion and White inclusion
Surface enhancement	Sample No. 8 and 20 having Red Slip on both the surfaces
Size range	Bowls measures 15 to 18 cms in rim diameter.

Bowls Type C Variety 1: straight-sided bowls having one variety and one rim variant:
(a) simple perpendicular, straight rounded rim
Sample No. 6 and 7

Fabric	Medium sandy with Mica, White inclusion and Organic matter
Surface enhancement	Sample No. 6 and 7 Unslip on both the surfaces
Size range	Bowls measures 15 to 16 cm in rim diameter.

Pots Type A Variety 1: small size pots with rim diameter 16 cm having one variety with short and ribbed neck and bulbous body and one rim variant:
(a) external projecting, short, narrow triangular rim
Sample No. 1 and 5

Fabric	Medium sandy with Mica, White inclusion and Organic matter
Surface enhancement	Sample No. 1 having Yellowish Red Slip on both the surfaces Sample No. 5 having Red Slip on the exterior surface and Unslip on the interior surface with incised horizontal line son the neck
Size range	Pot measures 16 cm in rim diameters.

Pots Type B Variety 1: medium size pots with rim diameter ranges between 18 to 22 cms having one variety with short and ribbed neck and bulbous body and one rim variant:
(a) external projecting, short, narrow rounded rim
Sample No. 2, 3 and 4

Fabric	Medium sandy with Mica, White inclusion and Organic matter
Surface enhancement	Sample No. 2 and 3 Unslip on both the surfaces Sample No. 4 Slip Worn-Out from both the surfaces
Size range	Pots measures 18 to 22 cms in rim diameters.

Basins Type A Variety 1: carinated basins having two varieties, first variety with carinated shoulder and one rim variant:
(a) external projecting, long, wide triangular rim
Sample No. 11

Fabric	Medium sandy with Mica, White inclusion and Organic matter
Surface enhancement	Sample No. 11 having Red Slip on both the surfaces
Size range	Basin measures 22 cm in rim diameter.

Basins Type A Variety 3: carinated basins having two varieties, second variety with angular shoulder and one rim variant:
(a) internal projecting, short, narrow triangular rim
Sample No. 12

Fabric	Medium sandy with White inclusion.
Surface enhancement	Sample No. 12 having Reddish Yellow Slip on both the surfaces
Size range	Basin measures 34 cm in rim diameter.

Undiagnostic Sherds: having rustication on the body with applied horizontal bands of sandy texture
Sample No. 13, 14 and 15

Fabric	Medium sandy with Mica, White inclusion and Organic matter
Surface enhancement	Sample No. 13, 14 and 15 Unslip on both the surfaces

Undiagnostic Sherd: having incised horizontal lines
Sample No. 17

Fabric	Medium sandy with Mica, White inclusion and Organic matter
Surface enhancement	Sample No. 17 having Light Red Slip on the exterior surface and Unslip o the interior surface

Undiagnostic Sherd: having incised horizontal lines
Sample No. 21

Fabric	Medium sandy with White inclusion
Surface enhancement	Sample No. 21 Unslip on both the surfaces

15. KAKRANA (Figure - 4.51)

Bowls Type B Variety 1: carinated bowls having three varieties, first variety with carinated shoulder and unmeasured rim variant
Sample No. 11

Fabric	Medium sandy with White inclusion
Surface enhancement	Sample No. 11 having Red Slip on both the surfaces with sherd having black color band on the neck
Size range	Bowl unmeasured in rim diameter

Bowls Type B Variety 3: carinated bowls having three varieties, second variety with blunt carinated shoulder and two rim variants:
(a) external projecting, short, narrow rounded rim
(b) external projecting, short, narrow triangular rim
Sample No. 5, 6 and 8

Fabric	Medium sandy with Mica, White inclusion and Organic matter
Surface enhancement	Sample No. 5 having Red Slip on both the surfaces Sample No. 6 and 8 having Light Red Slip on both the surfaces
Size range	Bowls measures 18 to 19 cms in rim diameter.

Bowls Type B Variety 4: carinated bowls having three varieties, third variety with angular shoulder and one rim variant:
(a) external projecting, long, wide triangular rim
Sample No. 7

Fabric	Medium sandy with Mica, White inclusion and Organic matter
Surface enhancement	Sample No. 7 having Light Red Slip on the interior surface and Unslip on the exterior surface
Size range	Bowl measures 16 cm in rim diameter.

Pots Type A Variety 1: small size pots with rim diameter ranges between 12 to 14 cms having one variety with short and ribbed neck and two rim variants:
(a) external projecting, short, narrow triangular rim
(b) external projecting, short, narrow tapered rim
Sample No. 3 and 4

Fabric	Medium sandy with Mica and White inclusion.
Surface enhancement	Sample No. 3 and 4 having Red Slip on both the surfaces
Size range	Pots measures 12 to 14 cms in rim diameters.

Pots Type B Variety 1: medium size pots with rim diameter ranges between 17 to 19 cms having one variety with short neck and bulbous body and two rim variants:
(a) external projecting, short, narrow rounded rim
(b) external projecting, long, wide rounded rim
Sample No. 1 and 2

Fabric	Medium sandy with Mica, White inclusion and Organic matter

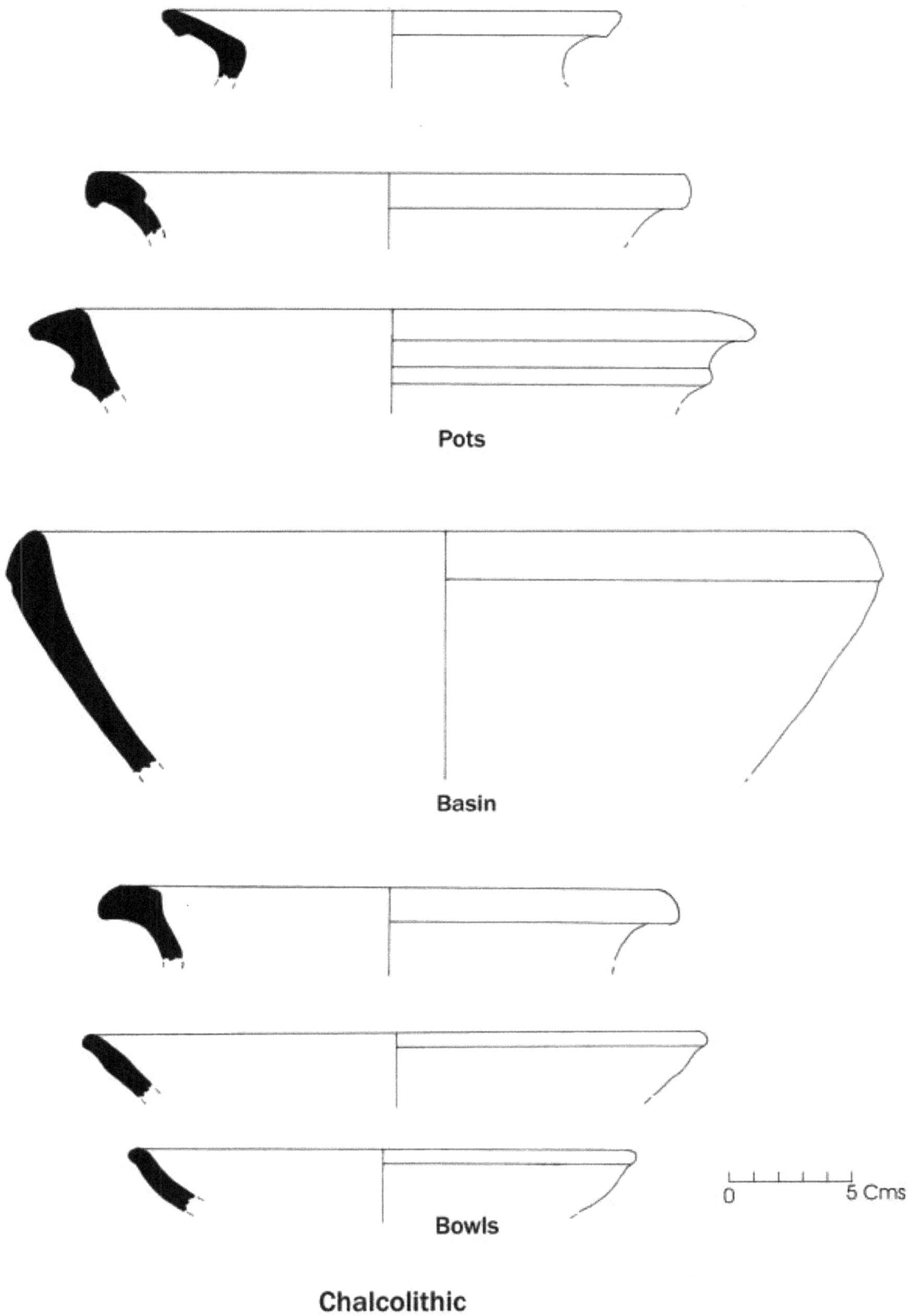

Figure 4.50 – Kakadiyo

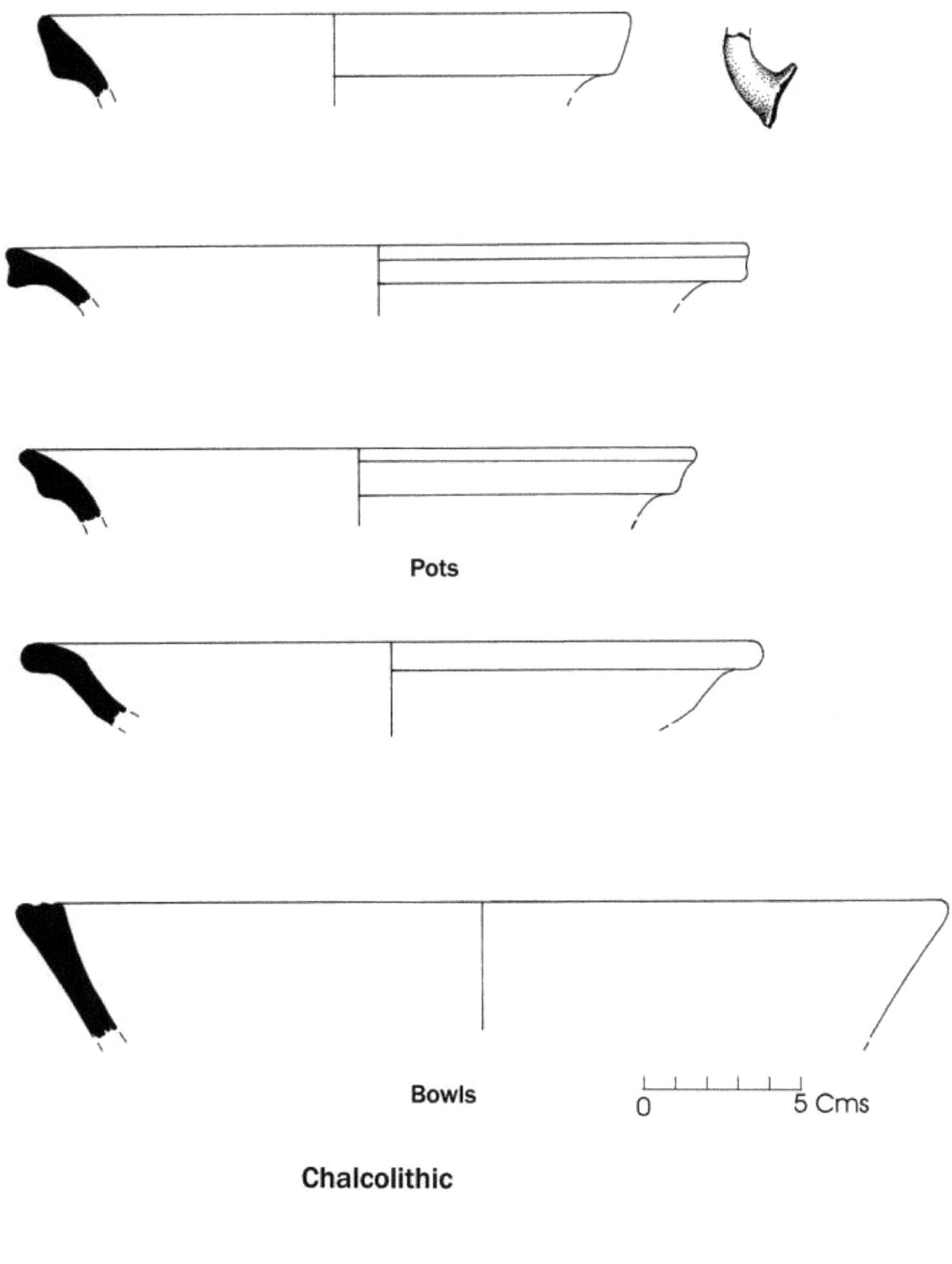

Figure 4.51 – Kakrana

Surface enhancement	Sample No. 1 having red Slip on both the surfaces Sample No. 2 Unslip on both the surfaces
Size range	Pots measures 17 to 19 cms in rim diameters.

16. KALI KANKARIA (Figure - 4.52)

Bowls Type A Variety 1: convex-sided bowls having one variety and one rim variant:
(a) simple internal projecting, extremely inverted rim
Sample No. 2

Fabric	Medium sandy with Tiny Mica and White inclusion.
Surface enhancement	Sample No. 2 Red Slip on both the surfaces
Size range	Bowl measures 27 cm in rim diameter.

Bowls Type B Variety 4: carinated bowls having one variety with angular shoulder and one rim variant:
(a) internal projecting, short, narrow triangular rim
Sample No. 4

Fabric	Medium sandy with Mica and White inclusion
Surface enhancement	Sample No. 4 having Red Slip on both the surfaces
Size range	Bowl measures 32 cm in rim diameter.

Bowls Type C Variety 1: straight-sided bowls having one variety and one rim variant:
(a) simple everted, slightly curved rounded rim
Sample No. 5

Fabric	Coarse sandy with Mica and White inclusion
Surface enhancement	Sample No. 5 having Light Red Slip on both the surfaces
Size range	Bowl measures 22 cm in rim diameter.

Pots Type B Variety 1: medium size pots with rim diameter 22 cm having one variety with bulbous body and one rim variant:
(a) simple everted, slightly curved rounded rim
Sample No. 1

Fabric	Medium sandy with Mica and White inclusion
Surface enhancement	Sample No. 1 having red Slip on both the surfaces
Size range	Pot measures 22 cm in rim diameters.

Basins Type A Variety 3: carinated basins having one variety with angular shoulder and one rim variant:
(a) internal projecting, short, narrow rounded rim
Sample No. 6

Fabric	Medium sandy with Mica and White inclusion.
Surface enhancement	Sample No. 6 having red slip on both the surfaces.
Size range	Basin measures 40 cm in rim diameter.

17. KANWAT (Figure - 4.53)

Bowls Type B Variety 3: carinated bowls having one variety with blunt carinated shoulder and one rim variant:
(a) external projecting, long, wide triangular rim
Sample No. 6

Fabric	Medium sandy with Mica and White inclusion.
Surface enhancement	Sample No. 6 Unslip on both the surfaces
Size range	Bowl measures 18 cm in rim diameter.

Bowls Type C Variety 1: straight-sided bowls having one variety and one rim variant:
(a) internal projecting, short, narrow triangular rim
Sample No. 7

Fabric	Medium sandy with Mica and White inclusion
Surface enhancement	Sample No. 7 having Very Pale Brownish Slip on the exterior surface and Unslip on the interior surface
Size range	Bowl measures 16 cm in rim diameter.

Pots Type A Variety 1: small size pots with rim diameter ranges between 14 to 16 cms having one variety with short neck and bulbous body and four rim variants:
(a) simple perpendicular, straight rounded rim
(b) external projecting, short, narrow quadrangular rim
(c) external projecting, long, wide rounded rim
(d) internal projecting, short, narrow triangular rim
Sample No. 1, 2, 3 and 4

Fabric	Medium with Mica, White inclusion and Organic matter
Surface enhancement	Sample No. 1 and 2 having Red Slip on both the surfaces Sample No. 3 having reddish Yellow Slip on both the surfaces Sample No. 4 having Light Red Slip on both the surfaces
Size range	Pots measures 14 to 16 cms in rim diameters.

Pot Type B Variety 1: medium size pots with rim diameter 22 cm having one variety with short neck and bulbous body and one rim variant:
(a) external projecting, short, narrow tapered rim
Sample No. 5

Fabric	Medium sandy with White inclusion.
Surface enhancement	Sample No. 5 having Dark Reddish Gray Slip on both the surfaces
Size range	Pot measures 22 cm in rim diameters.

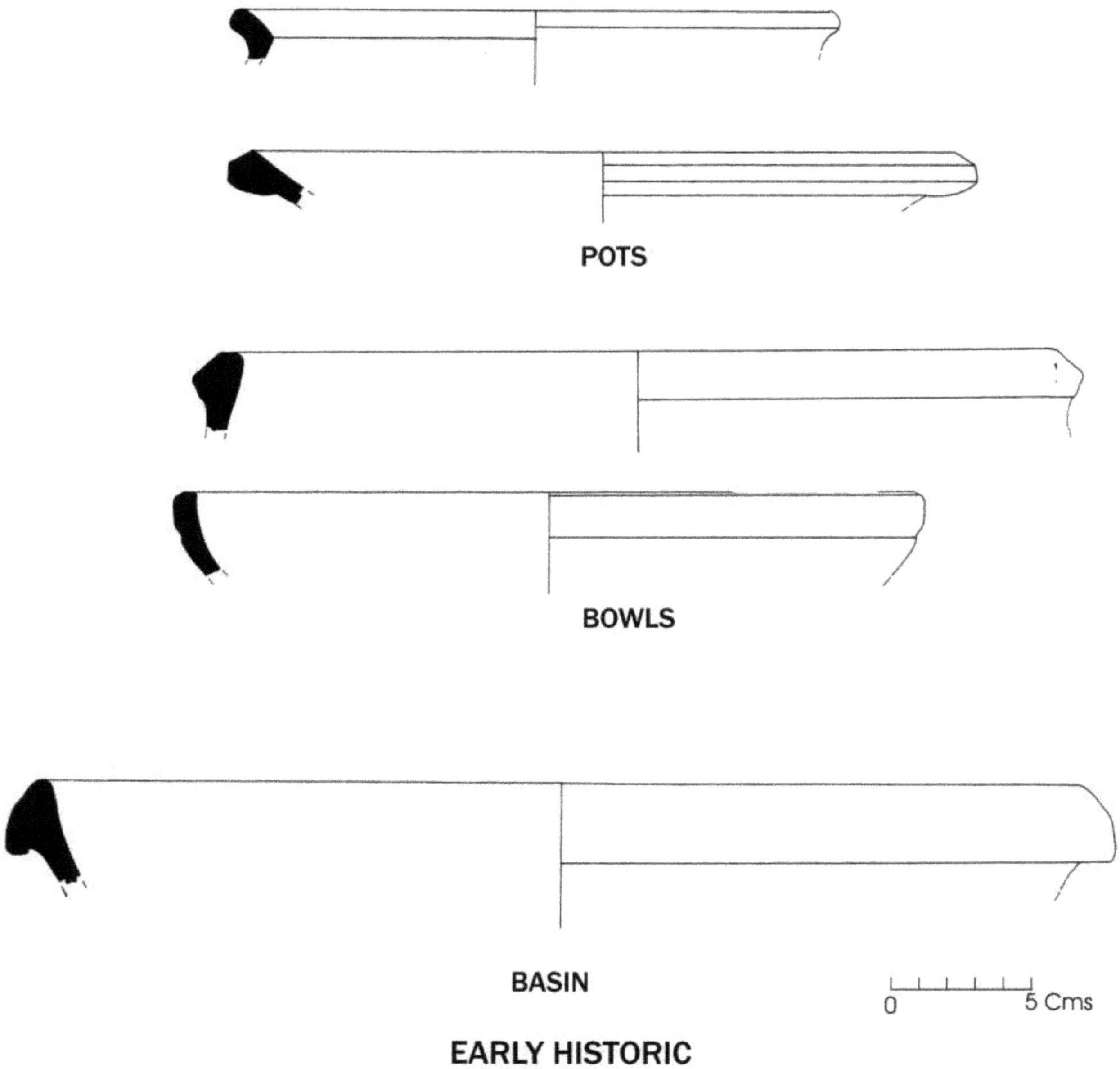

Figure 4.52 – Kali Kankaria

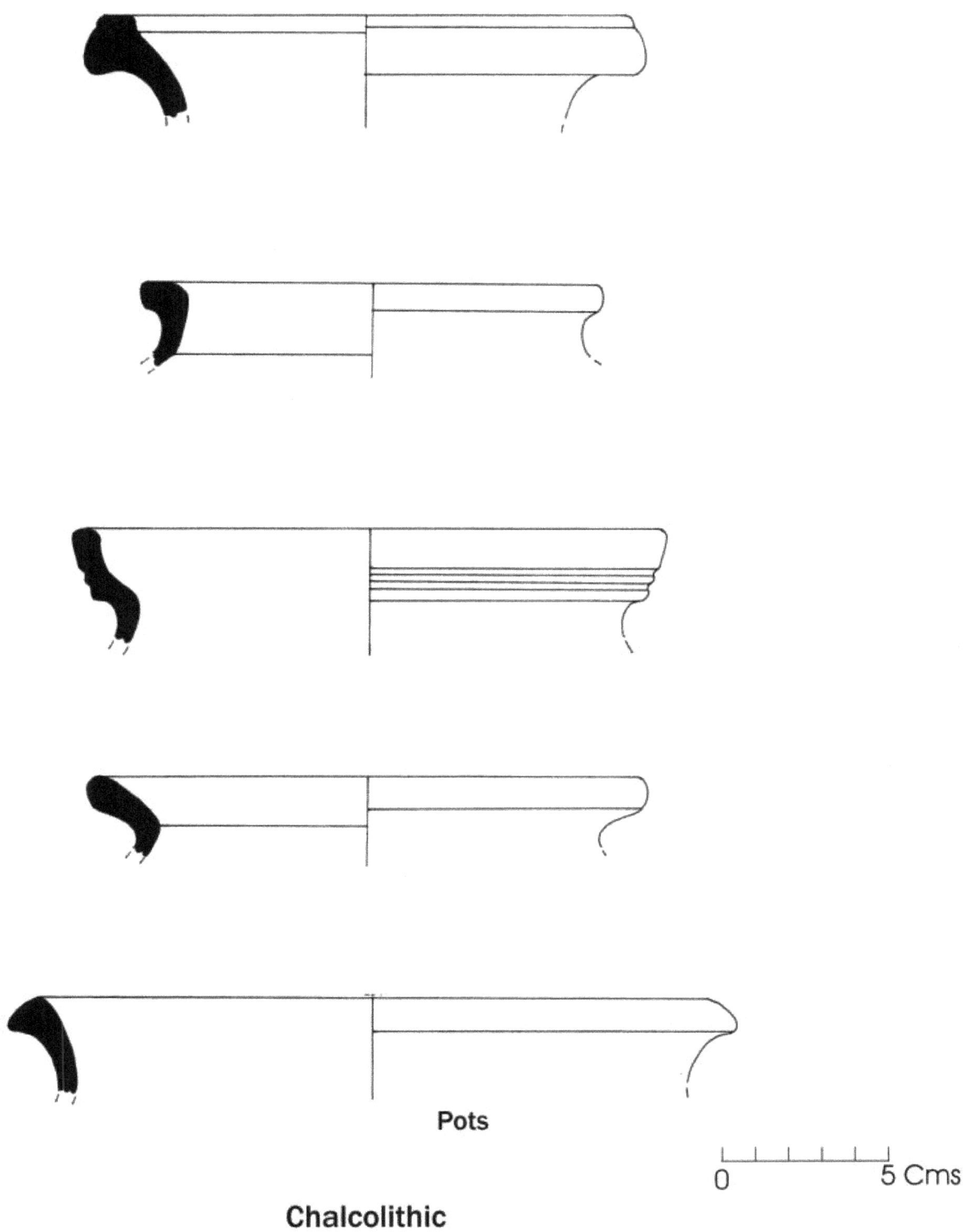

Figure 4.53 – Kanwat

18. KHARAGBINJPUR (Figure - 4.54)

Bowls Type C Variety 1: straight-sided bowls having one variety and two rim variants:
(a) external projecting, long, wide triangular rim
(b) internal projecting, short, narrow triangular rim
Sample No. 6 and 8

Fabric	Medium sandy with Mica and White inclusion.
Surface enhancement	Sample No. 6 having Light Red Slip on both the surfaces. Sample No. 8 having Red Slip on both the surfaces.
Size range	Bowls measures 12 to 16 cms in rim diameter.

Pots Type A Variety 2: small size pots with rim diameter ranges between 12 to 16 cms having one variety with long neck and one rim variant:
(a) external projecting, short, narrow, rounded rim
Sample No. 2 and 3

Fabric	Medium to Coarse sandy with Mica, White inclusion and Organic matter
Surface enhancement	Sample No. 2 and 3 having Slip Worn-Out from both the surfaces
Size range	Pots measures 12 to 16 cms in rim diameters

Pots Type B Variety 1: medium size pots with rim diameter 20 cms having one variety with short neck and bulbous body and one rim variant:
(a) external projecting, short, narrow tapered rim
Sample No. 1

Fabric	Medium sandy with Mica and White inclusion.
Surface enhancement	Sample No. 1 having Slip Worn-Out from both the surfaces
Size range	Pot measures 20 cm in rim diameters.

Basins Type A Variety 3: basin with carinated shoulder having one variety with angular shoulder and one rim variant:
(a) internal projecting, short, narrow triangular rim
Sample No. 9

Fabric	Medium sandy with Mica and White inclusion.
Surface enhancement	Sample No. 9 having Red Slip on both the surfaces.
Size range	Basin measures 24 cm in rim diameter.

19. KHO (Figure - 4.55)

Bowls Type A Variety 1: Convex-Sided Bowls having one variety and unmeasured rim variant
Sample No. 9

Fabric	Medium sandy with Mica, White inclusion and Organic matter
Surface enhancement	Sample No. 7 having Light Red Slip on the exterior surface and Unslip on the interior surface
Size range	Bowl unmeasured rim diameter

Bowls Type B Variety 3: carinated bowls having two varieties, first variety with blunt carinated shoulder and one rim variant:
(a) external projecting, short, narrow triangular rim
Sample No. 6

Fabric	Medium sandy with Mica, White inclusion and Organic matter
Surface enhancement	Sample No. 6 Unslip on both the surfaces with sherd having incised oblique lines on the shoulder
Size range	Bowl measures 18 cm in rim diameter.

Bowls Type B Variety 4: carinated bowls having two varieties, second variety with angular shoulder and two rim variants:
(a) external projecting, long, wide rounded rim
(b) bilateral projecting, equal projection rim
Sample No. 1 and 7

Fabric	Medium sandy with Mica, White inclusion and Organic matter
Surface enhancement	Sample No. 1 Unslip on both the surfaces Sample No. 7 Slip Worn-Out from both the surfaces
Size range	Bowl measures 22 to 26 cms in rim diameter.

Bowls Type C Variety 1: straight-sided bowls having one variety and two rim variants:
(a) simple perpendicular, straight rounded rim
(b) internal projecting, short, narrow triangular rim
Sample No. 4 and 8

Fabric	Medium sandy with Mica and White inclusion.
Surface enhancement	Sample No. 4 having Red Slip on both the surfaces Sample No. 8 having Light Red Slip on both the surfaces
Size range	Bowls measures 13 to 22 cms in rim diameter.

Pots Type A Variety 1: small size pots with rim diameter ranges between 11 to 16 cms having one variety with short neck and bulbous body and one rim variant:
(a) external projecting, short, narrow triangular rim
Sample No. 2, 3 and 5

Fabric	Medium sandy with Mica, White inclusion and Organic matter
Surface enhancement	Sample No. 2 having light Red Slip on both the surfaces Sample No. 3 having Reddish Yellow Slip on both the surfaces Sample No. 5 having Red Slip on both the surfaces
Size range	Pots measures 11 to 16 cms in rim diameters.

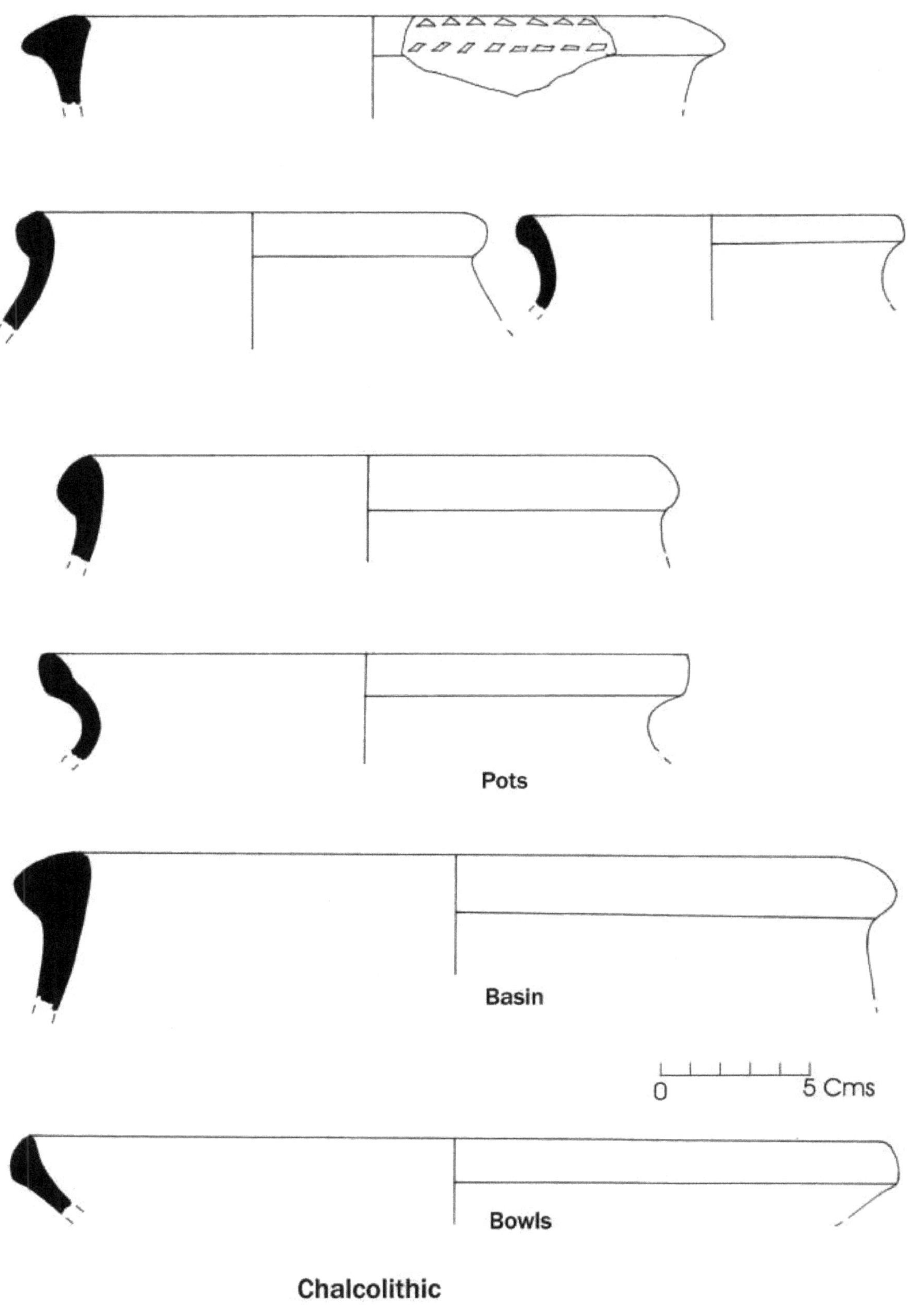

Figure 4.54 – Kharagbinjpur

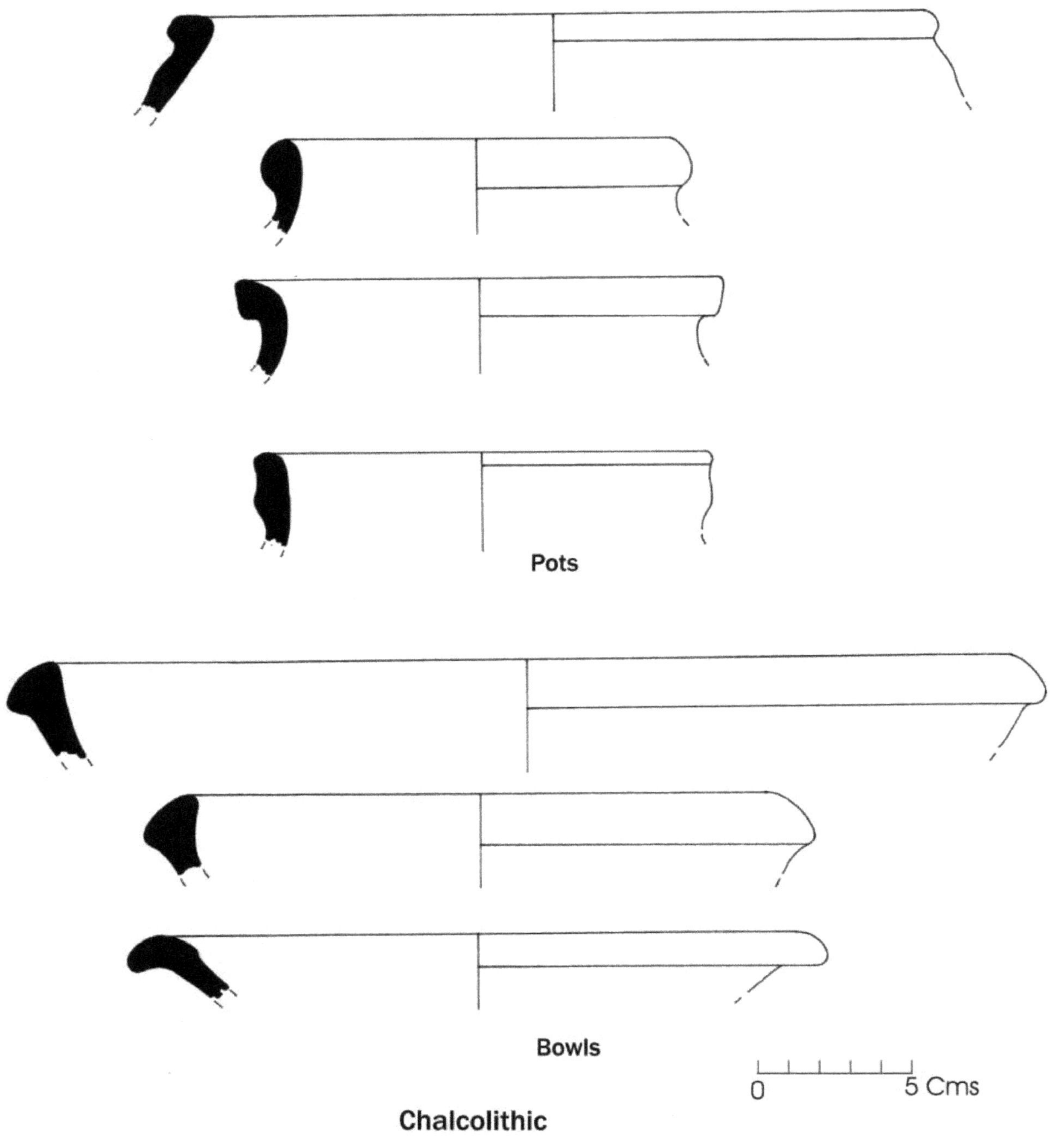

Figure 4.55 – Kho

20. KOT (SUNARI) (Figure - 4.56)

Bowls Type A Variety 1: convex-sided bowls have one variety and one rim variant:
(a) simple inverted, curved rim.
Sample No. 30 and 40

Fabric	Fine sandy with tiny mica and white inclusion.
Surface enhancement	Sample No. 30 Plain Gray Ware having Gray slip on both the surfaces with thin striation marks. Sample No. 40 Black Slip Ware having Black color slip on both the surfaces with thin striation marks and glossy surface.
Size range	Bowls measures 14 to 26 cms in rim diameter.

Bowls Type B Variety 1: carinated bowls have two varieties, first variety with sharp carination and two rim variants:
(a) simple inverted, curved tapered rim
(b) external projecting, long, wide triangular rim
Sample No. 20 and 22

Fabric	Fine to Medium sandy with tiny mica and white inclusion.
Surface enhancement	Sample No. 20 with long neck and ledge on the exterior surface and unslip on both the surfaces. Sample No. 22 with incised grooves on neck having red slip on both the surfaces.
Size range	Bowls measures 13 to 18 cms in rim diameter.

Bowls Type B Variety 2: carinated bowls have two varieties, second variety with blunt carination and three rim variants:
(a) simple perpendicular, straight rounded rim
(b) simple external projecting, extremely everted rounded rim
(c) external projecting, short, narrow triangular rim
Sample No. 19, 23 and 26

Fabric	Fine to Medium sandy with tiny mica and white inclusion.
Surface enhancement	Sample No. 19, bowl with corrugation on neck and body and unslip on both the surfaces. Sample No. 23, bowl with corrugation on neck and shoulder having red slip on the both the surfaces. Sample No. 26 with blunt carination on exterior surface and incised grooves on the neck.
Size range	Bowls measures 11 to 25 cms in rim diameter.

Bowls Type C Variety 1: straight-sided bowls have one variety and one rim variant.
(a) simple perpendicular, straight rounded rim.
Sample No. 31, 41 and 42

Fabric	Fine sandy with tiny mica and white inclusion.
Surface enhancement	Sample No. 31 plain gray ware having Gray slip on both the surfaces with thin striation marks. Sample No. 41 black slip ware having black slip on both the surfaces. Sample No. 42 black and red ware having black color slip on the interior surface and reddish yellow slip on the exterior surface.
Size range	Bowls measures 12 to 27 cms in rim diameter.

Pots Type A Variety 1: small size pots with rim diameter ranges between 12 to 18 cms having three varieties, first variety with short neck and bulbous body and three rim variants:
(a) simple external projecting, extremely everted rounded rim
(b) external projecting, long, wide rounded rim
(c) external ledge angular rim
Sample No. 5, 8 and 18

Fabric	Fine to Coarse sandy with tiny mica and white inclusion.
Surface enhancement	Sample No. 5 having light red slip on the exterior surface. Sample No. 8 slips worn-out from both the surfaces. Sample No. 18 gray ware pot having gray slip on both the surfaces.
Size range	Pots measures 12 to 17 cms in rim diameters.

Pots Type A Variety 2: small size pots with rim diameter ranges between 12 to 18 cms having three varieties, second variety with long neck and bulbous body and one rim variant:
(a) overturned, small, external projection rim.
Sample No. 7

Fabric	Medium sandy with tiny mica, white inclusion and organic matter.
Surface enhancement	Sample No. 7 having light red slip on the exterior surface.
Size range	Pot measures 15 cm in rim diameters.

Pots Type A Variety 3: small size pots with rim diameter ranges between 12 to 18 cms having three varieties, third variety with carination and two rim variants:
(a) simple perpendicular, straight rounded rim
(b) external projecting, long, wide rounded rim
Sample No. 12, 14 and 24

Fabric	Fine sandy with tiny mica and white inclusion.
Surface enhancement	Sample No. 12 and 24 having red slip on both the surfaces. Sample No. 14 slip worn-out from both the surfaces.

Size range	Pots measures 12 to 18 cms in rim diameters.

Pots Type B Variety 1: medium size pots with rim diameter ranges between 18 to 22 cms having three varieties, first variety with short neck and bulbous body and three rim variants:
(a) external everted rounded rim
(b) external projecting, short, narrow triangular rim
(c) external ledge angular rim
Sample No. 4, 9 and 15

Fabric	Fine sandy with tiny mica and white inclusion.
Surface enhancement	Sample No. 4 and 15 having red slip on both the surfaces. Sample No. 9 having light red slip on the exterior surface.
Size range	Pots measures 18 to 22 cms in rim diameters.

Pots Type B Variety 2: medium size pots with rim diameter ranges between 18 to 22 cms having three varieties, second variety with long neck and bulbous body and one rim variant:
(a) external projecting, long, wide rounded rim
Sample No. 10

Fabric	Medium sandy with tiny mica and white inclusion.
Surface enhancement	Sample No. 10 having red slip on the exterior surface.
Size range	Pot measures 20 cm in rim diameters.

Pots Type B Variety 3: medium size pots with rim diameter ranges between 18 to 22 cms having three varieties, third variety with carination and two rim variants:
(a) external projecting, long, wide beaked rim
(b) external projecting, long, wide quadrangular rim
Sample No. 1, 2, 3, 6, 11, 13 and 16

Fabric	Fine to Medium to Coarse sandy with tiny mica, white inclusion and organic matter.
Surface enhancement	Sample No. 1 and 6 slip worn-out from both the surfaces. Sample No. 2, 11, 13 and 16 having red slip on both the surfaces. Sample No. 3 having light red slip on both the surfaces.
Size range	Pots measures 19 to 22 cms in rim diameters.

Basins Type A Variety 1: thick sturdy basin having one variety with carination and incised notches and one rim variant:
(a) bilateral projecting, equal projection rim
Sample No. 28 and 29.

Fabric	Medium sandy with tiny mica and white inclusion.
Surface enhancement	Sample No. 28 having reddish brown slip on both the surfaces. Sample No. 29 having red slip on both the surfaces.
Size range	Basin measures 36 cms in rim diameter.

Lamp Variety 1: having one variety and one rim variant:
(a) simple perpendicular, straight rounded rim
Sample No. 43

Fabric	Fine sandy with tiny mica and white inclusion.
Surface enhancement	Sample No. 43 slip worn-out from both the surfaces.
Size range	Lamp measures 5 cm in rim diameter.

Undiagnostic Incised Sherds:
Sample No. 44, 45, 46 and 47

Fabric	Fine to Medium sandy with mica and white inclusion.
Surface enhancement	Sample No. 44, Sherd having incised stamp marks on the exterior surface. Sample No. 45 and 46, Sherd having horizontal incised lines on the exterior surface. Sample No. 47, Sherd having nail-headed incised notches on the exterior surface. Sample No. 48, Sherd having appliqué pattern on the exterior surface.

21. MATAJI NO TIMBO (GOVINDHPURA) (Figure - 4.57)

Bowls Type A Variety 1: convex-sided bowls having one variety and one rim variant:
(a) external ledge angular rim
Sample No. 11

Fabric	Coarse sandy with White inclusion and Organic matter
Surface enhancement	Sample No. 11 Unslip on both the surfaces with sherd having incised oblique lines on neck and shoulder
Size range	Bowl measures 32 cm in rim diameter.

Bowls Type B Variety 1: carinated bowls having one variety with carinated shoulder and three rim variants:
(a) external projecting, long, wide rounded rim
(b) external projecting, long, wide triangular rim
(c) bilateral projecting, equal projection rim
Sample No. 12, 14, 15 and 16

Fabric	Medium sandy with Mica, White inclusion and Organic matter
Surface enhancement	Sample No. 12 having Red Slip on both the surfaces Sample No. 14 having Light Red Slip on both the surfaces Sample No. 15 and 16 Unslip on both the surfaces
Size range	Bowls measures 19 to 26 cms in rim diameter.

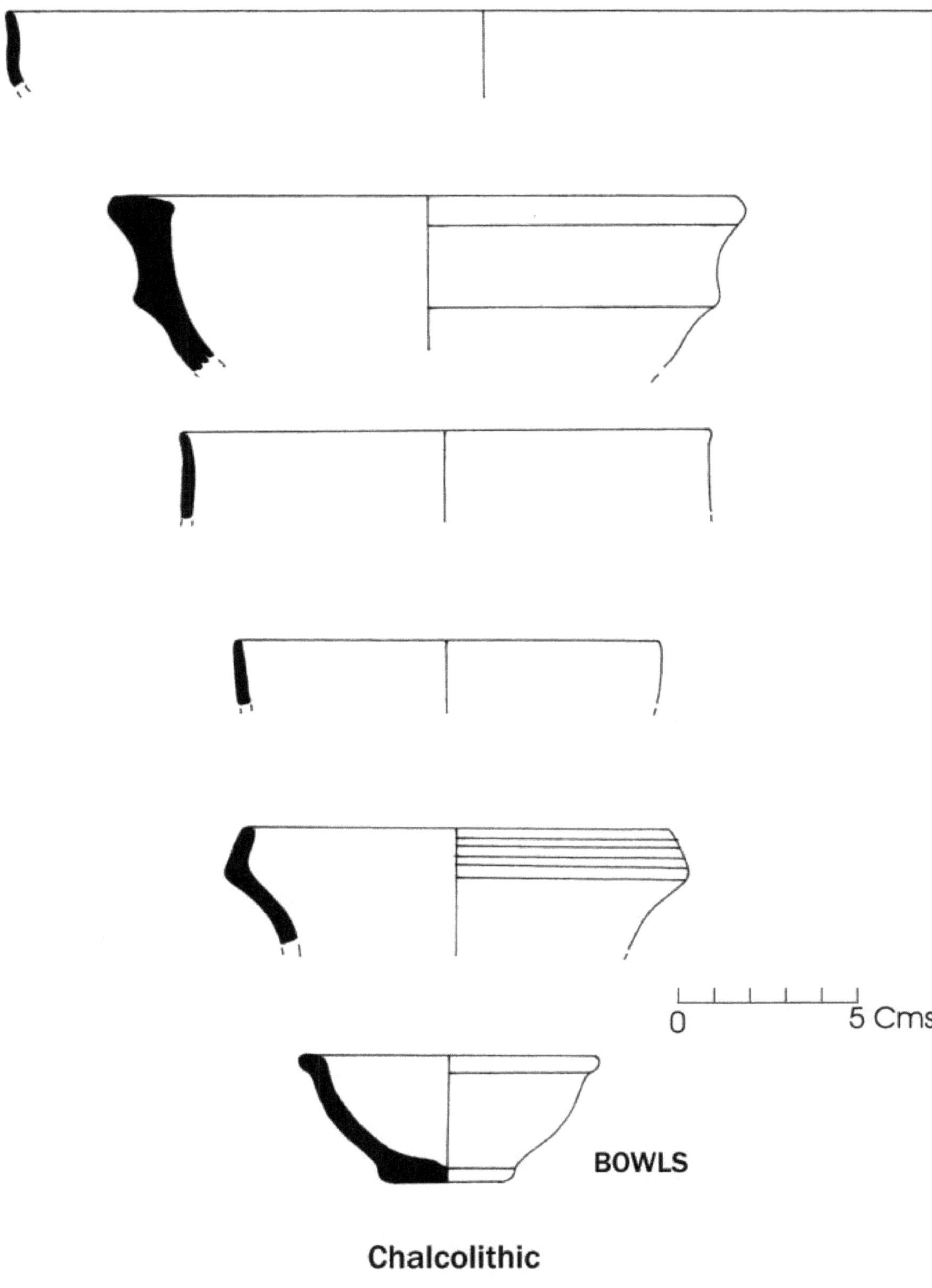

Figure 4.56 (a) – Kot (Sunari)

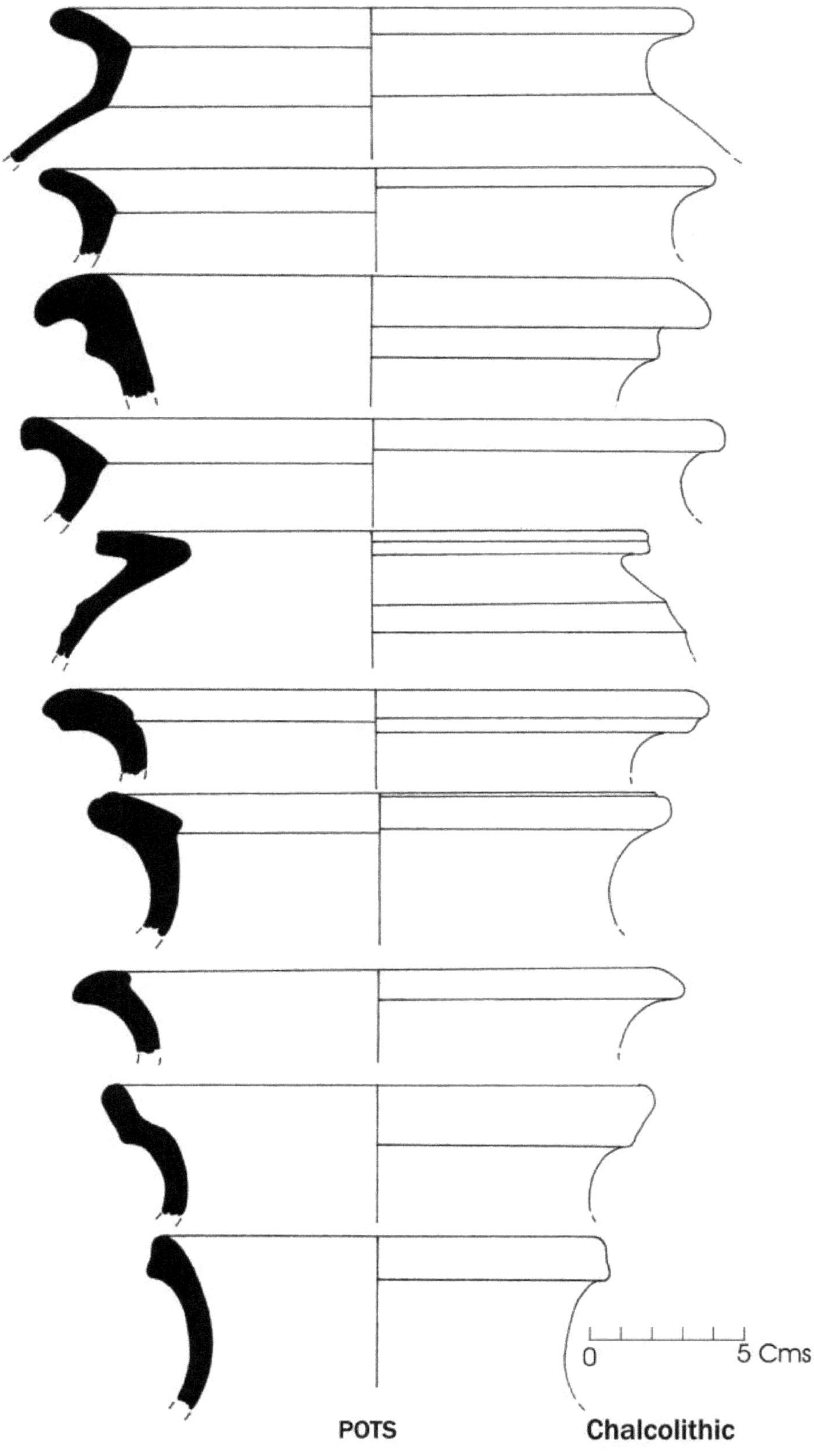

Figure 4.56 (b) – Kot (Sunari)

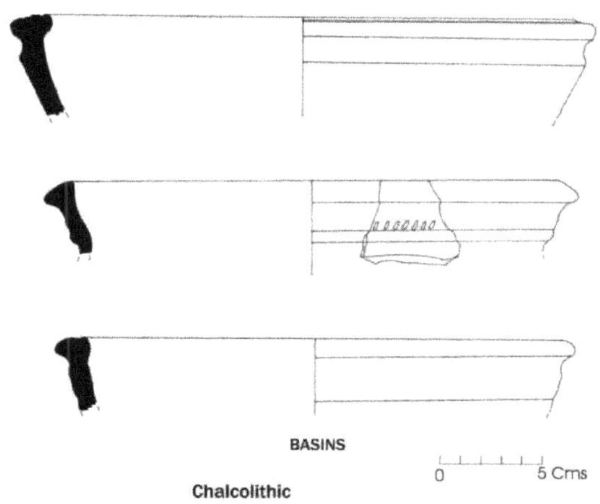

Figure 4.56 (c) – Kot (Sunari)

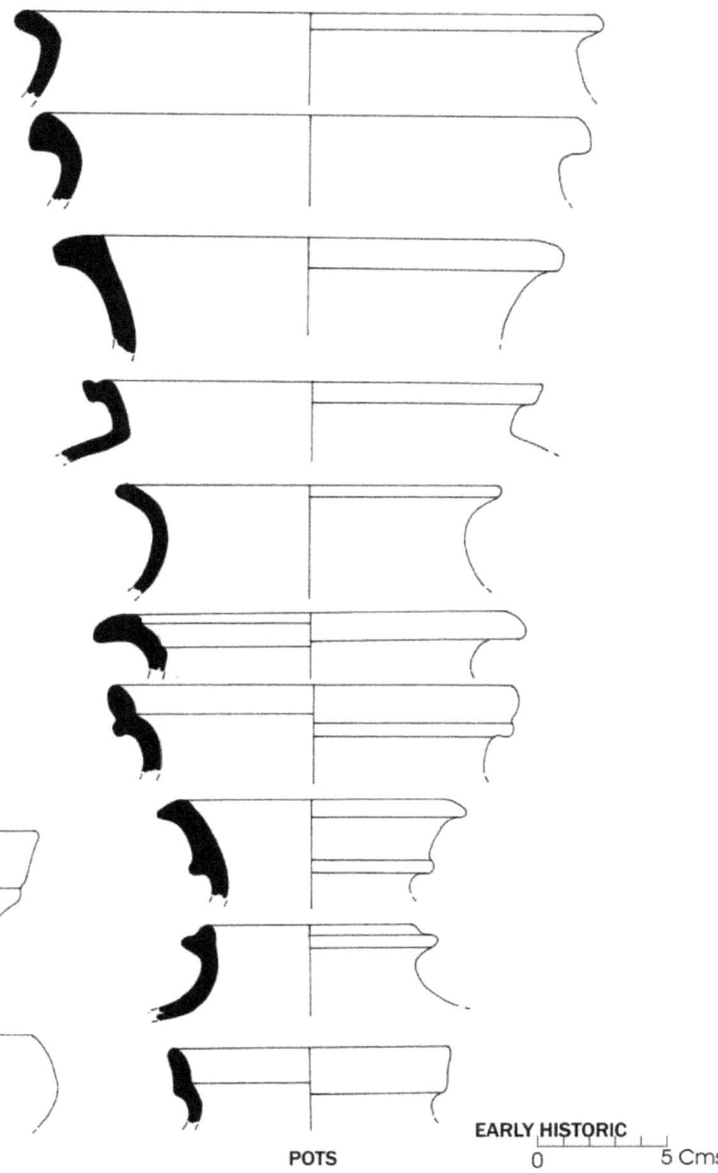

Figure 4.56 (d) – Kot (Sunari)

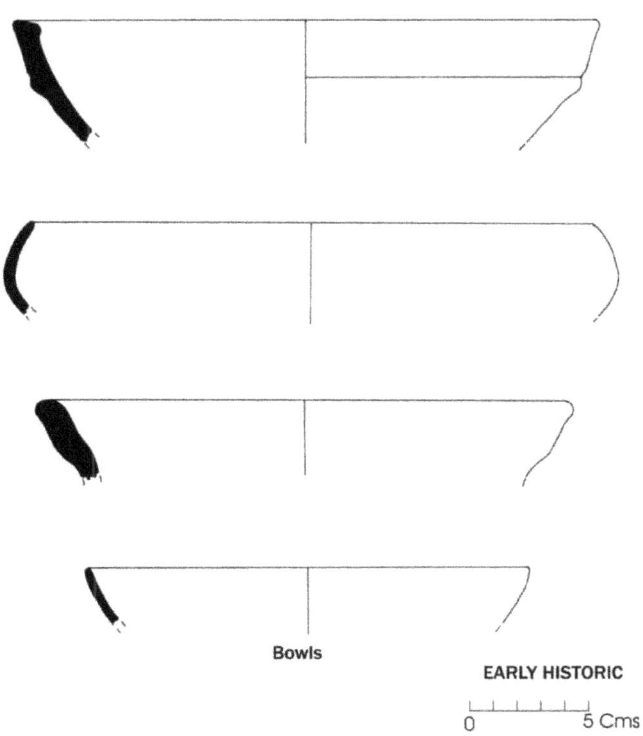

Figure 4.56 (e) – Kot (Sunari)

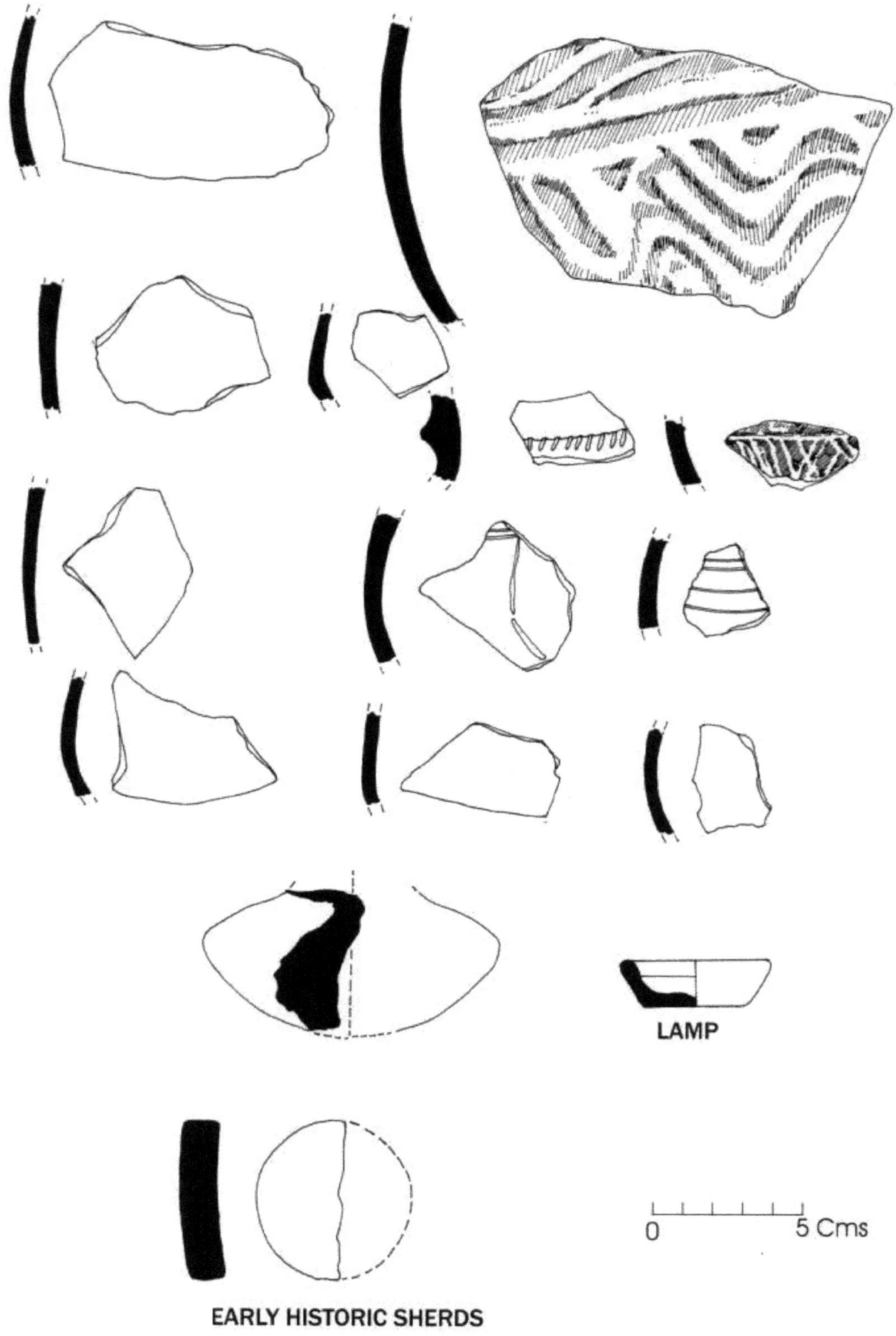

Figure 4.56 (f) – Kot (Sunari)

Bowls Type C Variety 1: *straight-sided bowls having one variety and two rim variants:*

(a) external projecting, short, narrow triangular rim
(b) internal projecting, short, narrow rounded rim
Sample No. 9 and 10

Fabric	Medium sandy with Mica and White inclusion
Surface enhancement	Sample No. 9 Unslip on both the surfaces Sample No. 10 having Light Red Slip on both the surfaces
Size range	Bowls measures 15 to 19 cms in rim diameter.

Pots Type A Variety 1: small size pots with rim diameter ranges between 11 to 18 cms having one variety with short neck and bulbous body and four rim variants:
(a) external projecting, short, narrow triangular rim
(b) external projecting, long, wide rounded rim
(c) external projecting, long, wide quadrangular rim
(d) external ledge angular rim
Sample No. 1, 4, 5 and 6

Fabric	Medium sandy with Mica, White inclusion and Organic matter
Surface enhancement	Sample No. 1 having Red Slip on both the surfaces Sample No. 4, 5 and 6 Unslip on both the surfaces
Size range	Pots measures 11 to 18 cms in rim diameters.

Pots Type B Variety 1: medium size pots with rim diameter ranges between 18 to 22 cms having one variety with short and ribbed neck and bulbous body and four rim variants:
(a) external projecting, short, narrow tapered rim
(b) external projecting, long, wide rounded rim
(c) internal projecting, short, narrow triangular rim
(d) bilateral projecting, equal projection rim
Sample No. 2, 3, 7 and 8

Fabric	Medium to Coarse sandy with Mica, White inclusion and Organic matter
Surface enhancement	Sample No. 2 having Red Slip on both the surfaces Sample No. 3, 7 and 8 Unslip on both the surfaces
Size range	Pots measures 16 to 22 cms in rim diameters.

Basins Type A Variety 3: carinated basins having one variety with angular shoulder and one rim variant:
(a) internal projecting, short, narrow rounded rim
Sample No. 19

Fabric	Medium sandy with Mica and White inclusion
Surface enhancement	Sample No. 19 Unslip on both the surfaces
Size range	Basin measures 36 cm in rim diameter

Dish Variety 1: dish have one variety with angular shoulder and one rim variant:
(a) simple everted, slightly rounded rim
Sample No. 13

Fabric	Medium sandy with Mica and White inclusion
Surface enhancement	Sample No. 13 Unslip on both the surfaces
Size range	Dish measures 13 cm in rim diameter.

Undiagnostic Incised Sherds:
Sample No. 20, 21 and 22

Fabric	Fine TO medium Too Coarse sandy with Mica, White inclusion and Organic matter
Surface enhancement	Sample No. 20 Unslip on both the surfaces with sherd having oblique lines with incised design between them Sample No. 21 having Gray Slip on both the surfaces with sherd having horizontal incised lines Sample No. 22 Unslip on both the surfaces with sherd having incised circular mark

22. MEENO KI DHANI (Figure - 4.58)

Bowls Type B Variety 3: carinated bowls having one variety with blunt carination on neck and one rim variant:
(a) simple external projecting, extremely everted rounded rim
Sample No. 13, 14 and 15

Fabric	Medium sandy with Mica, White inclusion and Organic matter.
Surface enhancement	Sample No. 13 and 15 unslip on both the surfaces. Sample No. 14 Reddish Yellow Slip on both the surfaces.
Size range	Bowls measures 19 to 20 cms in rim diameter.

Bowls Type C Variety 1: straight-sided bowls having one variety and three rim variants:
(a) simple perpendicular, straight rounded rim
(b) simple external projecting, extremely everted rounded rim
(c) internal projecting, short, narrow triangular rim
Sample No. 12, 16, 17, 19, 20 and 21

Fabric	Medium sandy with Mica, White inclusion and Organic matter.
Surface enhancement	Sample No. 12 having Light Red Slip on both the surfaces. Sample No. 16, 17, 19, 20 and 21 having Red Slip on both the surfaces.
Size range	Bowls measures 20 to 32 cms in rim diameter.

Pots Type A Variety 1: small size pots with rim diameter ranges 15 cm having one variety with short neck and bulbous body and four rim variants:
(a) simple everted, slightly, curved rounded rim
(b) simple perpendicular, straight rounded rim

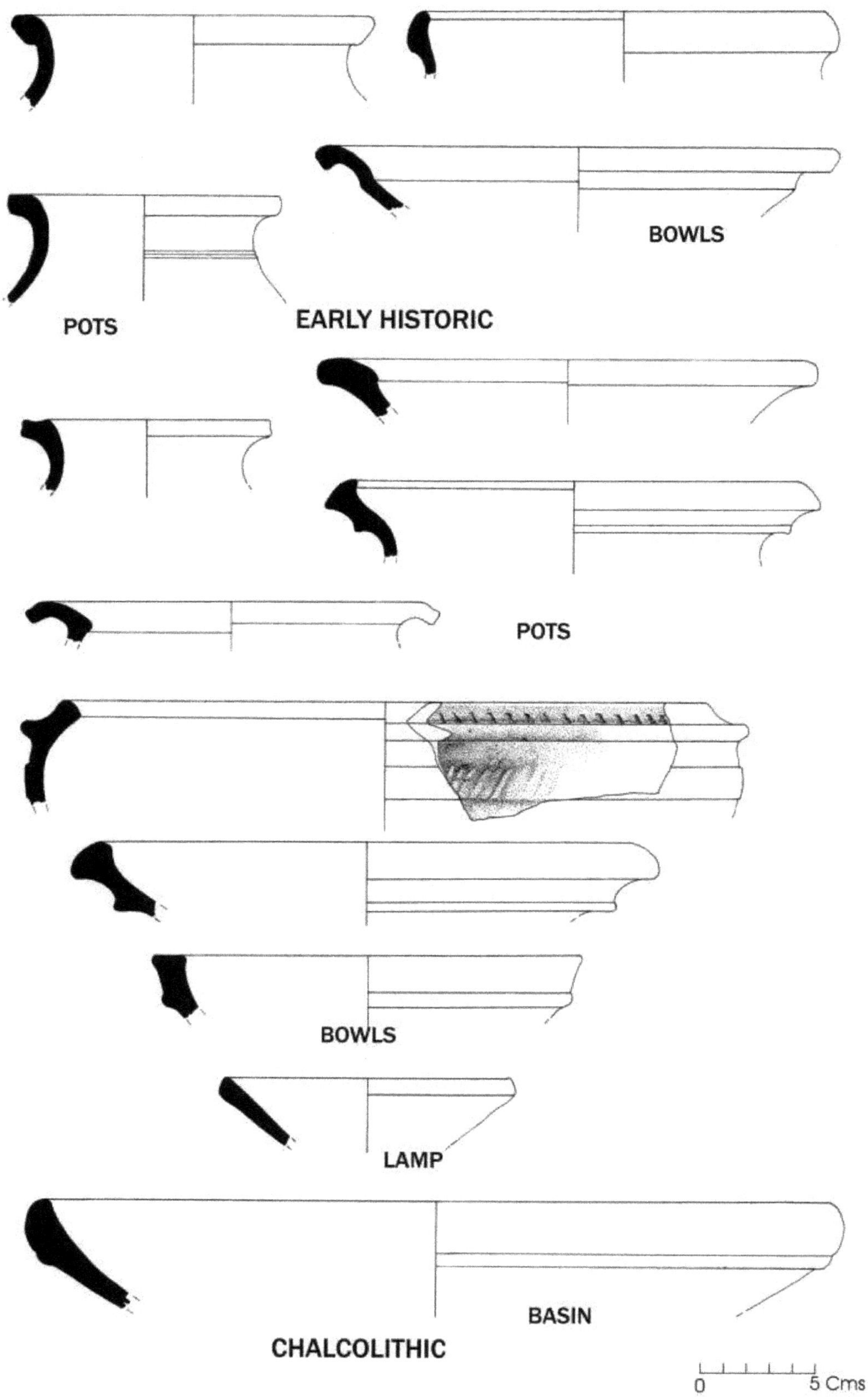

Figure 4.57 – Mataji no Timbo

(c) simple external projecting, extremely everted rounded rim
(d) external projecting, short, narrow triangular rim
Sample No. 4, 6, 9, 10 and 11

Fabric	Medium sandy with Mica, White inclusion and Organic matter.
Surface enhancement	Sample No. 4 having Red Slip on both the surfaces. Sample No. 6, 9 and 10 having Reddish Yellow Slip on both the surfaces. Sample No. 11 Unslip on both the surfaces.
Size range	Pots measures 15 cms in rim diameters.

Pots Type B Variety 1: medium size pots with rim diameter ranges between 18 to 20 cms having one variety with short neck and bulbous body and three rim variants:
(a) simple external projecting, extremely everted rounded rim
(b) simple external projecting, extremely everted tapered rim
(c) external projecting, short, narrow rounded rim
Sample No. 1, 2, 3, 5, 7 and 8

Fabric	Medium sandy with Tiny Mica, White inclusion and Organic matter.
Surface enhancement	Sample No. 1 Slip worn-out from both the surfaces. Sample No. 2 and 7 having Red Slip on both the surfaces. Sample No. 3 and 8 Unslip on both the surfaces. Sample No. 5 having Reddish Yellow Slip on the exterior surface and Weak Red Slip on the interior surface.
Size range	Pots measures 18 to 20 cms in rim diameters.

Turf Variety 1: turf having one variety and two rim variants:
(a) external projecting, short, narrow rounded rim
(b) overturned, large external projecting rim
Sample No. 22 and 23

Fabric	Medium sandy with Tiny Mica, White inclusion and Organic matter.
Surface enhancement	Sample No. 22 and 23 Unslip on both the surfaces.
Size range	Turf unmeasured

23. MOKULWAS (Figure - 4.59)

Bowls Type B Variety 3: carinated bowls having two varieties, first variety with blunt carinated shoulder and one rim variant:
(a) internal projecting, short, narrow triangular rim
Sample No. 13

Fabric	Medium sandy with Mica and White inclusion
Surface enhancement	Sample No. 13 having Red Slip on both the surfaces
Size range	Bowl measures 16 cm in rim diameter.

Bowls Type B Variety 4: carinated bowls having two varieties, second variety with angular shoulder and three rim variants:
(a) simple everted, straight rounded rim
(b) external projecting, short, narrow triangular rim
(c) bilateral projecting, equal projection rim
Sample No. 9, 10, 11, 12 and 14

Fabric	Medium to Coarse sandy with Mica, White inclusion and Organic matter
Surface enhancement	Sample No. 9, 10, 11 and 14 having Red Slip on both he surfaces Sample No. 12 having Light red Slip on both the surfaces
Size range	Bowls measures 18 to 24 cms in rim diameter.

Pots Type A Variety 1: small size pots with rim diameter ranges between 12 to 16 cms having one variety with short and ribbed neck and bulbous body and four rim variants:
(a) simple perpendicular, straight rounded rim
(b) simple external projecting, rounded rim
(c) external projecting, short, narrow rounded rim
(d) internal projecting, short, narrow triangular rim
Sample No. 1, 2, 3, 4, 5, 6, 7 and 8

Fabric	Medium to Coarse sandy with Mica and White inclusion
Surface enhancement	Sample No. 1, 2 5, 6 and 8 having Red Slip on both the surfaces with Sample No. 6 sherd having incised notches on the neck Sample No. 3 and 7 Unslip on both the surfaces Sample No. 4 Slip Worn – Out from both the surfaces
Size range	Pots measures 12 to 16 cms in rim diameters.

Dish Variety 1: dish have one variety with angular shoulder and one rim variant:
(a) simple everted, straight rounded rim
Sample No. 14

Fabric	Medium sandy with Mica and White inclusion
Surface enhancement	Sample No. 14 having Red Slip on both the surfaces
Size range	Dish measures 20 cm in rim diameter.

24. MURAJWALI No. 1 (Figure - 4.60)

Pots Type A Variety 4: small size pots with rim diameter ranges between 14 to 16 cms having one variety with short neck and corrugated neck and two rim variants:
(a) external projecting, long, wide quadrangular rim
(b) internal projecting, short, narrow rounded rim
Sample No. 1, 2 and 4

Fabric	Medium sandy with Mica, White inclusion and Organic matter

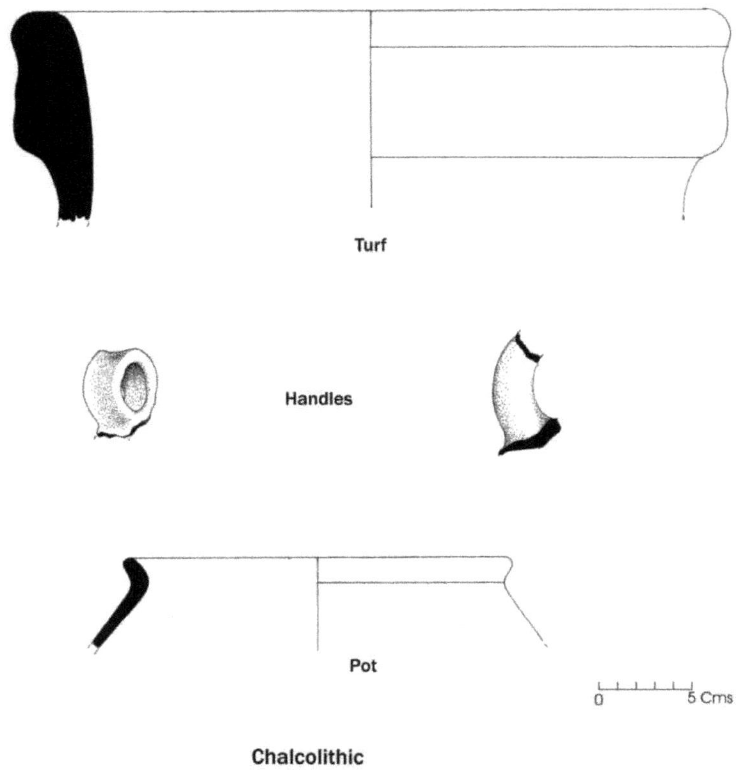

Figure 4.58 – Meeno ki Dhani

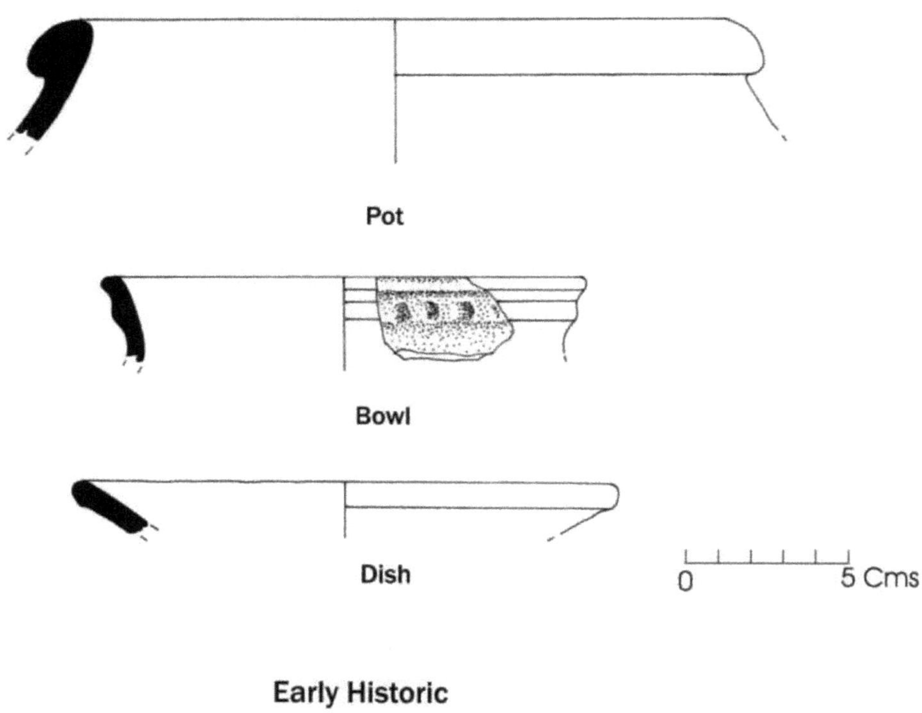

Figure 4.59 – Mokulwas

Surface enhancement	Sample No. 1 having light Red Slip on both the surfaces Sample No. 2 having red Slip on both the surfaces Sample No. 4 having Red Slip on the exterior surface and Light Red Slip on the interior surface
Size range	Pots measures 14 to 16 cms in rim diameters.

Pots Type B Variety 1: medium size pots with rim diameter 20 cm having one variety with short and ribbed neck and one rim variant:
(a) external projecting, short, narrow short beaked rim
Sample No. 3

Fabric	Medium sandy with Mica and White inclusion
Surface enhancement	Sample No. 3 having Red Slip on the both the surfaces
Size range	Pot measures 20 cm in rim diameters.

25. NALOT (Figure - 4.61)

Bowls Type A Variety 1: convex-sided bowls having one variety and one rim variant:
(a) simple internal projecting, extremely inverted rounded rim
Sample No. 8

Fabric	Medium sandy with White inclusion
Surface enhancement	Sample No. 8 having Light Red Slip on both the surface with sherd having incised horizontal lines on the body
Size range	Bowl measures 35 cm in rim diameter.

Bowls Type B Variety 1: carinated bowls having one variety with blunt carinated shoulder and two rim variants:
(a) simple external projecting, extremely everted rim
(b) external projecting, short, narrow tapered rim
Sample No. 2, 4, 6 and 10

Fabric	Medium sandy with Mica and White inclusion
Surface enhancement	Sample No. 2 having Red Slip on both the surfaces Sample No. 4 Unslip on both the surfaces Sample No. 6 Slip Worn-Out from both the surfaces Sample No. 10 having Red Slip on the exterior surface and Light Red Slip on the interior surfaces with sherd having incised mark on the interior surface
Size range	Bowls measures 14 to 29 cms in rim diameter.

Bowls Type C Variety 1: straight-sided bowls having one variety and one rim variant

rim variant
(a) bilateral projecting, equal projection rim
Sample No. 9

Fabric	Medium sandy with Mica
Surface enhancement	Sample No. 9 having Red Slip on both the surfaces
Size range	Bowl measures 35 cm in rim diameter

Pots Type A Variety 1: small size pots with rim diameter ranges between 14 to 15 cms having one variety with short ribbed neck and bulbous body and three rim variants:
(a) simple external projecting, extremely everted rim
(b) external projecting, short, narrow triangular rim
(c) bilateral projecting, equal projection rim
Sample No. 1, 3, 5 and 7

Fabric	Medium sandy with Mica and White inclusion
Surface enhancement	Sample No. 1 and 5 Slip Worn-Out from both the surfaces Sample No. 3 Unslip on both the surfaces Sample No. 7 having Reddish Brown Slip on both the surfaces
Size range	Pots measures 14 to 32 cms in rim diameters.

Basins Type A Variety 3: small size pots with rim diameter ranges between 14 to 15 cms having one variety with angular shoulder and two rim variants:
(a) external projecting, long, wide triangular rim
(b) bilateral projecting, equal projection rim
Sample No. 11 and 12

Fabric	Medium sandy with Mica, White inclusion and Organic matter
Surface enhancement	Sample No. 11 having Red Slip on both the surfaces Sample No. 12 Unslip on both the surfaces
Size range	Basins measures 30 to 32 cms in rim diameter

26. PAYGA KUNDALIA (Figure - 4.62)

Bowls Type B Variety 1: carinated bowls have two varieties, first variety with carinated shoulder, corrugated neck and incised notches and two rim variants:
(a) simple perpendicular, straight rounded rim
(b) internal projecting, short, narrow triangular rim
Sample No. 1, 2 and 3

Fabric	Coarse sandy with Abundant of Mica, White inclusion and Organic matter.

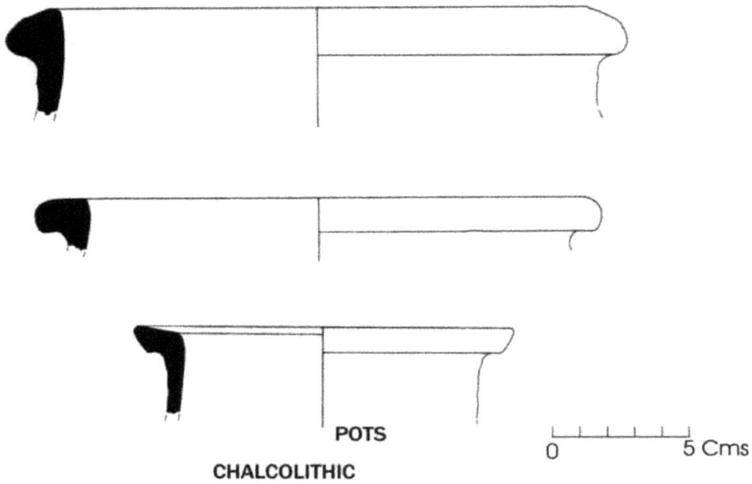

Figure 4.60 – Murajwali No 1

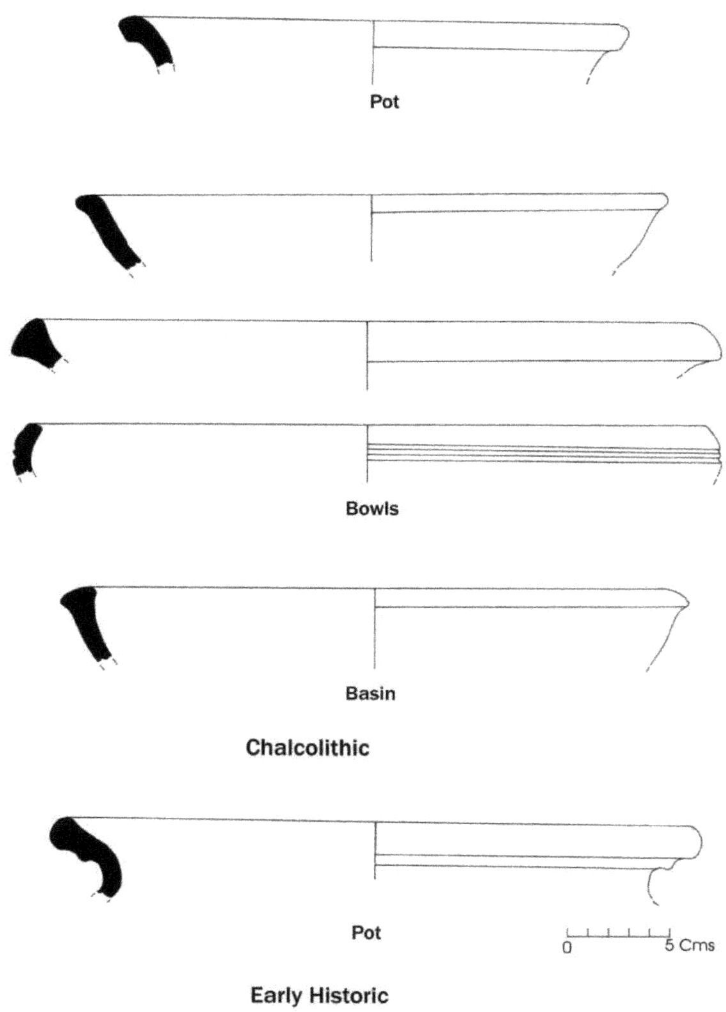

Figure 4.61 – Nalot

Surface enhancement	Sample No. 3 Unslip on both the surfaces with the sherd having incised triangular notches on the shoulder
Size range	Bowl measures 22 to 33 cms in rim diameter.

Bowls Type B Variety 3: carinated bowls have two varieties, second variety with blunt carinated shoulder and one rim variant:
(a) internal projecting, short, narrow triangular rim
Sample No. 4, 5 and 6

Fabric	Coarse sandy with Abundant of Mica, White inclusion and Organic matter.
Surface enhancement	Sample No. 4 Unslip on both the surfaces. Sample No. 5 and 6 having red slip on the interior surface.
Size range	Bowl measures 22 to 24 cms in rim diameter.

Pots Type A Variety 1: small size pots with rim diameter ranges between 15 to 18 cms having two varieties, first variety with short neck and bulbous body and one rim variant:
(a) external projecting, short, narrow tapered rim
Sample No. 12, 14 and 15

Fabric	Medium sandy with Tiny Mica and White inclusion.
Surface enhancement	Sample No. 12, 14 and 15 having red s lip on the exterior surface.
Size range	Pot measures 15 to 18 cms in rim diameters.

Post Type A Variety 4: small size pots with rim diameter ranges between 15 to 18 cms having two varieties, second variety with carinated shoulder and one rim variant:
(a) external projecting, short, narrow tapered rim
Sample No. 13

Fabric	Medium sandy with Tiny Mica and White inclusion.
Surface enhancement	Sample No. 13 having red s lip on the exterior surface.
Size range	Pot measures 13 cm in rim diameters.

Pots Type B Variety 1: medium size pots with rim diameter ranges between 16 to 19 cms having two varieties, first variety with short neck and bulbous body; second variety with short neck and bulbous body and two rim variants:
(a) external projecting, short, narrow, triangular rim
(b) external projecting, short, narrow tapered rim
Sample No. 8 and 11

Fabric	Medium sandy with Abundant Mica and White inclusion.
Surface enhancement	Sample No. 8 having red slip on both the surfaces with the sherd having incised hatches on the shoulder. Sample No. 11 slip worn-out on both the surfaces
Size range	Pot measures 16 to 19 cms in rim diameters.

Pots Type B Variety 3: medium size pots with rim diameter ranges between 16 to 19 cms having two varieties, second variety with long neck and bulbous body and one rim variant:
(a) external projecting, short, narrow, short quadrangular rim
Sample No. 9

Fabric	Medium sandy with Abundant Mica and White inclusion.
Surface enhancement	Sample No. 9 Unslip on both the surfaces.
Size range	Pot measures 16 cm in rim diameters.

Basins Type A Variety 3: carinated basins having one variety with angular shoulder and one rim variant:
(a) internal projecting, short, narrow triangular rim
Sample No. 20

Fabric	Medium sandy with Mica, White inclusion and Organic matter.
Surface enhancement	Sample No. 20 having Red Slip on both the surfaces.
Size range	Basin measures 22 cm in rim diameters.

Basins Type B Variety 1: thick sturdy basin having one variety and two rim variants:
(a) internal projecting, short, narrow triangular rim
(b) internal projecting, long, wide beaked rim
Sample No. 16, 17, 18, 19 and 22

Fabric	Medium sandy with Tiny Mica, White inclusion and Organic matter.
Surface enhancement	Sample No. 16 and 18 unslip on both the surfaces. Sample No. 17 slip worn-out from both the surfaces. Sample No. 19 and 22 having red slip on both the surfaces.
Size range	Basin measures 28 to 36 cms in rim diameters.

Basins Type C Variety 1: straight-sided basins having one variety and two rim variants:
(a) external projecting, short, narrow short quadrangular rim
(b) internal projecting, short, narrow triangular rim
Sample No. 21, 23, 24 and 25

Fabric	Medium sandy with Tiny Mica and White inclusion.
Surface enhancement	Sample No. 21 and 24 having Red Slip on both the surfaces. Sample No. 23 and 25 Unslip on both the surfaces.
Size range	Basin measures 22 to 28 cms in rim diameters.

Undiagnostic Sherds: having incised lines on the exterior surface
Sample No. 10

Fabric	Medium sandy with Mica and White inclusion
Surface enhancement	Sample No. 10 having Red Slip on both the surfaces

27. SADH MATA KA MAULA (GANWARI) (Figure – 4.63)

Bowls Type A Variety 1: convex–sided bowls having one variety and one rim variant:
(a) simple perpendicular, straight squared rim
Sample No. 15 and 27

Fabric	Medium sandy with Abundant Mica and White inclusion.
Surface enhancement	Sample No. 15 Unslip on both the surfaces. Sample No. 27 having Red Slip on both the surfaces.
Size range	Bowl measures 24 cm in rim diameter.

Bowls Type B Variety 4: carinated bowl having one variety with carinated shoulder and corrugated neck and one rim variant:
(a) simple perpendicular, straight rounded rim
Sample No. 18

Fabric	Medium sandy with Tiny Mica and White inclusion.
Surface enhancement	Sample No. 18 having red slip on both the surfaces. Sample No. 24 having red slip on both the surfaces with black color band on the interior surface.
Size range	Bowl measures 12 cm in rim diameter.

Pots Type A Variety 1: small size pots with rim diameter ranges between 12 to 16 cms having one variety with short neck and four rim variants:
(a) simple everted, slightly curved tapered rim
(b) external projecting, short, narrow triangular rim
(c) external projecting, long, wide rounded rim
(d) internal projecting, short, narrow triangular rim
Sample No. 3, 4, 5, 7, 8, 9, 11, 12 13 and 14

Fabric	Medium sandy with Mica and White inclusion.
Surface enhancement	Sample No. 3, 8, 10, 11, 12, 13 and 14 having Red Slip on the exterior surface. Sample No. 4 having Reddish Yellow Slip on both the surfaces. Sample No. 5 having Yellowish Red Slip on both the surfaces. Sample No. 7 and 9 having Light Red Slip on both the surfaces.
Size range	Pots measures 12 to 16 cms in rim diameters.

Pots Type B Variety 1: medium size pots with rim diameter ranges between 20 to 22 cms having one variety with short neck and bulbous body and four rim variants:
(a) simple everted, straight rounded rim
(b) external projecting, short, narrow triangular rim
(c) overturned small, external projection rim
(d) internal projecting, short, narrow triangular rim
Sample No. 1, 2, 6 and 10

Fabric	Medium sandy with Mica, White inclusion and Organic matter.
Surface enhancement	Sample 1 and 2 having Light Red Slip on both the surfaces. Sample No. 6 and 10 having Red Slip on both the surfaces.
Size range	Pots measures 20 to 22 cms in rim diameters.

Basins Type A Variety 2: carinated basins having two varieties, first variety with blunt carinated shoulder and one rim variant:
(a) internal projecting, short, narrow triangular rim
Sample No. 20

Fabric	Medium sandy with Mica, White inclusion and Organic matter.
Surface enhancement	Sample No. 20 having Red Slip on both the surfaces.
Size range	Basin measures 26 cm in rim diameter.

Basins Type A Variety 3: carinated basins having two varieties, second variety with angular shoulder and one rim variant:
(a) bilateral projecting, equal projection rim
Sample No. 21

Fabric	Medium sandy with Mica and White inclusion.
Surface enhancement	Sample No. 21 having Red Slip on both the surfaces.
Size range	Basin measures 32 cm in rim diameter.

Dish Variety 1: dishes having two varieties, first variety with carinated neck and one rim variant:
(a) simple perpendicular, straight rounded rim
Sample No. 24

Fabric	Medium sandy with Mica and White inclusion.
Surface enhancement	Sample 24 having Red Slip on both the surfaces.
Size range	Dish measures 11 cm in rim diameter.

Dishes Variety 2: dishes having two varieties, second variety and one rim variant:
(a) simple everted, straight rounded rim
Sample No. 22, 23 and 25

Fabric	Medium sandy with Mica and White inclusion.

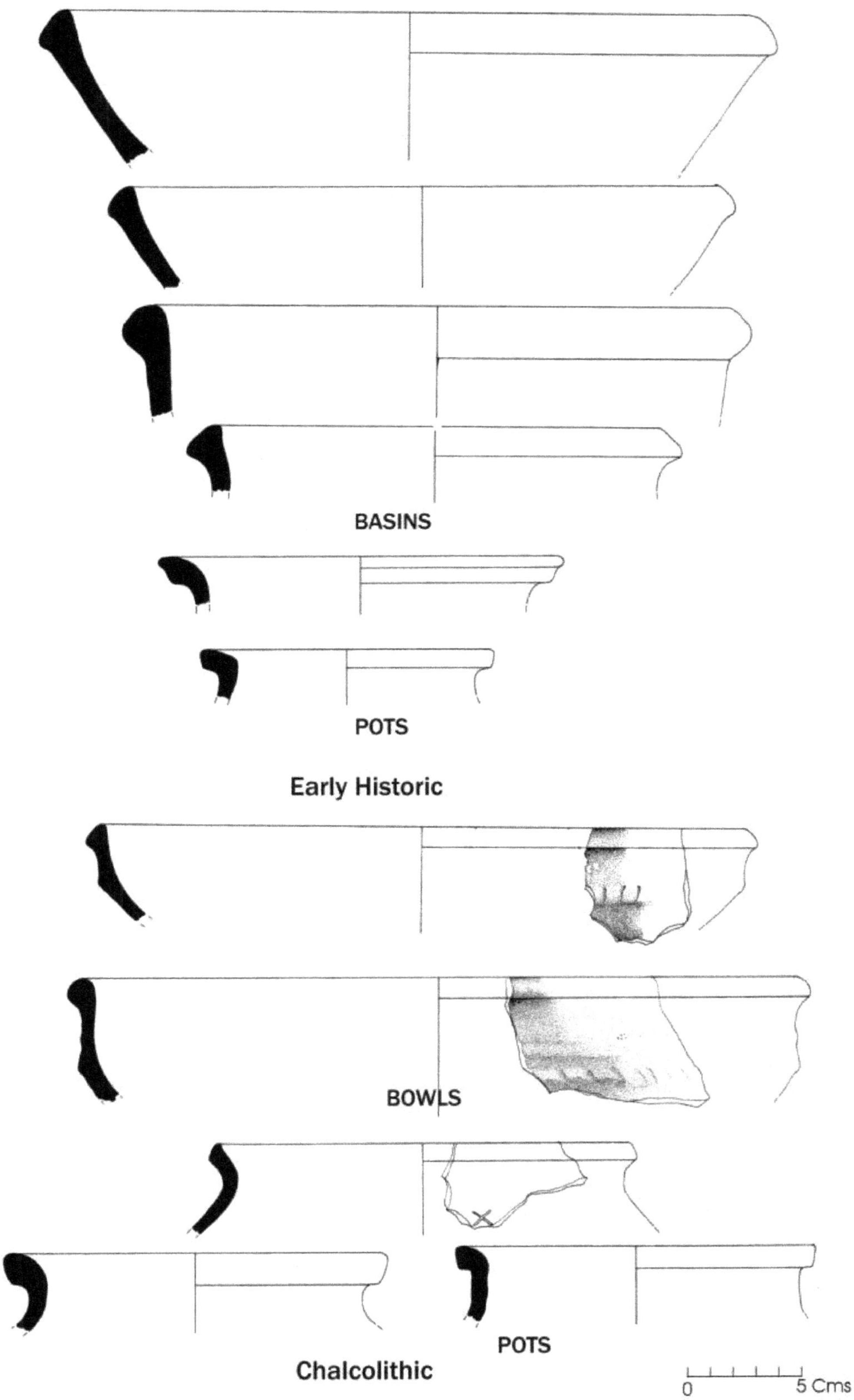

Figure 4.62 – Payga Kundalia

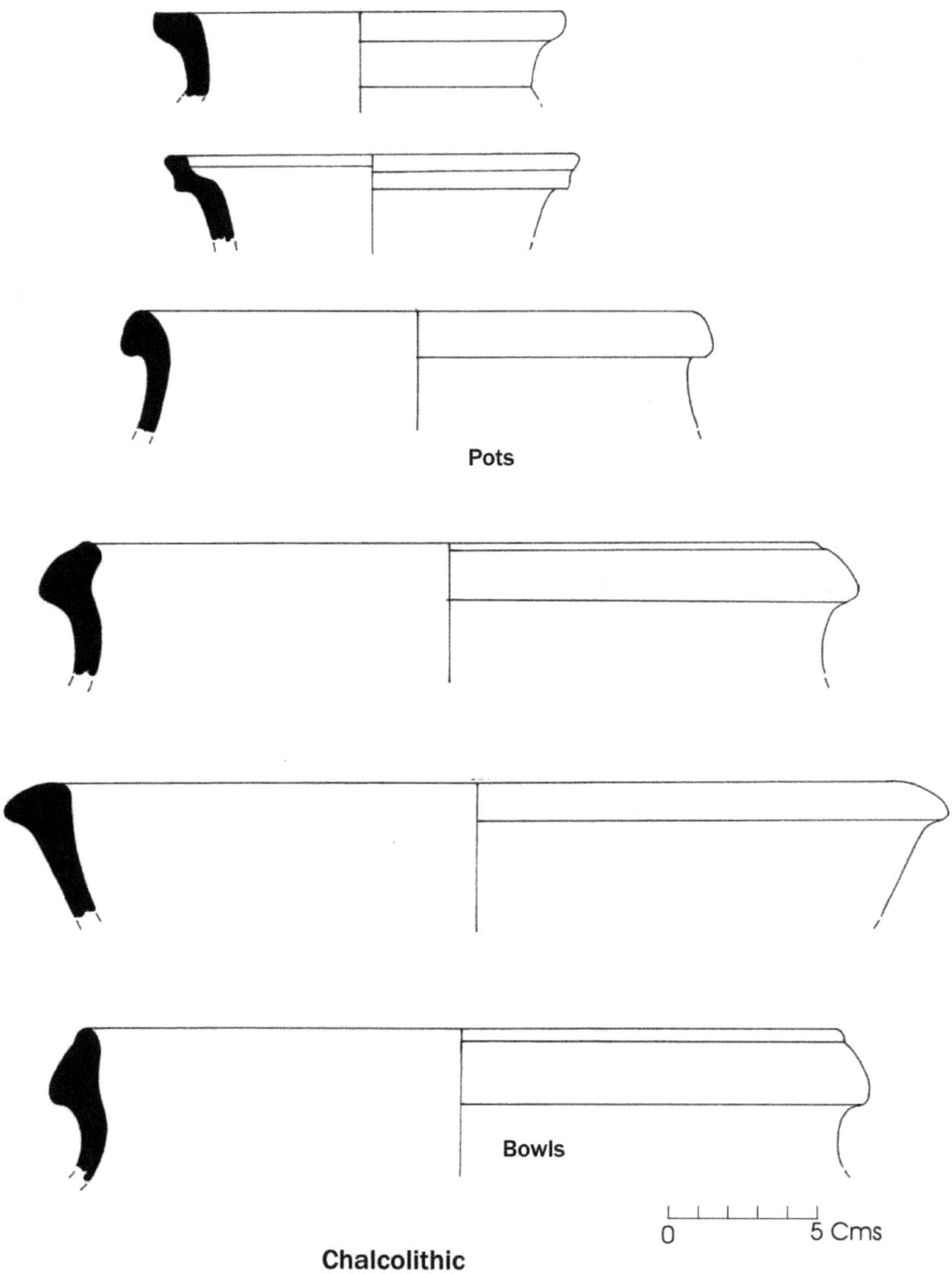

Figure 4.63 (a) – Sadmata ka Maula

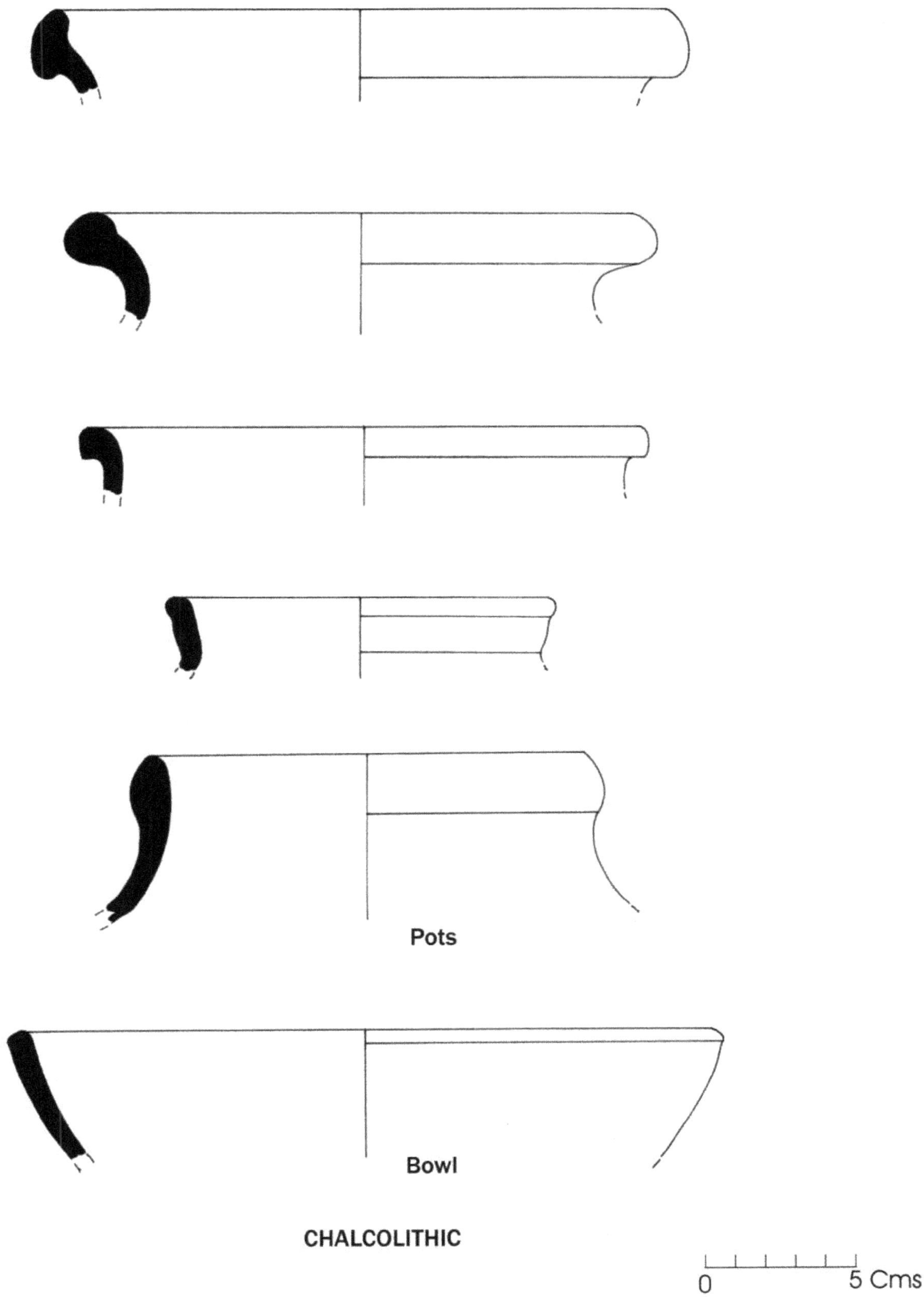

Figure 4.63 (b) – Sadmata ka Maula

Surface enhancement	Sample No. 22 and 23 Unslip on both the surfaces. Sample No. 25 having Light Red Slip on both the surfaces.
Size range	Dishes measures 16 to 21 cm in rim diameter.

28. SAINIYO KI DHANI (BEWA PATAN No. 2)
(Figure – 4.64_)

Bowl Type A Variety 1: convex–sided bowl having one variety and one rim variant:
(a) simple inverted, curved rounded rim.
Sample No. 12

Fabric	Medium sandy with White inclusion.
Surface enhancement	Sample No. 6 having Light Red Slip on both the surfaces
Size range	Bowl measures 21 cm in rim diameter.

Pots Type A Variety 1: small size pots with rim diameter ranges between 10 to 15 cms having two varieties, first variety with short and ribbed neck and one rim variant:
(a) internal projecting, short, narrow rounded rim
Sample No. 3

Fabric	Fine sandy with Mica and White inclusion
Surface enhancement	Sample No. 3 having Red Slip on both the surfaces.
Size range	Pot measures 15 cm in rim diameters.

Pots Type A Variety 4: small size pots with rim diameter ranges between 10 to 15 cms having two varieties, second variety with carinated shoulder and one rim variant:
(a) external ledge angular rim
Sample No. 4

Fabric	Medium sandy with Mica and White inclusion
Surface enhancement	Sample No. 4 with Slip Worn-Out from both the surfaces
Size range	Pot measures 10 cm in rim diameters.

Pots Type B Variety 1: medium size pots with rim diameter ranges between 20 to 22 cms having one variety with short neck and bulbous body and two rim variants:
(a) external projecting, short, narrow triangular rim
(b) external projecting, long, wide rounded rim
Sample No. 1, 2 and 5

Fabric	Medium sandy with Mica, White inclusion and organic matter
Surface enhancement	Sample No. 1 having Red Slip on both the surfaces with black color bad on lip of the rim on the interior surface and on neck on the exterior surface. Sample No. 2 having Red Slip on both the surfaces Sample No. 5 having Red Slip on the interior surface and Light Red Slip on the exterior surface
Size range	Pots measures 20 to 22 cms in rim diameters.

Basins Type A Variety 1: carinated basins having two varieties, first variety with corrugated neck and one rim variant:
(a) internal projecting, short, narrow triangular rim
Sample No. 7

Fabric	Medium sandy with Tiny Mica and White inclusion.
Surface enhancement	Sample No. 7 having Red Slip on the exterior surface and Light Red Slip on the interior surface
Size range	Basin measures 29 cms in rim diameter.

Basins Type A Variety 3: carinated basins having two varieties, second variety with angular shoulder and one rim variant:
(a) simple perpendicular, straight tapered rim
Sample No. 8

Fabric	Medium sandy with Mica, White inclusion and Organic matter
Surface enhancement	Sample No. 8 having light Red Slip on both the surfaces.
Size range	Basin measures 30cm in rim diameter.

Turf Variety 1: turf having one variety with corrugated neck and incised notches on neck and one rim variant:
(a) external projecting, long, wide triangular rim
Sample No. 26 and 27

Fabric	Medium sandy with Tiny Mica and White inclusion.
Surface enhancement	Sample No. 26 and 27 slip worn-out from both the surfaces.
Size range	Dishes measures 17 to 18 cms in rim diameter.

29. SALODARA - (Figure 4.65)

Bowls Type B Variety 3: carinated bowls having two varieties, first variety with blunt carinated shoulder and two rim variants:
(a) external projecting, long, wide triangular rim
(b) internal projecting, short, narrow triangular rim
Sample No. 6, 7 and 9

Fabric	Medium to Coarse sandy with Mica and White inclusion
Surface enhancement	Sample No. 6 Slip Worn-Out from both the surfaces Sample No. 7 Unslip on both the surfaces Sample No. 9 having Red Slip on both the surfaces
Size range	Bowls measures 15 to 23 cms in rim diameter.

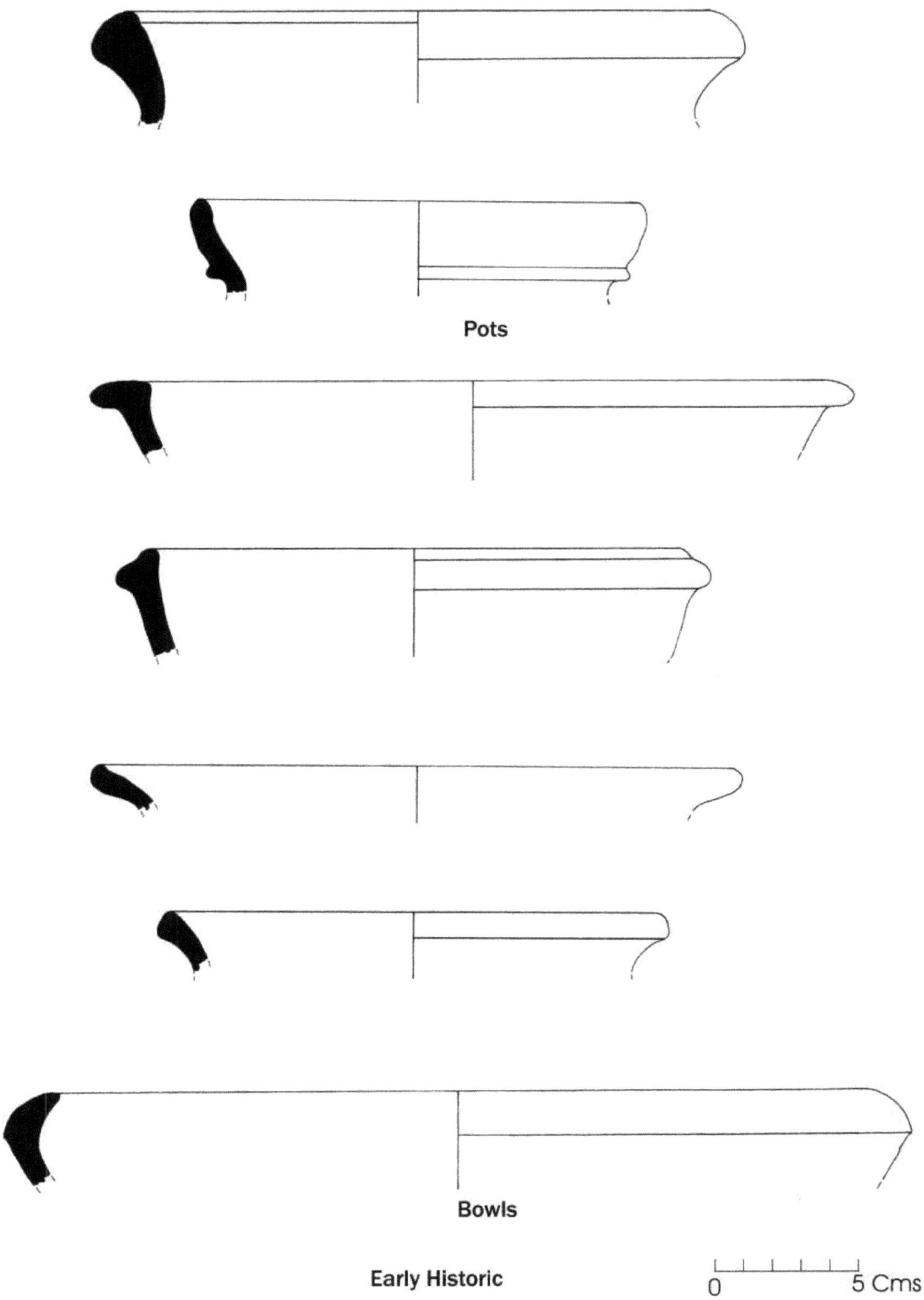

Figure 4.64 – Sainiyo ki Dhani

Bowls Type B Variety 4: carinated bowls having two varieties, second variety with angular shoulder and one rim variant:
(a) internal projecting, short, narrow triangular rim
Sample No. 8

Fabric	Coarse sandy with Mica, White inclusion and Organic matter
Surface enhancement	Sample No. 8 having Red Slip on both the surfaces
Size range	Bowl measures 16 cm in rim diameter

Pots Type A Variety 1: small size pots with rim diameter ranges between 11 to 16 cms having one variety with short neck and four rim variants:
(a) simple external projecting, everted rounded rim
(b) external projecting, short, narrow rounded rim
(c) internal projecting, short, narrow triangular rim
(d) internal projecting, short, narrow tapered rim
Sample No. 1, 2, 3, 4 and 5

Fabric	Medium sandy with Mica and White inclusion
Surface enhancement	Sample No. 1 having Light Red Slip on both the surfaces Sample No. 2 Unslip on both the surfaces Sample No. 3 and 5 having Red Slip on both the surfaces Sample No. 4 having Reddish Yellow Slip on both the surfaces
Size range	Pots measures 11 to 16 cms in rim diameters.

Undiagnostic Sherds:
Sample No. 10

Fabric	Medium sandy with Mica and White inclusion
Surface enhancement	Sample No. 10 Unslip on both the surfaces with the sherd having incised horizontal lines on the exterior surface

30. SWAMI KA KUHA (KOTRI) Figure (4.66)

Bowls Type A Variety 1: convex-sided bowls having one variety and two rim variants:
(a) simple inverted curved rounded rim
(b) simple internal projecting, extremely inverted rim
Sample No. 18 and 20

Fabric	Medium sandy with Mica, White inclusion and Organic matter
Surface enhancement	Sample No. 18 Unslip on both the surfaces Sample No. 20 having Light Red Slip on both the surfaces
Size range	Bowls measures 12 cms in rim diameter.

Bowls Type B Variety 1: carinated bowls having two varieties, first variety with carinated shoulder and unmeasured rim variant:
Sample No. 13

Fabric	Medium sandy with Mica and White inclusion
Surface enhancement	Sample No. 13 having Light Brownish Gray Slip on the exterior surface and Unslip on the interior surface with black color band on the neck
Size range	Bowl unmeasured rim diameter

Bowls Type B Variety 3: carinated bowls having two varieties, second variety with blunt carinated shoulder and one rim variant:
(a) simple perpendicular, straight rounded rim
Sample No. 19 and 21

Fabric	Medium sandy with Mica and White inclusion
Surface enhancement	Sample No. 19 Unslip on both the surfaces Sample No. 21 having Light Red Slip on the interior surface and Reddish Yellow Slip on the exterior surface
Size range	Bowls measures 17 to 18 cms in rim diameter.

Bowls Type C Variety 1: straight-sided bowls having one variety and two rim variants:
(a) internal projecting, short, marrow triangular rim
(b) bilateral projecting, equal projection rim
Sample No. 14,15,16 and 17

Fabric	Medium sandy with Mica and White inclusion
Surface enhancement	Sample No. 14 having Weal Red Slip on both the surfaces Sample No. 15 having Light Reddish Brown Slip on both the surfaces Sample No. 16 having Red Slip on the interior surface and Gray Slip on the exterior surface Sample No. 17 having Light Red Slip on both the surface
Size range	Bowls measures 10 to 21 cms in rim diameter.

Pots Type A Variety 1: small size pots with rim diameter ranges between 11 to 17 cms having one variety with short and ribbed neck and bulbous body and six rim variants:
(a) simple external projecting, extremely everted rim
(b) external projecting, short, narrow rounded rim
(c) external projecting, short, narrow triangular rim
(d) external projecting, short, narrow tapered rim
(e) external projecting, long, wide, beaked rim
(f) internal projecting, short, narrow triangular rim
Sample No. 1, 2, 3, 4, 5, 6, 8, 9, 10 and 11

Fabric	Medium to Coarse sandy with Mica, White inclusion and Organic matter
Surface enhancement	Sample No. 1, 2 and 5 having Red Slip on both the surfaces. Sample No. 1, Sherd having black color bands on the neck.

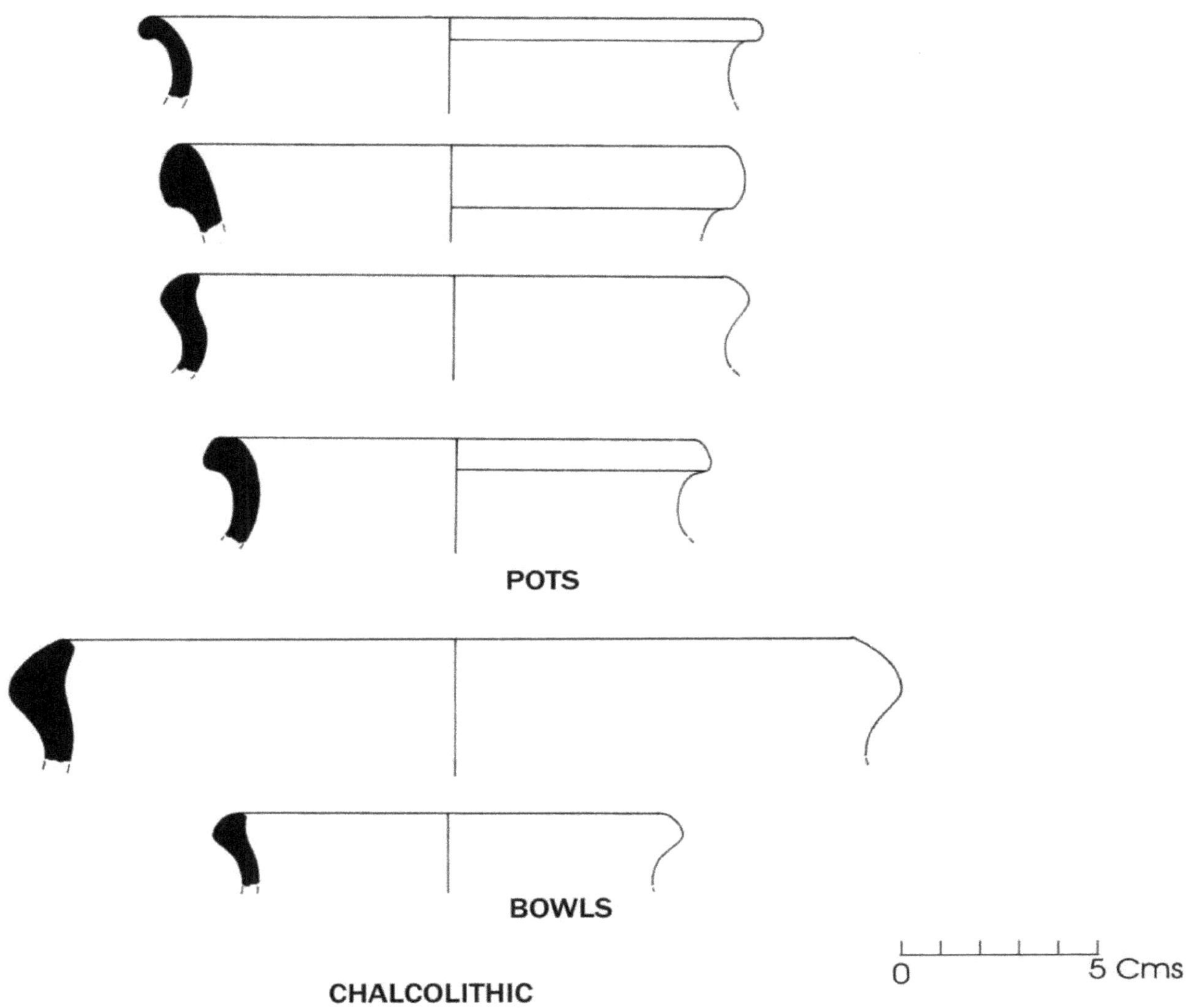

Figure 4.65 – Salodara

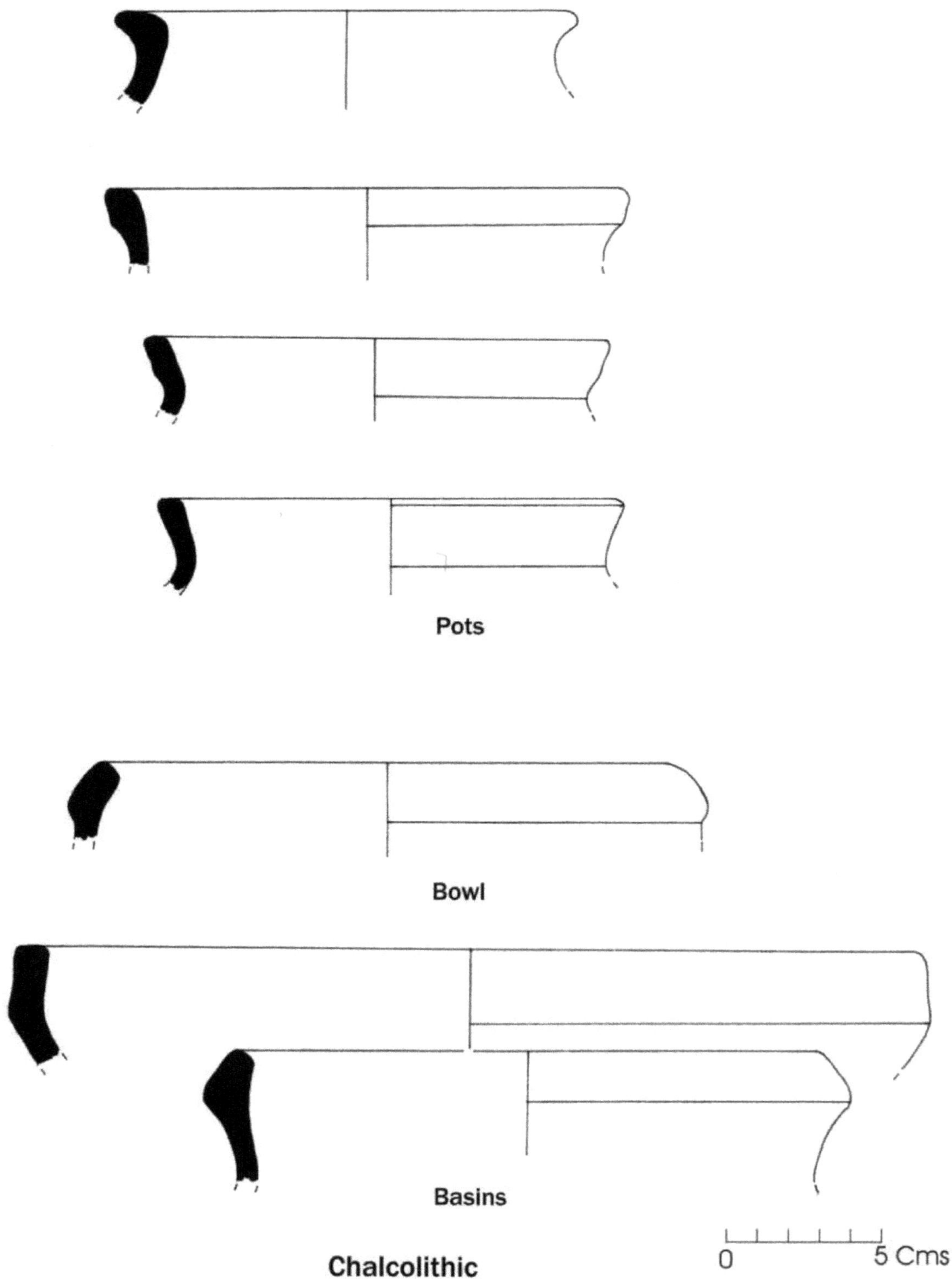

Figure 4.66 – Swamiyo ka Kuha

	Sample No. 3 having Light Red Slip on both the surfaces Sample No. 4 and 9 having Light Red Slip on the exterior surface and Unslip on the interior surface Sample No. 6, 8 and 11 Unslip o both the surfaces Sample
Size range	Pots measures 11 to 17 cms in rim diameters.

**** Sample No. 22, Bowl with External Ledge having Simple Perpendicular, Straight Rounded Rim. Sherd having series of incised design on the neck and shoulder*

Basins Type A Variety 1: carinated basins having three varieties, first variety with carinated shoulder and one rim variant:
(a) external projecting, short, narrow tapered rim
Sample No. 24

Fabric	Medium sandy with Mica and White inclusion
Surface enhancement	Sample No. 24 having Light Red Slip on both the surfaces
Size range	Basin measures 25 cm in rim diameter

Basins Type A Variety 2: carinated basins having three varieties, second variety with blunt carinated shoulder and one rim variant:
(a) simple perpendicular, straight rounded rim
Sample No. 25

Fabric	Medium sandy with White inclusion
Surface enhancement	Sample No. 25 having Red Slip on both the surfaces
Size range	Basin measures 30 cm in rim diameter

Basins Type A Variety 3: carinated basins having three varieties, third variety with angular shoulder and two rim variants:
(a) internal projecting, short, narrow triangular rim
(b) bilateral projecting, equal projection rim
Sample No. 23, 26 and 27

Fabric	Medium sandy with Mica, White inclusion and Organic matter
Surface enhancement	Sample No. 23 and 26 having Red Slip on both the surfaces Sample No. 27 having Light Red Slip on both the surfaces
Size range	Basins measures 20 to 21cms in rim diameter

31. TAPKESHWAR (Figure – 4.67)

Bowls Type B Variety 3: carinated bowls having two varieties, first variety with blunt carinated shoulder and one rim variant:
(a) external projecting, short, narrow, short beaked rim
Sample No. 4

Fabric	Coarse sandy with Mica, White inclusion and Organic matter
Surface enhancement	Sample No. 4 Unslip on both the surfaces
Size range	Bowl measures 19 cm in rim diameter.

Bowls Type B Variety 4: carinated bowls having two varieties, second variety with angular shoulder and two rim variants:
(a) external projecting, long, wide rounded rim
(b) external projecting, long, wide beaked rim
Sample No. 2 and 3

Fabric	Medium to Coarse sandy with Mica, White inclusion and Organic matter
Surface enhancement	Sample No. 2 and 3 Unslip on both the surfaces
Size range	Bowls measures 16 to 19 cms in rim diameter.

Pots Type A Variety 1: small size pots with rim diameter ranges between 14 to 17 cms having one variety with short neck and bulbous body and two rim variants:
(a) external projecting, long, wide rounded rim
(b) external projecting, short, narrow triangular rim
Sample No. 1, 5 and 7

Fabric	Medium to Coarse sandy with Mica, White inclusion and organic matter
Surface enhancement	Sample No. 1, 5 and 7 Unslip on both the surfaces
Size range	Pots measures 14 to 17 cms in rim diameters.

Basins Type A Variety 3: carinated basins having one variety with angular shoulder and one rim variant:
(a) external projecting, long, wide rounded rim
Sample No. 9

Fabric	Medium sandy with Tiny Mica, White inclusion and Organic matter
Surface enhancement	Sample No. 9 having red Slip on both the surfaces
Size range	Basin measures 36 cm in rim diameter.

Undiagnostic Sherds: having rustication on the body with applied bands of sandy texture
Sample No. 11 and 12

Fabric	Medium sandy with Tiny Mica, White inclusion and organic matter
Surface enhancement	Sample No. 11 and 12 having rustication on the exterior surface and Unslip on the both the surfaces

32. THANESHWAR (Figure – 4.68)

Bowls Type A Variety 1: convex – sided bowl having one variety and two rim variants:
(a) simple inverted, curved tapered rim
(b) simple inverted, curved rounded rim
Sample No. 8, 11 and 13

Fabric	Medium to Coarse sandy with Mica, White inclusion and Organic matter.
Surface enhancement	Sample No. 8, 11 and 13 Unslip on both the surfaces.
Size range	Bowls measures 14 to 20 cms in rim diameter.

Bowls Type B Variety 3: carinated bowls having one variety with blunt carinated shoulder and two rim variants:
(a) simple everted, slightly curved rounded rim
(b) simple perpendicular, straight rounded rim
Sample No. 9, 10 and 12

Fabric	Medium to Coarse sandy with Mica, White inclusion and Organic matter.
Surface enhancement	Sample No. 9, 10 and 12 Unslip on both the surfaces.
Size range	Bowls measures 10 to 12 cms in rim diameter.

Pots Type A Variety 1: small size pots with rim diameter ranges between 10 to 15 cms having three varieties, first variety with short neck and one rim variant:
(a) internal projecting, short, narrow triangular rim
Sample No. 3

Fabric	Coarse sandy with Mica, White inclusion and Organic matter.
Surface enhancement	Sample No. 3 Unslip on both the surfaces.
Size range	Pot measures 15 cm in rim diameters.

Pots Type A Variety 3: small size pots with rim diameter ranges between 10 to 15 cms having three varieties, second variety with long neck and bulbous body and one rim variant:
(a) simple perpendicular, straight rounded rim
Sample No. 1

Fabric	Medium sandy with Mica, White inclusion and Organic matter.
Surface enhancement	Sample No. 1 having Red Slip on both the surfaces.
Size range	Pot measures 10 cm in rim diameters.

Pots Type A Variety 4: small size pots with rim diameter ranges between 10 to 15 cms having three varieties, third variety with carinated neck and one rim variant:
(a) simple perpendicular, straight rounded rim
Sample No. 4

Fabric	Coarse sandy with Mica, White inclusion and Organic matter.
Surface enhancement	Sample No. 4 Unslip on both the surfaces.
Size range	Pot unmeasured.

Basins Type A Variety 1: carinated basin having two varieties, first variety with carinated shoulder and corrugated neck and one rim variant:
(a) simple perpendicular, straight rounded rim
Sample No. 7

Fabric	Coarse sandy with White inclusion and Organic matter.
Surface enhancement	Sample No. 7 Unslip on both the surfaces.
Size range	Basin measures 25 cm in rim diameter.

Basins Type A Variety 3: carinated basin having two varieties, second variety with angular shoulder and one rim variant:
(a) simple perpendicular, straight rounded rim
Sample No. 5 and 6

Fabric	Coarse sandy with Mica, White inclusion and Organic matter.
Surface enhancement	Sample 5 and 6 Unslip on both the surfaces.
Size range	Basins measures 24 to 25 cms in rim diameter.

Undiagnostic Incised Sherds:
Sample No. 14, 15, 16 and 17

Fabric	Medium to Coarse sandy with Abundant Mica, White inclusion and Organic matter.
Surface enhancement	Sample No. 14, Sherd having rows of incised triangular notches. Sample No. 15, 16 and 17, Sherds having rows of incised horizontal lines.

33. TONDA (Figure – 4.69)

Bowls Type B Variety 1: carinated bowls having two varieties, first variety with carinated shoulder and one rim variant:
(a) internal projecting, short, narrow triangular rim
Sample No. 8 and 9

Fabric	Medium sandy with Mica, White inclusion and Organic matter
Surface enhancement	Sample No. 8 having Light Red Slip on both the surfaces Sample No. 9 having Light Red Slip on both the surfaces
Size range	Bowls measures 23 to 34 cms in rim diameter.

Bowls Type B Variety 3: carinated bowls having two varieties, second variety with blunt carinated shoulder and one rim variant:
(a) bilateral projecting, equal projection rim
Sample No. 11

Fabric	Medium sandy with Mica and White inclusion
Surface enhancement	Sample No. 11 having red Slip on both the surfaces
Size range	Bowl measures 20 cm in rim diameter.

Bowls Type C Variety 1: straight-sided bowls having one variety and one rim variant:
(a) internal projecting, long, wide rounded rim

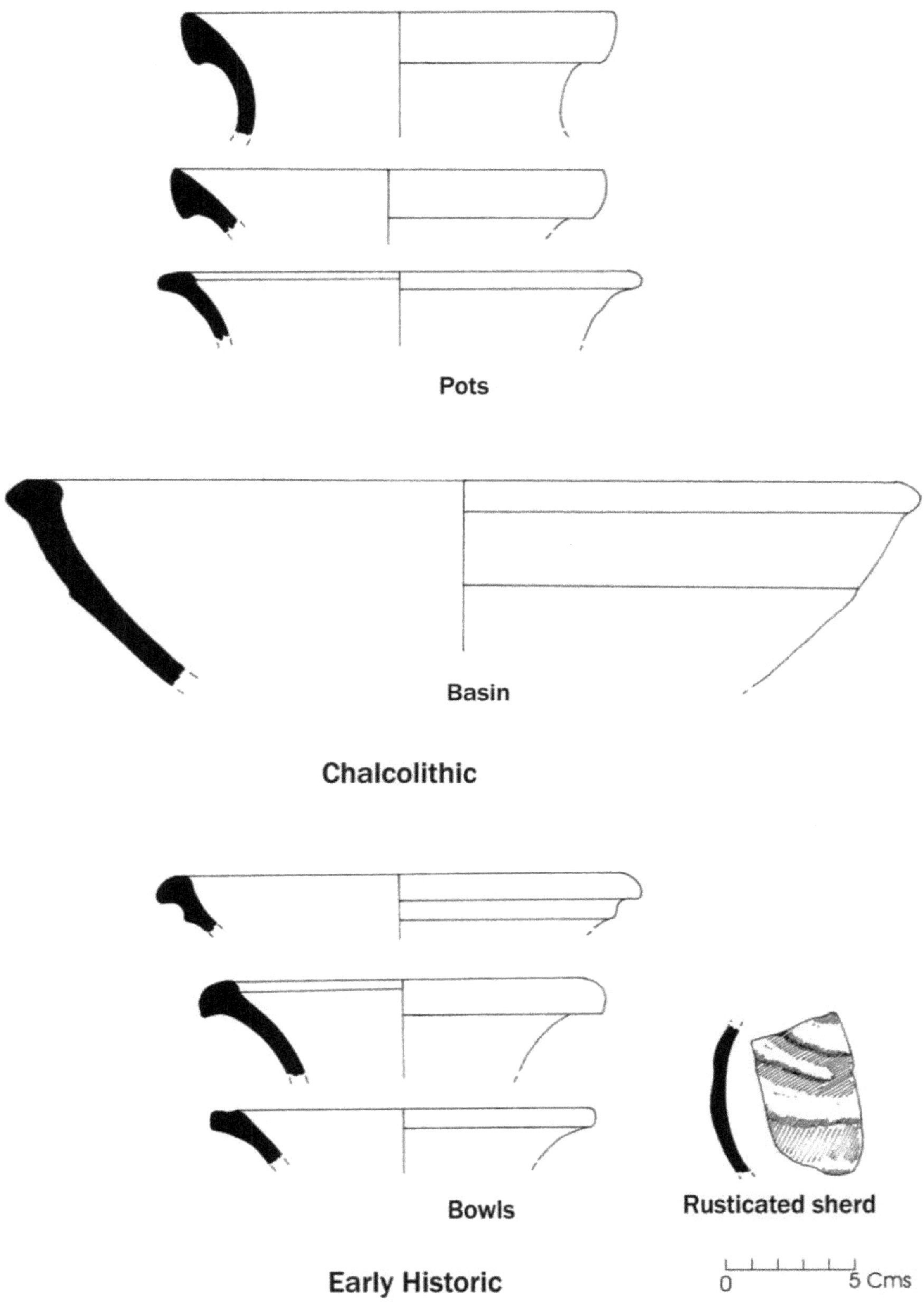

Figure 4.67 – Tapkeshwar

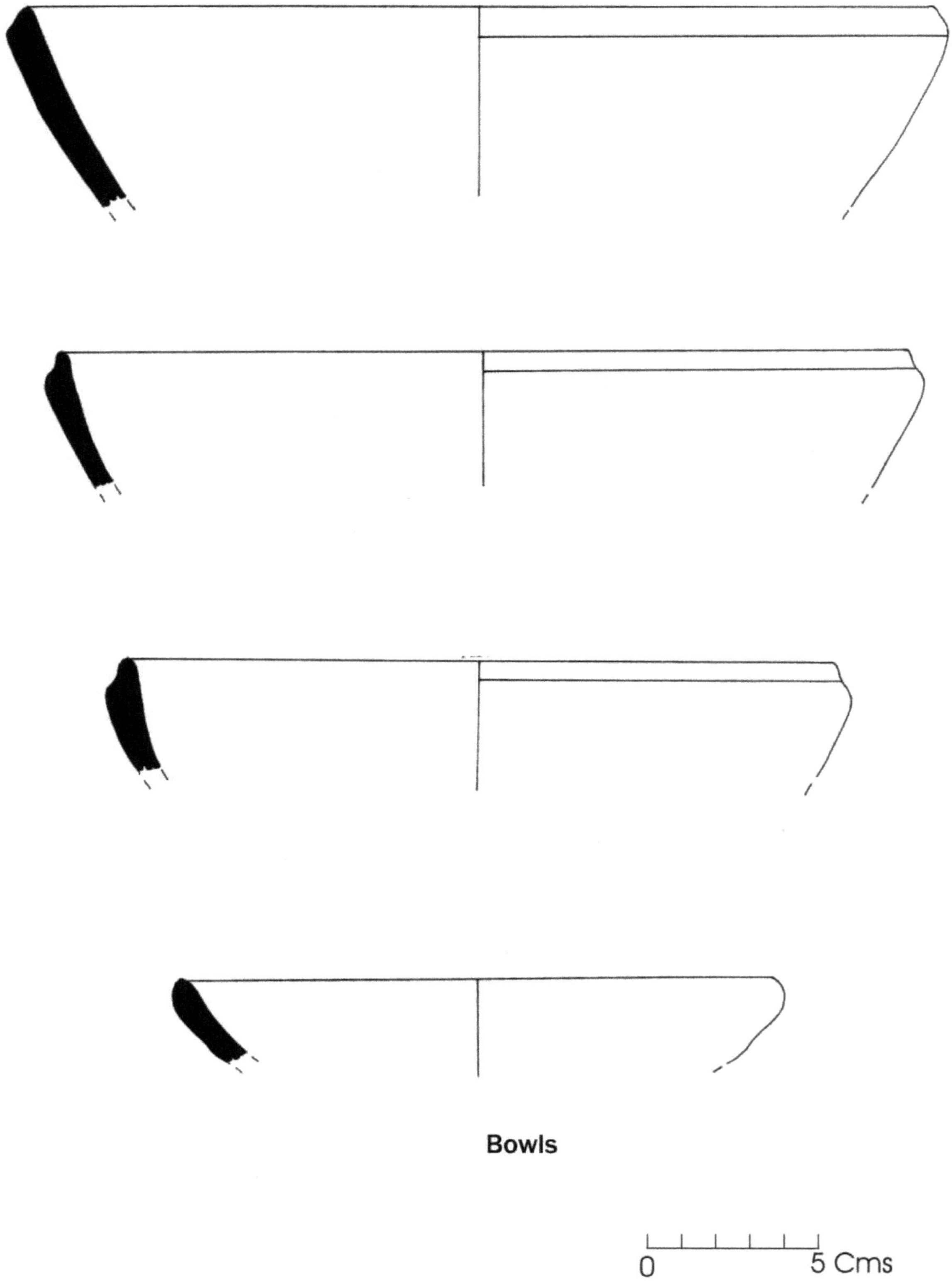

Figure 4.68– Thaneshwar

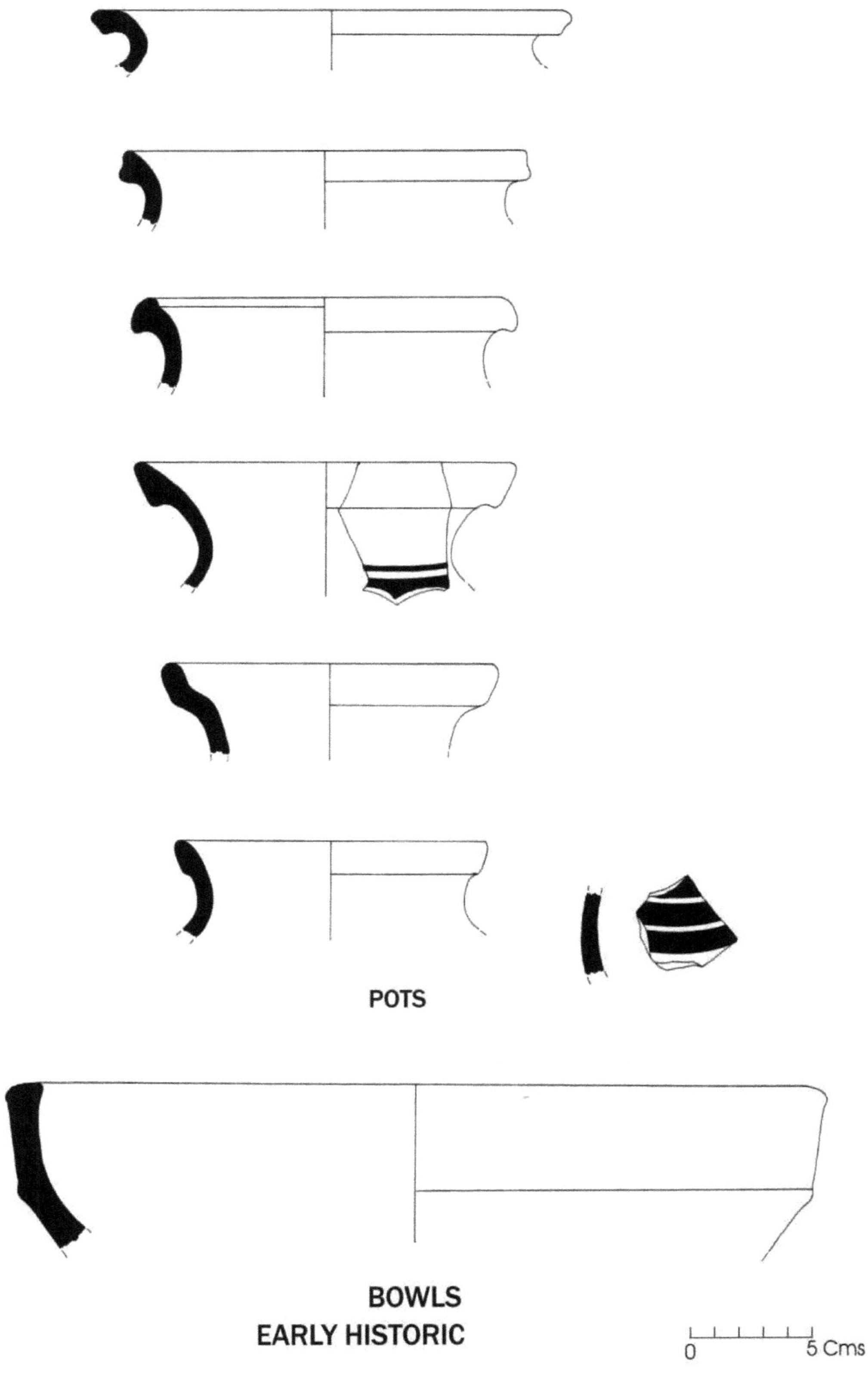

Figure 4.69 – Tonda

Sample No. 10

Fabric	Medium sandy with Mica and White inclusion
Surface enhancement	Sample No. 10 having Light Red Slip on both the surfaces
Size range	Bowl measures 15 cm in rim diameter.

Pots Type A Variety 1: small size pots with rim diameter ranges between 12 to 17 cms having two varieties, first variety with short neck and bulbous body and four rim variants:

(a) external projecting, short, narrow triangular rim
(b) internal projecting, short, narrow triangular rim
(c) external ledge, angular rim
(d) bilateral projecting, equal projection rim

Sample No. 1, 2, 4, 6 and 7

Fabric	Medium sandy with Mica, White inclusion and organic matter
Surface enhancement	Sample No. 1, 4 having Light Red Slip on both the surfaces Sample No. 2, 6 having Red Slip on both the surfaces Sample No. 7 Unslip on both the surfaces
Size range	Pots measures 12 to 17 cms in rim diameters.

Pots Type A Variety 3: small size pots with rim diameter ranges between 12 to 17 cms having two varieties, second variety with long neck and bulbous body and one rim variant:

(a) external projecting, short, narrow triangular rim

Sample No. 3

Fabric	Medium sandy with Mica and White inclusion
Surface enhancement	Sample No. 3 having Red Slip on both the surfaces
Size range	Pot measures 14 cm in rim diameters.

Pots Type B Variety 1: medium size pots with rim diameter 18 cm having one variety with short neck and bulbous body and two rim variants:

(a) external projecting, short, narrow triangular rim
(b) internal projecting, long, wide rounded rim

Sample No. 5 and 12

Fabric	Medium sandy with White inclusion and Organic matter
Surface enhancement	Sample No. 5 Unslip on both the surfaces Sample No. 12 having Red Slip on both the surfaces
Size range	Pots measures 18 cms in rim diameters

Undiagnostic Sherds:

Sample No. 13, 14, 15, 16, 17, 18, 19 and 20

Fabric	Medium sandy with Mica and White inclusion
Surface enhancement	Sample No. 13 having Red Slip on both the surfaces with incised horizontal lines on the exterior surface Sample No. 14 having Reddish Yellow Slip on the exterior surface and Unslip on the interior surface with black color horizontal bands on the exterior surface Sample No. 15 having Red Slip on the exterior surface and Unslip on the interior surface with black color horizontal bands on the exterior surface Sample No. 16 having Brown Slip on the exterior surface and Unslip on the interior surface with black color horizontal bands on the exterior surface Sample No. 17 having Light Red Slip on the exterior surface and Unslip on the interior surface with sherd having incised and black color horizontal lines on the exterior surface Sample No. 18, 19 and 20 having Red Slip on the exterior surface and Unslip on the interior surface with black color horizontal bands on the exterior surface

4.4 Classification of Ancient Mines

In the present survey, 19 ancient mining activities has been identified and typologically classified into open mines, pit mines, shallow mines and deep mines.

1. **Abrak ki Khan** - comprises open mine having ore deposits in a form of quartz-mica schist minerals, which was extensively worked by the ancient miners (Figure - 4.70).
2. **Akawali Area** – four localities have been identified: Akawali Mine, Akawali Mine No.1, Akawali Mine No. 2 and Suradhi (Figure - 4.71). There are ample of evidence, which shows that these areas were extensively worked by the ancient miners in the form open mine, deep mine and pit mines.
3. **Bavan Kuhi** – has ancient mines in a form of 52 open pits along with drainage channels, which was used to extract water from the pits (Figure - 4.72). There are narrow shafts leads into big stopes and some of the stopes are more than 35m wide.
4. **Beed Ki Dhani** - comprise open mine, which is around 3 m wide and 15 m deep having stains of chalcopyrite in quartzite stone.
5. **Chandmari Mine** – At Chandmari Hindustan Copper Limited was drilling out the deep mine at the depth of 434 feet. On the footsteps of open mine, there are opening for old working on its wall, which shows heaps of mine spoil, gossan bands and malachite staining (Figure - 4.73).

6. **Gura – Ponkh** - has given evidence of ancient soapstone mine, which has revealed good quality of black and dark green soap stone mines (Figure - 4.74).
7. **Kakrana** - has shown evidence of ancient opened mine in a form of gossans.
8. **Khetri Mine No. 1 and 2** - are the part of Khetri copper complex, which extends 3 km by 1 km. The surface manifestations of the sulphides mineralization in this area are quite prominent and present in the form of gossans, old workings and huge heaps of metal slags (Figure - 4.75).
9. **Kho** – has ancient copper mine in a form of an open mine. This area also gives evidence of copper ore nodules lying scattered all around the mine area. As a part of living tradition, shepherds in this area collect these copper nodules and sell it to the local craftsmen or buyers. This site also given evidence of soapstone mine within the vicinity (Figure - 4.76).
10. **Kolihan – Chandmari** - comprises open pit mine. The old workings with dumps of mine spoil show that the ancient miners have worked in this area, which extend to a depth of 15 - 30 m.
11. **Kolihan Mine No. 486** – is a deep mine lies on the northern part of Khetri Copper Belt, which is now closed for mining by HCL (Figure - 4.77).
12. **Ponkh ki Dhani** - The ancient mining area comprises open mine in the form spoil dumps (Figure - 4.78).
13. **Samiyo ki Dhani** - is an open mine comprises evidence of ancient mining in a form of mine spoil.
14. **Sandhawala Kuha** – is an open mine, which gives evidence of ancient mine in a form of stope (Figure - 4.79.
15. **Sawalpur** – the site gives evidence of ancient mine in a form of open pit having quartzite-mica minerals.

4.5 Classification of Metal Processing Activity Areas

In the present survey 14 metal processing activity areas were identified and classified the metal processing indicators in the form of slag deposits, metallurgical slags, terra cotta tuyeres, terra cotta crucibles and vitrified clay lumps.

1. **Agri Ki Dhani** - has revealed vitrified clay lumps, patches of metallurgical slags and terracotta tuyeres in the form of metal processing indicators (Figure 4.80).
2. **Baleshwar** - has revealed metallurgical slags spread in the area of about half kilometre on the slope of a hillock (Figure 4.81).
3. **Ganeshwar No. 1** - is a single culture site comprising Chalcolithic material with metal processing indicators in a form of vitrified clay lumps, charred wood and metallurgical slags.
4. **Ganeshwar No. 6** - comprises metallurgical slags lying scattered on the surface of a hillock. There are stone alignments in circular fashion on the surface, presumed to be used by the ancient miners. These stones has also shown grooves and cut marks formed during metal processing activities (Figure 4.82).
5. **Kharagbinjpur** - is a single culture site, which has revealed Chalcolithic materials along with vitrified clay lumps and metallurgical slags. The site is located on the foothills in an open scrub area of about 1 Km along with the structural remains and slag deposits. The huge slag deposit scattered on the surface indicates that the area was intensively used for metal processing (Figure 4.83).
6. **Kot** - is a multi-cultural site, which reveals Chalcolithic, Early Historic – I and Early Historic – II cultural materials. Along with this, it also reveals metal processing indicators in a form of terra cotta tuyeres and vitrified clay lumps (Figure 4.84).
7. **Murajawali No. 1** - is a single culture site of Chalcolithic period identified as a metal processing activity area with the findings of vitrified clay lumps, terra cotta tuyeres and crucibles (Figure – 4.85).
8. **Murajawali No. 2** – comprises vitrified clay lumps and terra cotta tuyeres. The site also gives evidence of fire activity patches along with the metal processing indicators (Figure – 4.86).
9. **Naka ki Dhani** – is an important site, which has revealed thick deposit of metal processing activities with the findings of vitrified clay lumps, terra cotta tuyeres, terra cotta crucibles and metallurgical slags (Figure – 4.87). The site has also given evidence of fire activity patches.
10. **Pitrawali Dongri** – is another important site, which has also revealed thick deposit of metal processing activities. The site has revealed terra cotta tuyeres and crucibles lying scattered on a large area with patches of fired activities along with the huge deposit of metallurgical slags (Figure – 4.88 a and 4.88 b).
11. **Salodara** - is a single culture site. Along with Chalcolithic material the site also shows metal processing indicators in a form of vitrified clay lumps and metallurgical slags along with the drainage channel.
12. **Shivnaryan ki Dhani** – The site has high lighten with the findings of terra cotta tuyeres in different sizes and terra cotta crucibles (Figure – 4.89).
13. **Singhana** - The area was intensively used for metal processing or dumping ground by the ancient miners. The present village inhabits over the slag heaps. There are three huge slag heaps in the village. The site falls in the Khetri Copper Belt (KBC), which is around 80 km copper belt from Singhana to Raghunathgarh. The area forms the northern most continuation of KCB, under cover of blown sand/alluvium and strike of about 11 km (Figure – 4.90).
14. **Thaneshwar** - is a single culture site. Along with Chalcolithic material the site also revealed metal processing indicators in a form of vitrified clay lumps, terra cotta tuyeres, terra cotta crucibles, stone muller, and circular cut stone (Figure – 4.91). The important feature of this site is that most of the metal processing indicators were in insitu position.

Figure 4.70 Ancient Mine at Arbak ki Khan

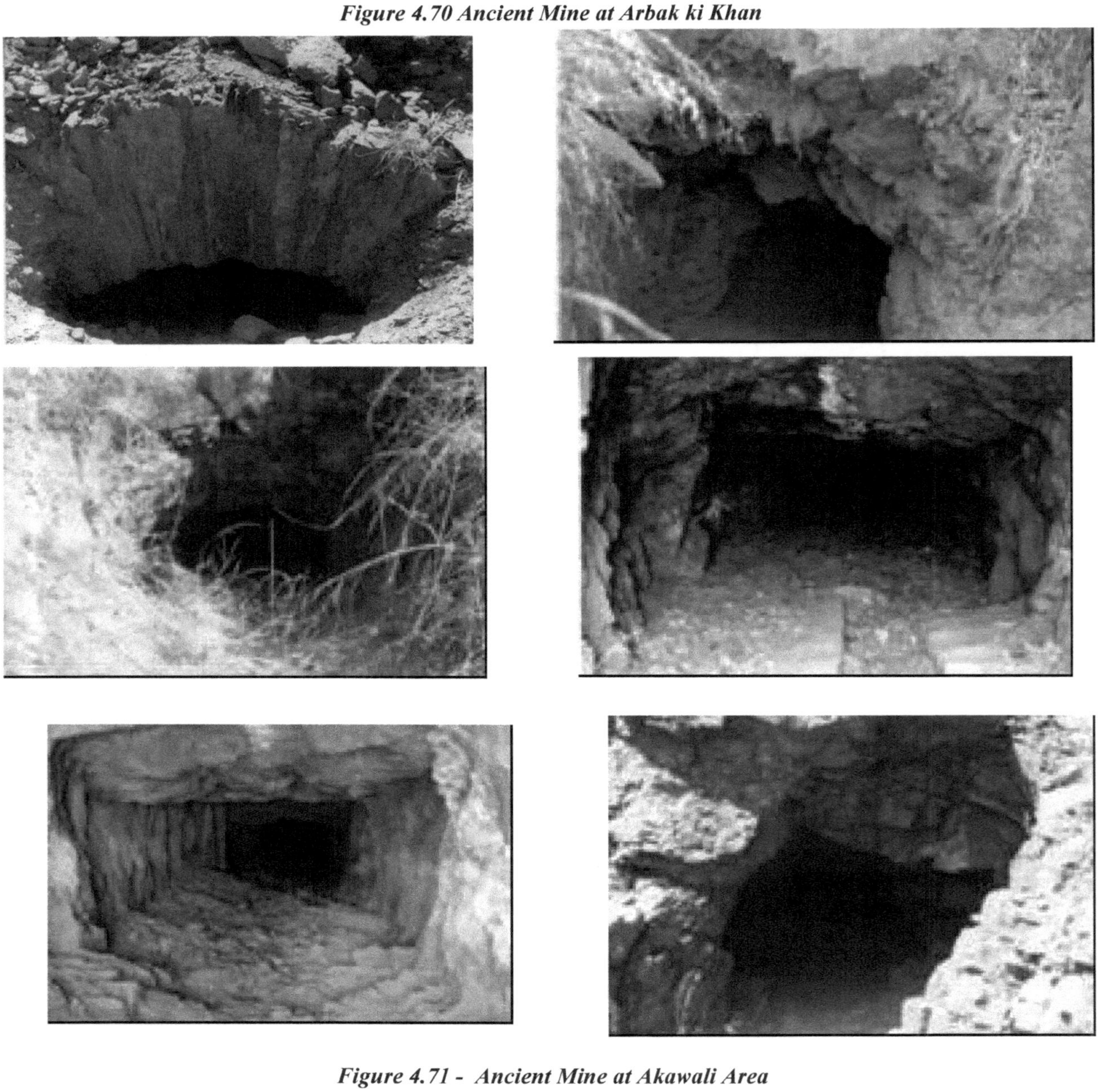

Figure 4.71 - Ancient Mine at Akawali Area

(a)

(b)

(c)

(d)

(e)

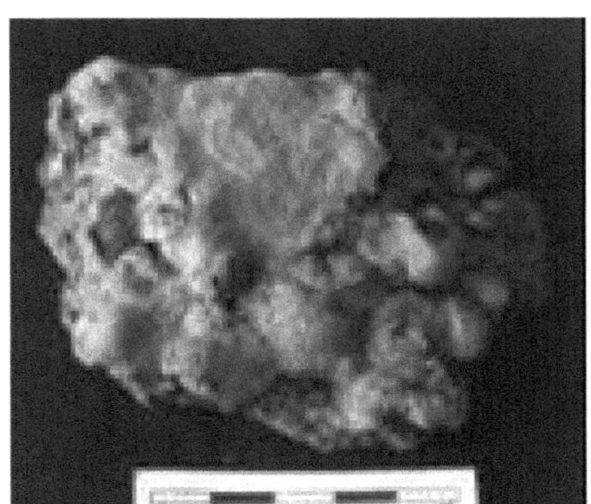

Figure 4.72 – Ancient Mine at Bavan Kuhi

(a)

(b)

(c)

Figure 4.73 – Ancient Mine at Chandmari

(a) (b) (c)

Figure 4.74 – Ancient Mine at Gura Ponkh

Figure 4.75 – Ancient Mine at Khetri Mine # 1 & 2

(a)

(b)

(c)

(d)

Figure 4.76 – Ancient Mine at Kho

Figure 4.77 – Ancient Mine at Kolihan Mine # 486

(a) **(b)** **(c)**

Figure 4.78 – Ancient Mine at Ponkh Ki Dhani

Figure 4.79 – Ancient Mine at Sandhawala Kuha

terra cotta tuyeres, vitrified clay lumps and metallurgical slag

metallurgical slag

Figure 4.81 – Remnants of Metal Processing Activity at Baleshwar

Figure 4.80 – Remnants of Metal Processing Activity at Agri ki Dhani

Slag heap

groove and cut marks

stone alignment

stone alignment

Figure 4.82 – Remnants of Metal Processing Activity at Ganeshwar No 6

metallic Slag Deposit

Figure 4.83 – Remnants of Metal Processing Activity at Kharagbinjpur

(a) terra cotta tuyeres

(b) **vitrified clay lumps**

Figure 4.84 – Remnants of Metal Processing Activity at Kot

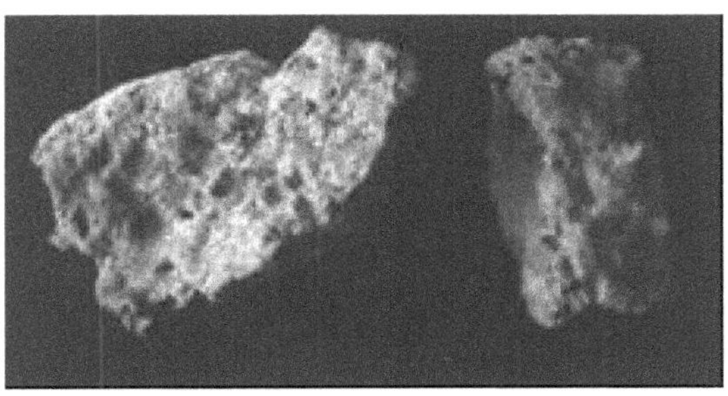

(a) **vitrified clay lumps**

(c) **terra cotta tuyeres**

Figure 4.85 – Remnants of Metal Processing Activity at Murajawali # 1

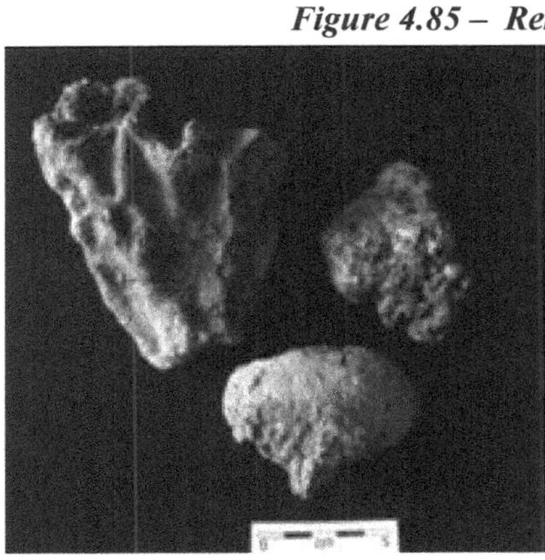

terra cotta tuyees and vitrified clay lumps

Figure 4.86 – Remnants of Metal Processing Activity at Murajawali # 2

(a) smelting activity area

(b) smelting activity area

(c) smelting activity area

(d) terra cotta tuyeres

(e) metallurgical slags

Figure 4.87 – Remnants of Metal Processing Activity at Naka ki Dhani

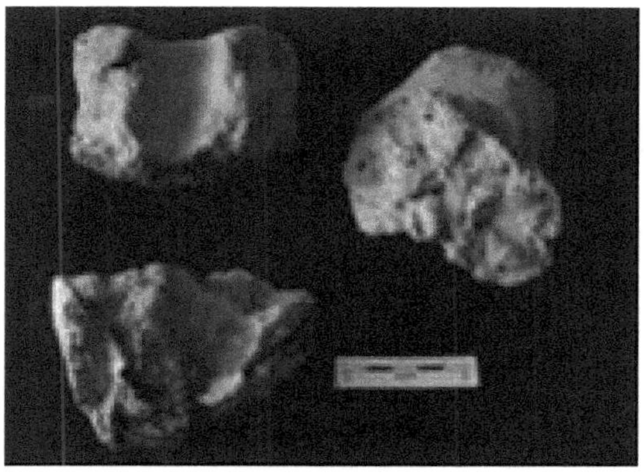

terra cotta tuyeres and crucible

terra cotta tuyeres

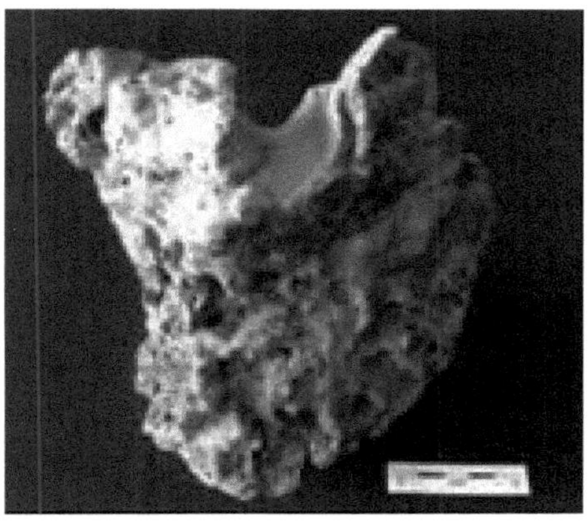

terra cotta tuyeres

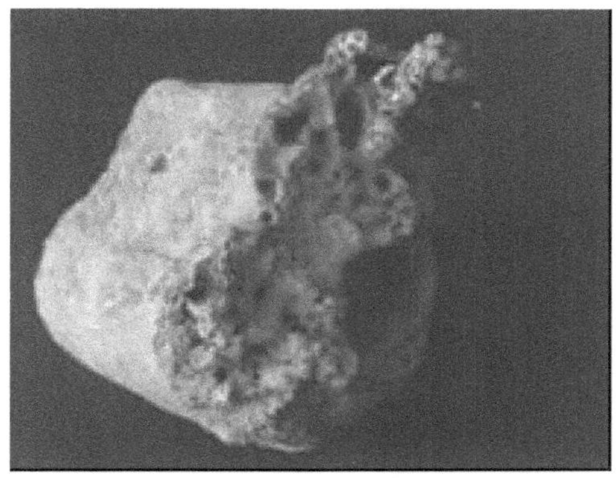

terra cotta crucible

smelting activity area

Figure 4.88 (a) – Remnants of Metal Processing Activity at Pitrawali Dongri

smelting activity area

smelting activity area

terra cotta tuyeres with furnace wall

terra cotta tuyeres

Figure 4.88 (b) – Pitrawali Dongri

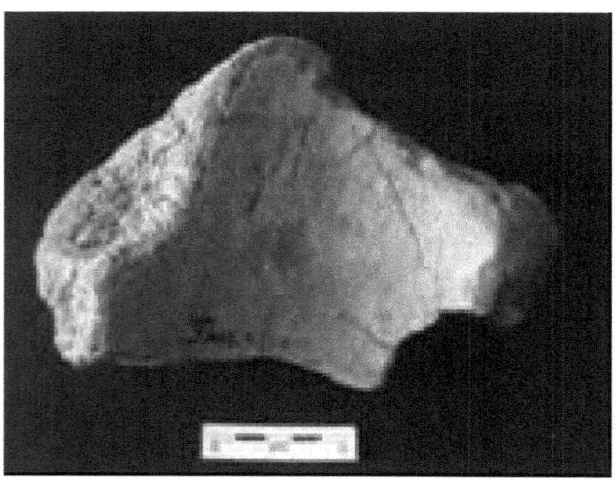

terra cotta tuyeres

terra cotta tuyeres and crucibles

Figure 4.89 – Remnants of Metal Processing Activity at Shivnarayan ki Dhani

Huge slag deposits

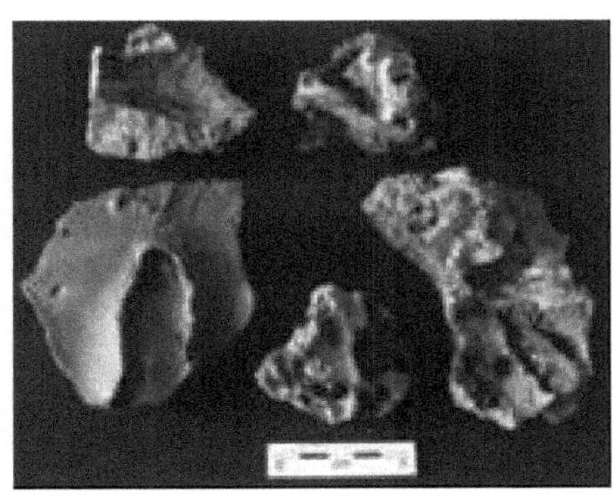

Metal slags

Huge slag deposit

Figure 4.90 - Remnants of Metal Processing Activity at Singhana

terra cotta tuyeres

terra cotta crucibles

insitu terra cotta tuyeres

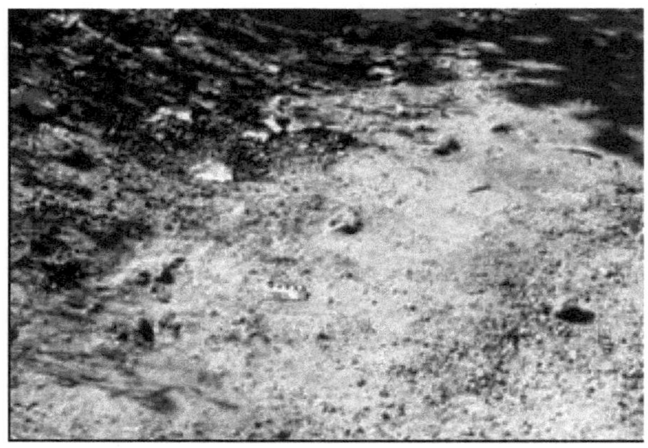

insitu terra cotta tuyeres

insitu terra cotta tuyeres

Figure 4.91 - Remnants of Metal Processing Activity at Thaneshwar

CHAPTER 5 - SETTLEMENT PATTERN STUDIES

5.1 Settlement Pattern Studies

Settlement archaeology refers to that subset of archaeological research that is aimed at entire regions rather than a single site. While settlement patterns is a term used to regional approaches where settlement data are typically collected in the context of regional surveys, and the problems addressed are likewise regional in scope.

Settlement Pattern in a particular region has increased archaeological awareness of regional diversity and complexity. The concept of settlement patterns consists of spatial distribution of human activities and occupation, which is ranging from locating an activity from a single site to the arrangement of sites in the region.

The development of settlement pattern studies can be traced to the work of L.H. Morgan carried out in 1881. The study mainly concern on the aboriginal residential architecture of North America, which reflects the social organisation of the Prehistoric people. At the end of 19th century, Mindeleff investigate sociology of architectural remains. The study reveals settlement accretion and growth and presented a method for reconstructing occupation chronology and settlement composition in archaeological remains. Julian Steward in 1937 carried out research on aboriginal social-organisation in North America. Where he has recognised the pattern; how location of residence could be understand as a product of interaction between the environment and culture and the factor of technology and social organisation. Cyril Fox's publication "Personality of Britain" in 1940s brought a new understanding in archaeological studies. In which he relates the distribution of archaeological sites to the distribution of environmental features. Subsequently, more studies were carried out on similar lines with the work of G. Childe in 1934, Hoggs in 1943 and Grimes in 1945. G.R. Willey's publication on "Prehistoric Settlement Patterns in Viru Valley" in 1953 brought a major break-through in the field of archaeology. The study aimed at inferring cultural process from regional patterning of settlements. Willey not only stressed the need of settlement pattern studies, but also defined their scope and potential utility in archaeology. He defined settlement pattern as the way in which human disposed themselves over the landscape on which they dwelled. It refers to the dwelling, to their arrangement and to the nature and disposition of other building pertaining to community life. The study has reflect natural environment, level of technology on which the builder operates and various institutions of social interaction and control, which the culture maintained. Vogt's in 1956, formulate the scope of settlement pattern studies as description of; a) nature of individual, domestic house type. b) spatial arrangement of this domestic house types with respect to one another within a village or community unit. c) relationship of domestic house types to other spatial architectural features. d) overall, village or community plans. e) and spatial relationship of villages or communities to one another, over as large area as feasible. Sanders in 1956, introduce the concept of symbiotic region in archaeological literature. He gave more emphasis on analysing the distribution of human settlement in the context of agricultural system, local specialisation and inter-regional exchanges. He made a clear distinction between Community Settlement Patterns and Zonal Settlement Patterns. Community Settlement Pattern comprises individual units of population and Zonal Settlement Pattern is concerned with the distribution of community size, distance between the communities, density of the population and the symbiotic inter-relationship between the communities. Chang in 1958, made the first serious attempt to generate and test hypothesis concerning the relationship between social organisation and settlement patterns in simple Neolithic societies. He made an important distinction between the term "Settlement Pattern" and "Community Pattern". Settlement Pattern as the manner in which human settlements are arranged over the landscape in relation to geographic environment and Community Pattern as the manner in which the inhabitants arrange their various structures within the community and their communities within the aggregates. Winter in 1969, made a distinction between Settlement Pattern and Settlement System. According to him, Settlement Pattern is the geographic and physiographic relationship of contemporaneous group of site within a single culture; whereas Settlement System refers to the functional relationship among the contemporaneous group of sites within a single culture. Bruce_Trigger in 1968 isolated two approaches in settlement pattern studies:

(i) Ecological Approach, it is based on the assumption that the settlement pattern was a product of the simple interaction of two variables environment and technology.

(ii) Sociological approach, in which settlement data were used as a basis for making inferences about the social, political and religious organisation of a culture.

Trigger suggested that settlement study could be more profitably done if we study them in terms of:

a) Individual building or structure.
b) Manner in which these buildings or structures are arranged within a single community.
c) And the manner in which the communities are distributes over the landscape.

In this book, an attempt has been made to know geomorphic features and their relationship with the archaeological sites to understand and reconstruct the dynamics of settlement patterns from Prehistoric to Early Historic periods in Shekhawati region of Rajasthan and also to get a light on geomorphic adaptation of archaeological site and their possible interaction with mining and metal processing activity areas.

5.2 Distribution and Area of Sites

There are 104 sites identified in the research area, which includes 3 Palaeolithic sites, 12 Mesolithic sites, 33 Ganeshwar related Chalcolithic sites, 23 Early Historic sites, 19 Ancient mining areas and 14 Metal processing activity areas. Where archaeological culture sites are categories into Single Culture Site (25) and Multi Culture Site (46). Further Single Culture Site classified into

Palaeolithic (2), Mesolithic (3), Chalcolithic (15) and Early Historic (5). Whereas in Multi Culture Site Palaeolithic (1), Mesolithic (9), Chalcolithic (18) and Early Historic (18) (Figure 5.1).

Area of Sites – The sites are categorized into six types:

Type (1) - Area of sites > 100 Sq. m

Sr. No.	Name of Sites	Culture	Period	Remarks
1.	Bhudoli	Single Culture Site	Mesolithic	
2.	Ganeshwar No. 3	Single Culture Site	Mesolithic	
3.	Murajwali No. 1	Single Culture Site	Chalcolithic	Metal processing activity area

Type (2) - Area of sites < 100 > 500 Sq. m

Sr. No.	Name of Sites	Culture	Period	Remarks
1.	Bavadi ki Jhamin	Single Culture Site	Chalcolithic	
2.	Beed ki Dhani	Multi Culture Site	Mesolithic Chalcolithic	Ancient Mining Area
3.	Bhadwari	Multi Culture Site	Mesolithic Chalcolithic Early Historic	
4.	Gidhali No. 1	Single Culture Site	Mesolithic	
5.	Gidhali No. 2	Multi Culture Site	Mesolithic Chalcolithic Early Historic	
6.	Gogajiwali Dongri	Multi Culture Site	Chalcolithic Early Historic	
7.	Jodhpura	Multi Culture Site	Chalcolithic Early Historic	
8.	Kakrana	Multi Culture Site	Mesolithic Chalcolithic	Ancient Mining Area
9.	Mataji No Timbo	Multi Culture Site	Mesolithic Chalcolithic Early Historic	
10.	Meeno ki Dhani	Single Culture Site	Chalcolithic	
11.	Murajwali No. 2	Single Culture Site	Chalcolithic	Metal processin g activity area
12.	Naka ki Dhani			Metal processin g activity area

Type (3) - Area of sites < 500 0 > 1000 Sq. m

Sr. No.	Name of Sites	Culture	Period	Remarks
1.	Biharipura	Multi Culture Site	Mesolithic Chalcolithic Early Historic	
2.	Gadhka	Multi Culture Site	Chalcolithic Early Historic	
3.	Jasi ka Bas	Single Culture Site	Chalcolithic	
4.	Kakadiyo	Multi Culture Site	Chalcolithic Early Historic	
5.	Kanwat	Single Culture Site	Chalcolithic	
6.	Kharagbinjp ur	Single Culture Site	Chalcolithic	Metal processing activity area
7.	Nalot	Multi Culture Site	Chalcolithic Early Historic	
8.	Pir Baba ki Bani	Single Culture Site	Palaeolithic	
9.	Thaneshwar	Single Culture Site	Chalcolithic	Metal processing activity area

Type (4) - Area of sites < 1000 - 5000 Sq. m

Sr. No.	Name of Sites	Culture	Period	Remarks
1.	Bhagegawal a Tila	Single Culture Site	Early Historic	
2.	Gujjaro ki Dhani	Multi Culture Site	Chalcolithic Early Historic	
3.	Payga Kundalia	Multi Culture Site	Palaeolithic Mesolithic Chalcolithic Early Historic	
4.	Sadmata ka Maula	Multi Culture Site	Mesolithic Chalcolithic	
5.	Shivnaryan ki Dhani			Metal processing activity area
6.	Swamiyo ka Kuha	Single Culture	Chalcolithic	

		Site		
7.	Tonda	Single Culture Site	Early Historic	

Type (5) - Area of sites < 5000 - > 10000 Sq. m

Sr. No.	Name of Sites	Culture	Period	Remarks
1.	Jogi Badh	Multi Culture Site	Chalcolithic Early Historic	
2.	Mokulwas	Single Culture Site	Early Historic	
3.	Narshingpur i	Single Culture Site	Chalcolithic	
4.	Sainiyo ki Dhani	Single Culture Site	Early Historic	
5.	Ganeshwar No. 1,2,4 and 5	Multi Culture Site	Chalcolithic Early Historic	

Type (6) - Area of sites < 10000 Sq. m

Sr. No.	Name of Sites	Culture	Period	Remarks
1.	Kot (Sunari)	Multi Culture Site	Chalcolithic Early Historic	Metal processing activity area

5.3 Distribution of Sites in Geological Formation

There are four geological deposits identified in research area (Figure 5.2):

1. **Deposit 1** - sand, alluvium etc. covers around 6.41% of the research area and no sites have been found in this deposit.
2. **Deposit 2** - schist, gneisses, calc-gneisses, dolmitic marble, amphibiolite and phyllite, which covers around 24.22% of the research area and around 44 sites were found. Out of which Palaeolithic (1), Mesolithic (5), Chalcolithic (14), Early Historic (11), Ancient Mining Areas (10) and Metal Processing Activity Areas (3).
3. **Deposit 3** - intrusive and extrusive rocks covers around 0.96% of the research area and here also no sites have been found in this deposit.
4. **Deposit 4** - quartz, schist and conglomerate, which covers around 68.41% of the research area and around 59 sites were found. Out of which Palaeolithic (2), Mesolithic (7), Chalcolithic (19), Early Historic (12), Ancient Mining Areas (9) and Metal Processing Activity Areas (11).

5.4 Distribution of Sites in Geomorphological Features

There are five geomorphic features identified in research area (Figure 5.3):

1. **Denudational Hills / Ridges** – covers around 20.92 % of the area. Around 51 sites were found. Out of which Palaeolithic (3), Mesolithic (2), Chalcolithic (14), Early Historic (10), Ancient Mining Areas (13) and Metal Processing Activity Areas (9).
2. **Rocky Pediments** – covers around 8.98 % of the area. Around 11 sites were found out of which Mesolithic (2), Chalcolithic (5) and Early Historic (4).
3. **Alluvial Plain** – covers around 25.45 % of the area. Around 8 sites were found. Out of which Mesolithic (2), Chalcolithic (3), Early Historic (2) and Ancient Mining Areas (1).
4. **Aeolian Plain** – covers around 40.74 % of the area. Around 33 sites were found out of which Mesolithic (6), Chalcolithic (11), Early Historic (7), Ancient Mining Areas (5) and Metal Processing Activity Areas (5).
5. **Buried Rocky Peneplain** – covers around 3.81 % of the research area and there are no sites found in this feature.

5.5 Distribution of Sites in Dune Types

There are dunes have been notice in research area (Figure 5.4):

1. **Sand Sheets / Mounds** - covers around 16.23 % of the area. Around 22 sites were found. Out of which Mesolithic (2), Chalcolithic (7), Early Historic (5), Ancient Mining Areas (6) and Metal Processing Activity Areas (2).
2. **Rocky Peneplains** – covers around 29.16 % of the area. Around 31 sites were found out of which Mesolithic (3), Chalcolithic (15), Early Historic (6), Ancient Mining Areas (1) and Metal Processing Activity Areas (6).
3. **Hills / Rocky ridges** – covers around 20.19 % of the area. Around 50 of sites were found out of which Palaeolithic (3), Mesolithic (7), Chalcolithic (11), Early Historic (12), Ancient Mining Areas (12) and Metal Processing Activity Areas (6).
4. **Transverse Dunes** – covers around 13.82 % of the research area and there are no sites found in this feature.
5. **Alluvial Plain** - covers 18.94 % of the research area and no sites have been found.
6. **Sand Sheets and Longitudinal Dunes** – covers around 1.63 % of the research area and there also no sites have been found.

5.6 Distribution of Sites in Drainage System

The present drainage system in research area is supported by Kantli, Rupangarh, Dohan, Dongar, Sabi (Sahibi), Chandrawati, Udaipur-Lohargarh ki Nadi, Madhobini Nadi, Sota Nadi, Dhudhala Nadi, Bandi Nadi, Sukh Nadi (Singhana Nadi), and Khariya Nadi. It is observed in the research area that archaeological sites are located on the small streams and nallas rather than on the main rivers and their tributaries. This imply that prehistoric to early-historic people have preferred water bodies where water is stored for longer period than the rivers and their tributaries where water is available seasonally (Figure 5.5).

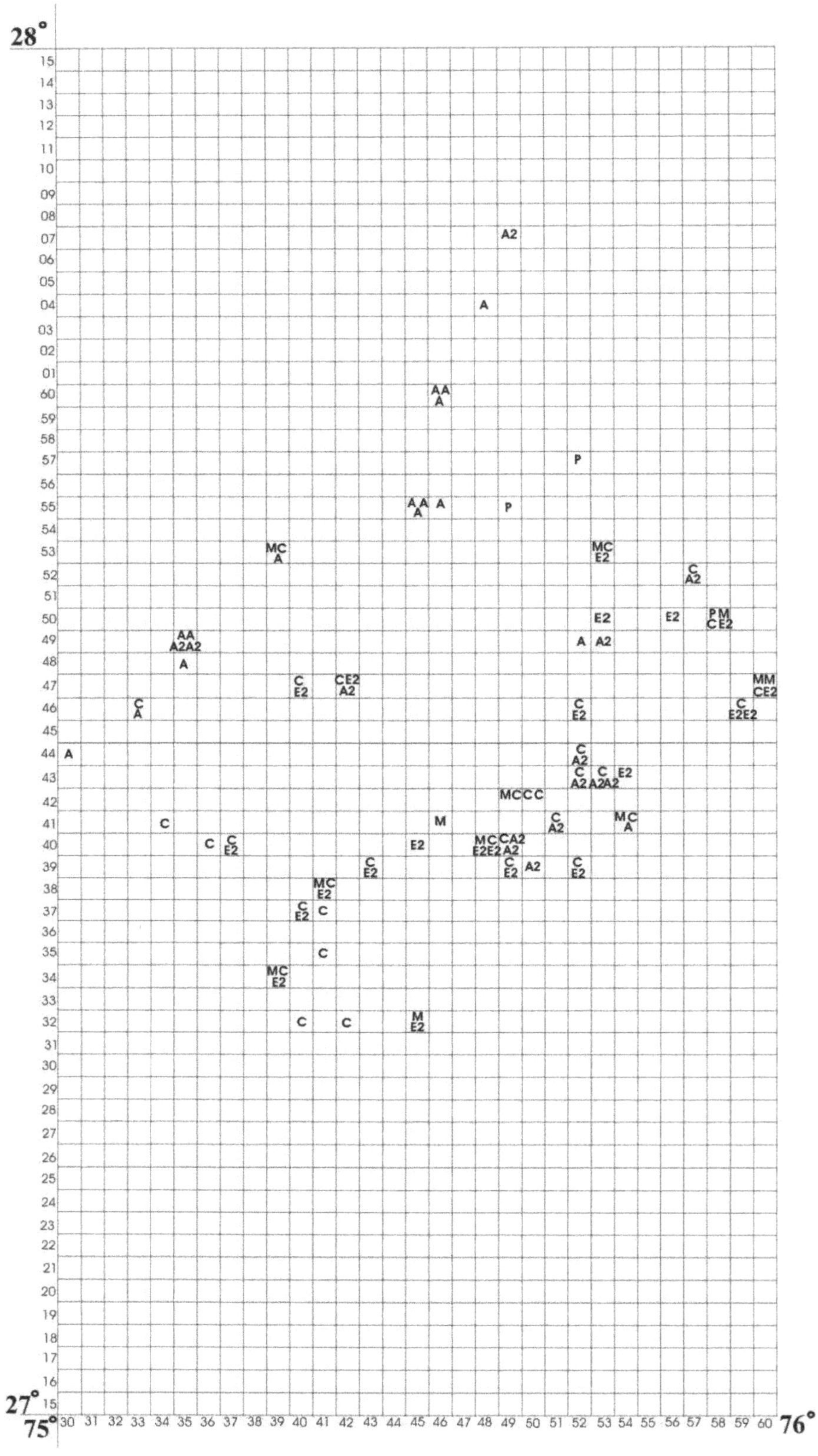

Figure – 5.1 Distributions of Sites in Research Area

Symbols: P – Palaeolithic; M – Mesolithic; C – Chalcolithic; E2 – Early Historic; A – Ancient Mining Area and A2 – Metal Processing Activity Area.

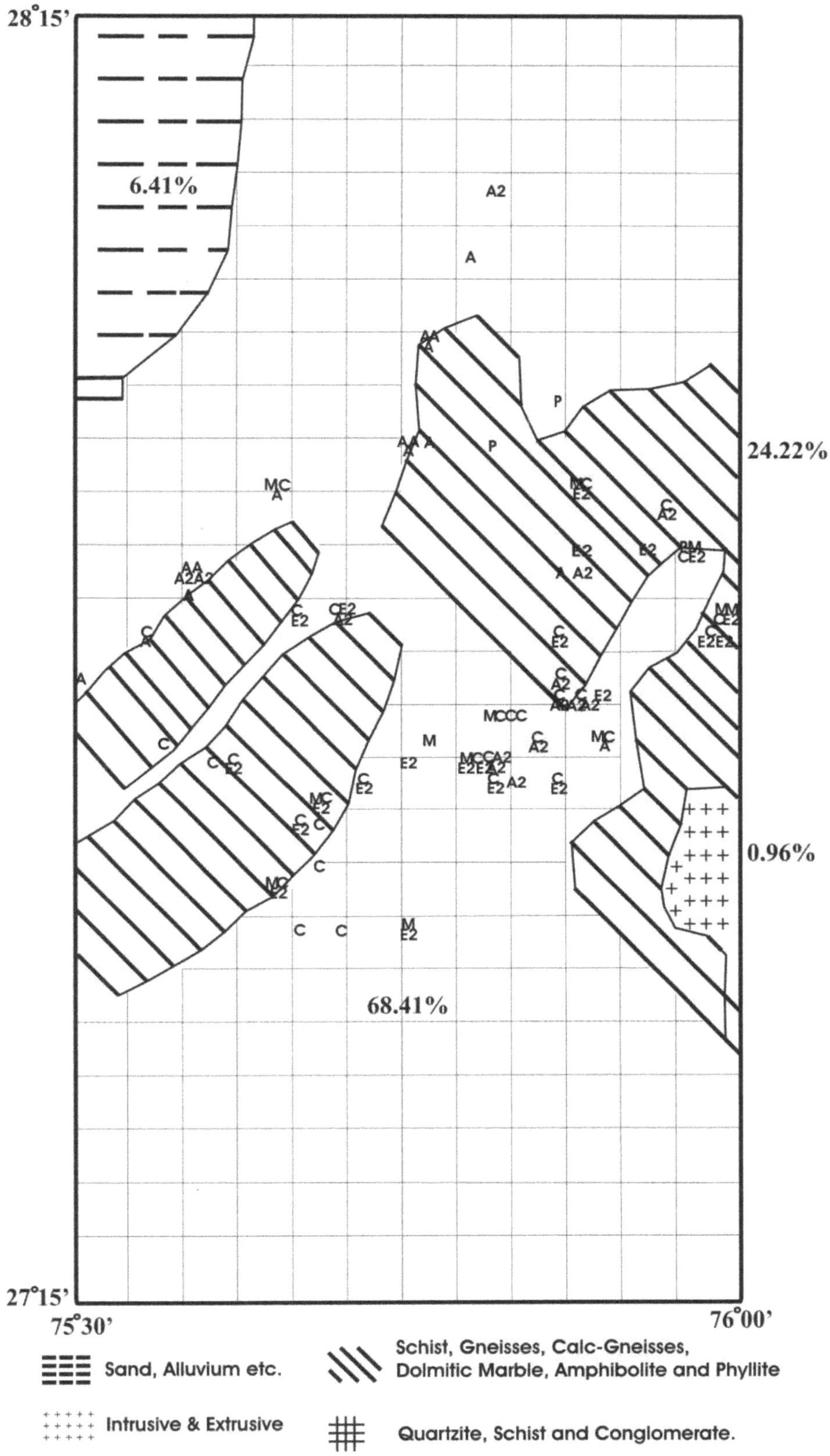

Figure – 5.2 Distributions of Sites in Geological Formation

Symbols: P – Palaeolithic; M – Mesolithic; C – Chalcolithic; E2 – Early Historic; A – Ancient Mining Area and A2 – Metal Processing Activity Area.

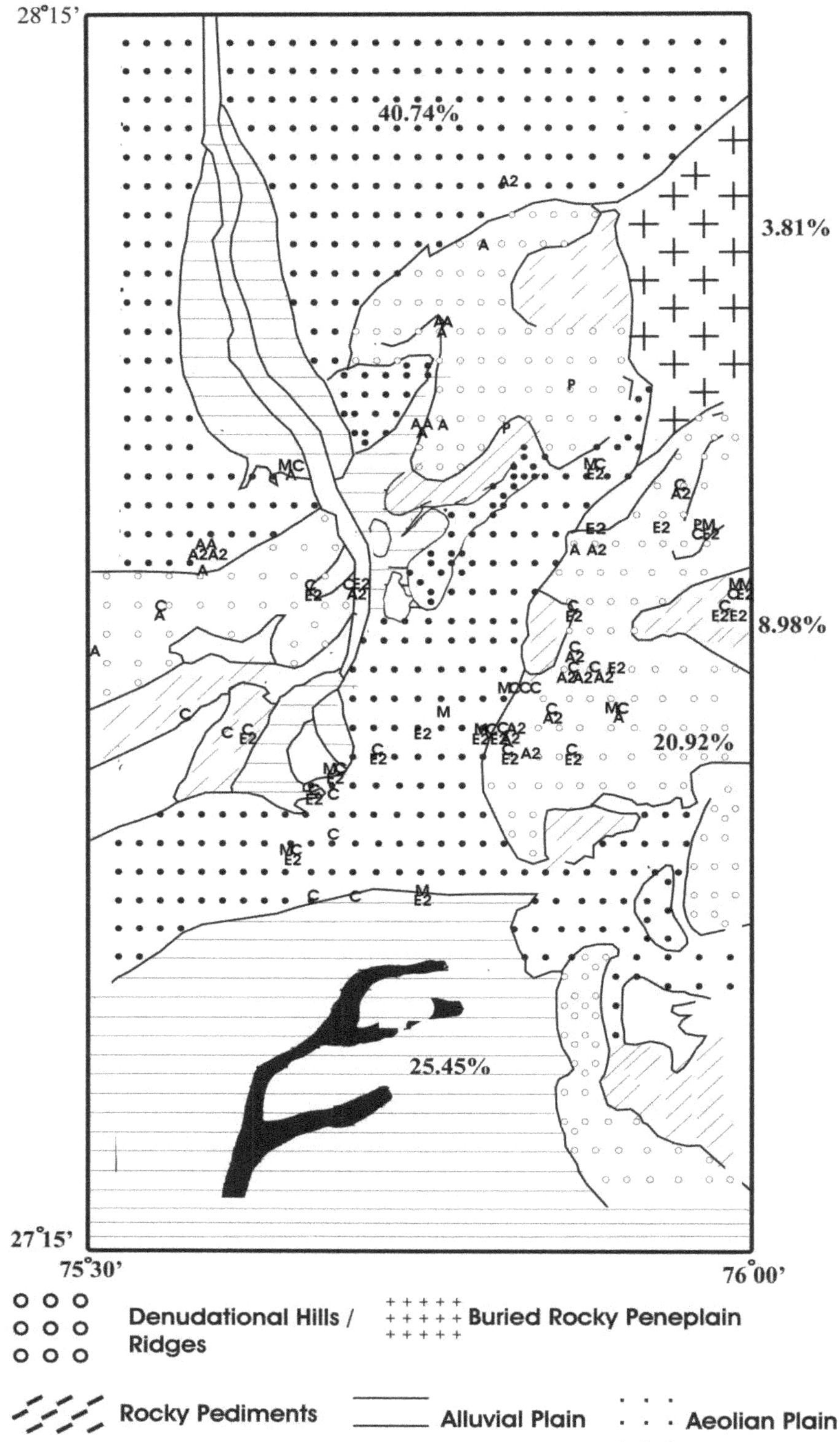

Figure – 5.3 Distributions of Sites in Geomorphological Features

Symbols: P – Palaeolithic; M – Mesolithic; C – Chalcolithic; E2 – Early Historic; A – Ancient Mining Area and A2 – Metal Processing Activity Area.

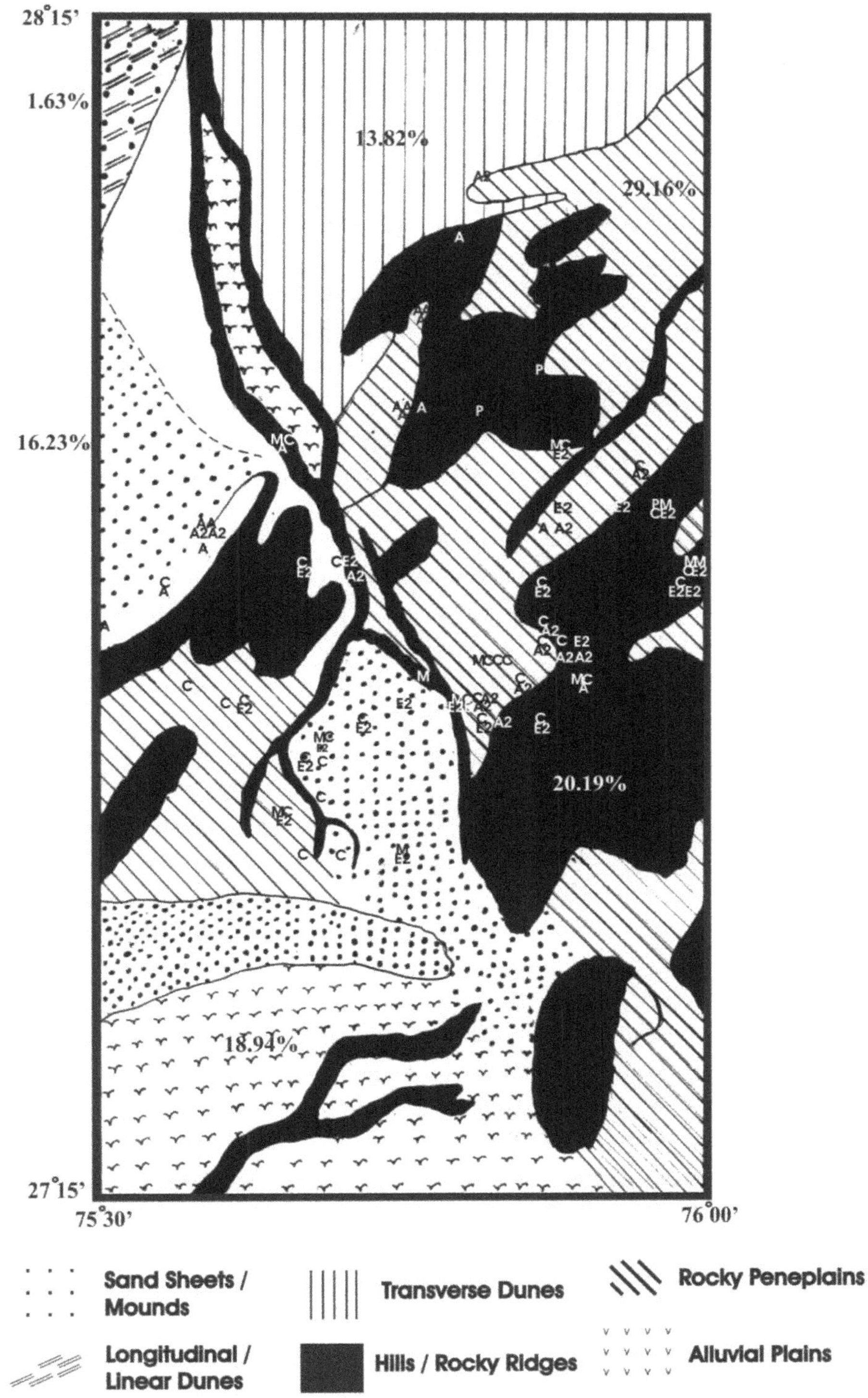

Figure - 5.4 Distributions of Sites in Dune Types

Symbols: P – Palaeolithic; M – Mesolithic; C – Chalcolithic; E2 – Early Historic; A – Ancient Mining Area and A2 – Metal Processing Activity Area.

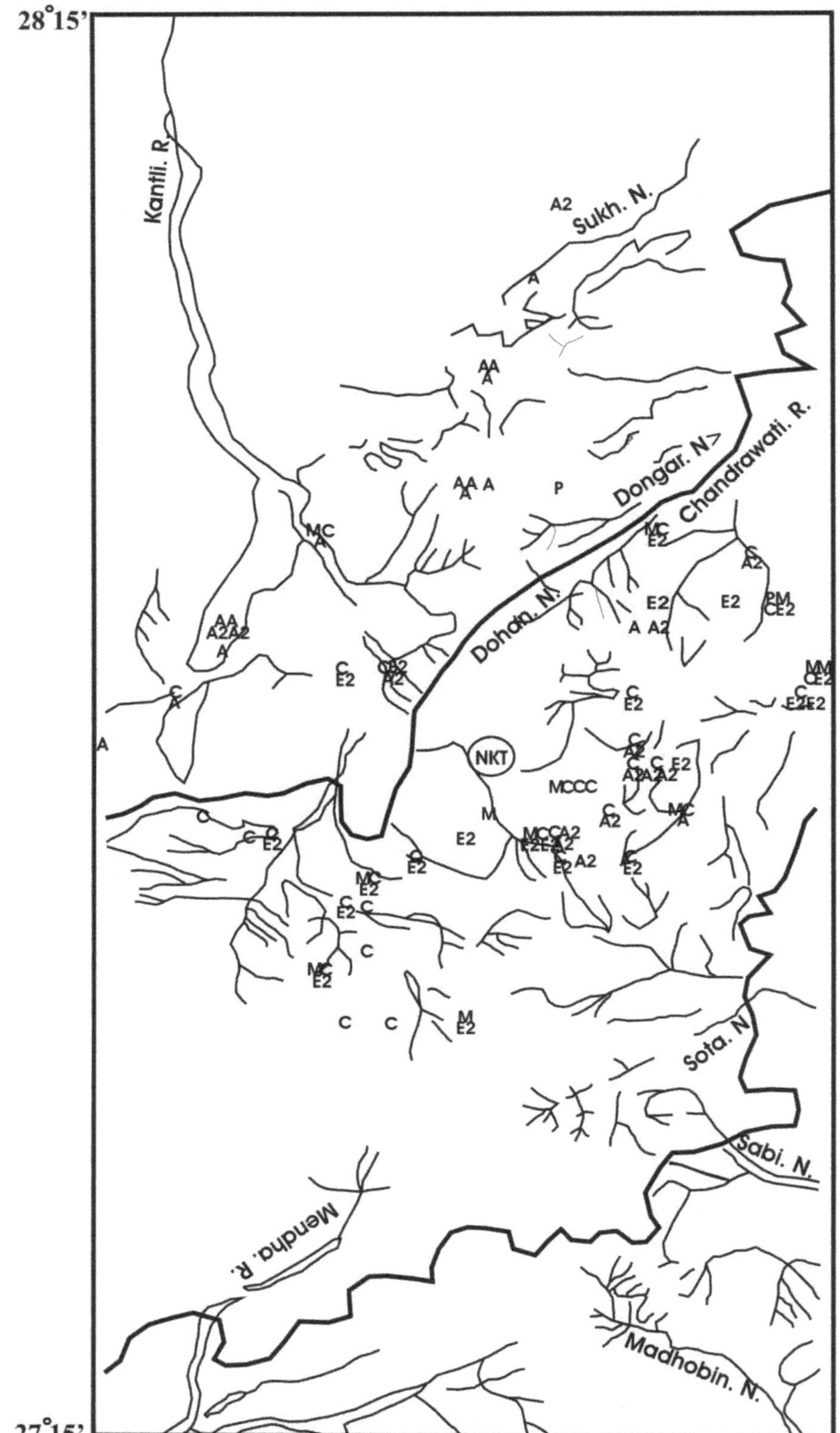

Figure – 5.5 Distributions of Sites in Drainage System

Symbols: P – Palaeolithic; M – Mesolithic; C – Chalcolithic; E2 – Early Historic; A – Ancient Mining Area and A2 – Metal Processing Activity Area.

CHAPTER 6 CONCLUSION

The systematic studies of settlement patterns in different parts have increased archaeological awareness of regional diversity and complexity. The concept of settlement-pattern-studies in archaeology consists of spatial distribution of human activities and occupation, which ranges from location of activities from single site to the arrangement of sites in the region. An understanding of settlement-patterns requires a proper understanding of the geological, geomorphic and environmental features of the region. It is an attempt to provide a comprehensive geological and ecological framework for delineating settlement-patterns in the research area.

To the best of my understanding, none of the previous studies in the research area has given sufficient information on the geomorphic features and their relationships with the archaeological evidence. Thus, the present study enunciates the position of prehistoric to early historic settlement-patterns in the research area. The research brings forth structure-inference concerning settlement location, function, distribution and trend in settlement density at regional scale with a view to understand ecological adaptation and cultural changes through prehistoric to early historic periods. The method of regional analysis has developed models for explaining economic and functional relations between settlements. Economic development is understood through analysing variations in style and technologies used for certain artefacts like ceramics, lithics and metals. Functional differences in terms of raw material resources, smelting sites, processing sites and possible interaction between these are adequately looked into.

Palaeolithic Culture

Rajasthan saw the beginning of Palaeolithic research in 1950s, which has revealed numerous sites that provide us with a broader and better picture of this culture. The distribution of Palaeolithic sites in Rajasthan is mostly concentrated on the southern, central and western Rajasthan. The Lower Palaeolithic sites are largely confined to southern and central Rajasthan and partly to the western Rajasthan whereas Middle Palaeolithic sites are largely confined to central and partly to the southern and western Rajasthan. The Upper Palaeolithic sites are largely confined to central and partly to the western and northeastern Rajasthan.

In spite of all these collective works done by different agencies, the present research area was shown devoid of any Palaeolithic findings. In the light of above resolution, a fresh survey was conducted, which has revealed three Lower Palaeolithic secondary sites discovered in the research area at Laluda ki Dhani, Payga Kundalia and Pir Baba ki Bani. The tools collected from these sites comprise handaxes, cleavers scrapers and hammer stones made from quartz and quartzite raw materials. Typo-technologically, these tools have been designated as Lower Palaeolithic tools. All the above mention sites are located on the denudational hills and rocky ridges on deposit 2 of the research area consists of quartzite, schist, conglomerate, gneisses, dolmitic marble and phyllite deposits. The sites are located on the nearby streams and nallas. This infers that Palaeolithic people have preferred hilly and rocky ridges for their settlements near to the small streams and nallas instead of main rivers. The findings of these sites has given lead for the future research in the Shekhawati region where Middle Palaeolithic and Upper Palaeolithic sites are still not discernible.

Mesolithic Culture

Generally, lithic assemblages are taken to predate the use of metal. So the Mesolithic of Rajasthan is believed to be necessarily earlier than the Chalcolithic. However, there is some, as yet insufficient evidence to suggest that sometimes the technological transition from lithic to metal was not as total as one had previously been led to believe, nor sudden and universal. However, the Mesolithic site of Bagor indicates that microlithic using hunter-gatherers inhabited Bagor even when Chalcolithic stage had been firmly established in parts of the region. This implies that some degree of hunter-gatherers' lifestyle continued to be in existence even when a more advanced technological skill had come into use. "So it is assumed that Agriculture neither dies nor begins everywhere at the same time and hence Mesolithic phase also expands or shrinks in a particular region depending on how late or early agriculture economy begins".

Following the above-mentioned direction, present survey on Mesolithic research was carried out to find sites showing single culture having microliths with or without pottery and multi-cultural site having microliths with other cultural materials. In the light of these assumptions, 12 Mesolithic sites reported from the research area, which has given single culture site with microliths and lithic debitages, multi-cultural sites with Chalcolithic and Early historic materials and few sites near to the mining areas. The significant feature observed in the research area is the earliest Mesolithic site at Payga Kundalia in Sohanpura village. The site revealed Mesolithic rock shelters with paintings, which represents in ochre color depicting animal figures, hunting scenes and group rituals. The site has also revealed cup marks on the boulders, which were used for preparing colors.

The finding of shell bangles as extraneous material from Kot in Sunari village and Biharipur indicate trade contacts with nearby coastal people i.e. from Gujarat coast or these shells may have came in exchange from the regional Chalcolithic people. This has to be looked in more detail in future.

The research area also gives evidence for primary Mesolithic sites from Gidhali No. 1 and 2 in Haripura village. The site is located on the stabilized sand dunes and reveals microlithic cores, blades and lithic debitage in large quantity, which is made from different raw materials.

The other important feature observed at Beed ki Dhani in Dariba village and in Kakrana that these sites are located near to the ancient mining areas, which indicate that Mesolithic people were also involved in mining that has been established in the research area.

The Mesolithic sites in the research area are mainly distributed on the aeolian plains and few of them are spread on the denudational hills, rocky pediments and alluvial plains. This point out that Mesolithic people has preferred their seasonal settlements on sand dunes, which lie in-between the hills and rocky ridges.

It is observe from the above facts that Mesolithic people in the research area have developed two phases of culture development and changes. The first phase demonstrates general lithic assemblages of microliths and lithic debitage whereas second phase shows continuation of lithic assemblages of the previous phase along with the Chalcolithic materials. This indicates that Mesolithic people in the research area were semi–nomadic microlithic using hunter-gatherer. Later they came in contact with copper using people from southern and northern Rajasthan and providing copper ore, which was readily available in the research area. Simultaneously they also involve in smelting copper ores and providing copper ingots. Subsequently this phase was continued for longer period in the research area, which was reflected in their material culture. With the bygone of time, their economy has also change to copper using semi-nomadic culture having hunter-gatherer life style and develope to copper using Chalcolithic culture.

Chalcolithic Culture

A discussion on Chalcolithic stage in India as a chrono-cultural phase becomes difficult because of the acute incongruencies recorded between various regions within the country.

The cultural distribution of Chalcolithic sites in Rajasthan on northern region along the Ghaggar – Hakra river and its associate water courses recognizes as Harappan culture sites whereas southeastern zone beyond south of Luni River identified as Ahar culture sites while central and northeastern part on the Shekhawati region known as Ganeshwar–Jodhpura culture sites. In the present survey, 33 Chalcolithic (Ganeshwar related sites) were discovered in the Shekhawati region of Rajasthan, i.e. Ganeshwar culture sites, which is known from a typical kind of pottery called as "OCP" (Ganeshwar), which emanate as an archaeological entity in its earlier stages of development. The term "OCP" for Ganeshwar pottery was a misnomer, which has also been misunderstood. Ganeshwar pottery is defined as red slipped painted pottery with profuse incised design, executed with the help of a sharp, comb like instrument. Due to the sketchy availability of information on Ganeshwar pottery, hitherto nothing much has been said or understood about the culture as a whole. In the present study, an attempt has been made to examine the explored pottery with the aim to present corpus of Ganeshwar ceramics types.

The distribution of Chalcolithic related Ganeshwar sites in the research area are by and large spread on denudational hills and rocky ridges and on aeolian plains and partly on the rocky pediments and on alluvial plains. The research area is paramount by Aravalli ranges running diagonally from northeast to southwest with the intervention of Kantli River flowing in south–north direction in the gap between the Aravalli ranges. The available data from research area indicates that Chalcolithic people have preferred their settlements mainly on hills and rocky ridges adjacent to the Aravalli ranges on the east. The other areas opt by Chalcolithic people were the aeolian plains, which lies in between the hills and rocky ridges and few settlements are rarely scattered on the rocky pediments and alluvial plains. This also imply that Ganeshwar people have preferred the nearby area where there is also availability of raw material for mining and smelting the ores into ingots

From the above data, it shows that Ganeshwar culture form an individual entity in the central and northeastern region of Rajasthan that is known as Shekhawati region, which marked the influence from both the Chalcolithic Culture from northern Rajasthan and Semi-Nomadic Copper using culture from southeastern Rajasthan. In the light of all these notions, it shows that Ganeshwar people have some trade affinities with the Harappans from north Rajasthan, which is reflected in the copper objects found during the excavation at Ganeshwar. While the site of Bagor and Gilund from southeast Rajasthan and Kalibangan from north Rajasthan shows resemblance in ceramic tradition with the Ganeshwar incised ware e.g. the incised pottery from Bagor phase 2 (Misra, 1973, Fig: 22-23), incised pottery from Gilund in Chalcolithic level (IAR 1959-60, plate: XLIV) and fabric D from Kalibangan (Madhu Bala, 1997, plate: 13.11). Along with this, the finding of graffiti sherds and reserved slip wares from Ganeshwar during exploration has thrown more light on contact with the Harappans or regional Chalcolithic people from southern Rajasthan.

Thus Ganeshwar culture appear as an individual entity in Shekhawati region as a regional Chalcolithic culture, whose main economy is based on metallurgy – mining and extracting copper and supply this material to the neighbouring regional Chalcolithic people. This has been supported by lead isotope analysis on Ganeshwar and Singhana copper samples, which have further, strengthen the argument of Ganeshwar contacts with the Harappans. Hence, Ganeshwar people identified as regional Chalcolithic people using more advance technological skill of metal technology that had come in use.

Early Historic Culture

After Painted Gray Ware culture, all inhabitants in northern Rajasthan seem to have deserted perhaps due to the desiccation of Saraswati River. Then in the early centuries of the Common Era saw the rise of flourishing Rang Mahal Culture in northern Rajasthan. The scenario of eastern Rajasthan during Early Historic period seems a continuation of Painted Gray Ware Culture as exemplified at Noh and Bairat as at many sites in northern India.

An Early Historic site of Noh, Bairat, Rairh, Jodhpura, Nagar, Sambhar, Nagari, Rang Mahal and Ahar from Rajasthan form four stages, which furnishes some ideas of the material culture of the Early Historic period. In the research area, Early Historic period is designated into two phases; Early Historic – I represent unpainted PGW, unpainted black and red ware and NBPW and Early Historic – II represents Rang Mahal ware sites.

The only sites in the research area Kakadiyo in Thikriya village and Kot in Sunari village have shown Early Historic – I culture material. The remaining 21 early historic sites identified as Rang Mahal ware. The significance of Rang Mahal sites from Ganeshwar shows human head in black schist stone, which is dateable to Sunga – Kushana period around 3rd century BCE to 2nd – 3rd century CE. The important finding of terra cotta seal with *Brahmi* letters from Kot has also established the early date to the site. The site has also revealed terra cotta anvil and iron bangles. The important findings of stamped pottery from Jogi Badh; terra cotta beads from Payga Kundalia and Jasi ka Bas; and iron bangles from Gidhali No. 2 has further enhance the material culture of the Early Historic period in the research area .

The research in early historic period was carried out to identify the distribution of sites in the research area. The distribution of Early Historic sites formed in the same fashion as formed during the Chalcolithic period. Most of the early historic sites were scattered on the denudational hills and rocky ridges and on the aeolian plains and remaining few sites are found on the rocky pediments and alluvial plains.

The available data from research area indicates that Early Historic people have preferred their settlements mainly on hills and rocky ridges adjacent to the Aravalli ranges and few of them scattered on the Aravalli ranges. The other areas prefer by the Early Historic people were the aeolian plains in between the hills and rocky ridges and few of them are scattered on the rocky pediments and alluvial plain. This implies that Early Historic people have preferred the same area, which was previous occupied by the Chalcolithic people.

Ancient Mining and Metal Processing Activity Area

Ancient mining and metallurgical research was carried out to probe its relation with the archaeological sites. In the present survey 19 ancient mining-areas and 14 metal-processing activity areas have been identified from the research area. Hence, this region gives the evidence of exploitation of metals by the inhabitants in different periods. Most of the mining and processing areas have been found on the denudational hills and rocky ridges and few have been scattered on the aeolian plains.

In the Aravalli's, Chalcopyrite is the main source of copper ore deposit, which is spread over the entire terrain from Singhana in the north to Ragunathgarh in the south at the stretch on 80 Kms. It is in this stretch most of the ancient mining areas are located and exploited for the copper ores. In the adjacent hills on the east of the Aravallis there are also few deposits, which were exploited for copper ores. It is in these adjacent hills nearby the small streams and nallas, there are number of metal processing activity areas were identified.

A preliminary investigation of lead isotope analysis was carried out by Hoffman, Randal and Raghubans (2005) to determine the provenance of copper ore during Harappan period. The copper sample were derived from Harappa and were compared with seven regional copper source areas i.e. Ganeshwar and Singhana in Rajasthan, Chagai hills in Baluchistan, Shin Kai copper deposit in Waziristan, North Gujarat, Himachal Pradesh, Uttaranchal and Oman. The analysis resulted that Harappans perhaps acquired most of copper ore from the west of Harappa and some of its copper ore was also obtained from Rajasthan. This information has given a lead for further metallurgical research in the Shekhawati region.

Majority of the archaeological sites in the research area are distributed on the denudational hills and rocky ridges and few of them are on the rocky pediments and on the alluvial plains. Similarly ancient mining and processing areas have been found on the same geomorphological settings. This implies that settlement patterns in the research area have preferred denudational hills, rocky ridges and aeolian plains for their settlements and for their metal processing activities. Another important feature observed in the research area is the archaeological sites, which are spread on the nearby small streams and nallas rather than on the main rivers and their tributaries. This shows that prehistoric to early historic people have preferred water bodies where water is stored for longer period then the rivers and their tributaries where water is available seasonally because all the rivers in the research area are seasonal. Another important feature observed is about Aravalli ranges, which has formed as a major geomorphological feature in the research area. Aravallis has also played major role in human colonization and dispersion, which is reflected in the archaeological sites. Aravalli has also formed as a natural barrier for cultural and ecological dispersion. For instance the rainfalls in Rajasthan rely on the depression form in the Central India by southwest and northeast monsoon and moves towards Rajasthan. So the Aravallis running diagonally in Rajasthan block the major rainfall on the western Rajasthan that Marwar region or the west of the Aravalli. This barrier is reflected in archaeological sites also, where major archaeological sites are flourishing on the east of the Aravallis.

BIBLIOGRAPHY

- Agarwal, B.D. 1978. **Rajasthan District Gazetteers, Sikar**. Directorate of District Gazetteers, Government of Rajasthan, Jaipur.
- Agrawal, D.P. 1971. **The Copper Bronze of India.** New Delhi: Munishiram Manoharlal.
- Agrawal, D.P. 1982. **The Archaeology of India.** Scandinavian Institute of Asian Studies Monograph Series # 40. Curzon Press. London.
- Agrawal, D.P. 1985. **The Archaeology of India.** Scandinavian Institute of Asian Studies Monograph Series # 46. Curzon Press.
- Agrawal, R.C., Datta, P.S., Hussain, Z., Krishnamurthy, R.V., Misra, V.N., Rajaguru, S.N. and Thomas, K. 1980. *Palaeolithic, Stratigraphy and Prehistory in north and west Rajasthan.* Proc. Ind. Acad. Sci. 89 (1): pp. 51-56.
- Agrawala, R.C. 1978. *Archaeological Discoveries at Ganeshwar, Rajasthan.* Archaeological Studies Vol. #3. Mysore.
- Agrawala, R.C. 1979. *More Copper finds from Rajasthan,* Man and Environment III: 91-92.
- Agrawala, R.C. 1981a. *Protohistoric Copper Objects from Rajasthan: New Parameters.* (Ed) V.S. Srivastava. **Cultural Contours of India,** pp. 25-31. New Delhi.
- Agrawala, R.C. 1981b. *Recent Explorations in Rajasthan.* Man & Environment, Vol. V. ISPQS, Pune.
- Agrawala, R.C. 1984. *Ganeshwar Culture: A Review.* Journal of Oriental Institute. Baroda 34(1-2): 89-95.
- Agrawala, R.C. and Vijay Kumar. 1982. *Ganeshwar – Jodhpur Culture. A New Trait in Indian Archaeology.* (Ed.) G.L. Possehl. **Harappan Civilization. A Contemporary Perspective**, pp. 125 – 134. New Delhi: Oxford IBH. 1982.
- Allchin, B. and Allchin, R. 1982. **The Rise of Civilization in India and Pakistan.** Cambridge.
- Allchin, B. and Goudie, A.S. 1971. *Dunes, Aridity and Early Man in Gujarat, Western India.* Man 6: pp. 248-265.
- **Allchin, B., A. Goudie and K.T.M. Hegde. 1978.** The Prehistory and Palaeography of the Great Indian Desert. **Academic Press. London. New York. San Francisco.**
- Antonava, K., G. Bongard-Levin and G. Kotovsky. 1979. **A History of India.** Book # 1. Progress Publication. Moscow.
- Asthana, S. 1982. *Harappan Trade in Metals and Minerals – A Regional Approachable.* (Ed) G.L.Possehl. **Harappan Civilization – A Contemporary Perspective** New Delhi: Oxford IBH. 1982.
- Bachmann, H.G. 1982. **The Identification of Slags from Archaeological Sites.** Institute of Archaeology, London.
- Bailey, A.R. 1964. **A Text Book of Metallurgy.** London
- Bala, M. 1997. *Some Unique Antiquities And Pottery From Kalibangan.* (Ed) J. P. Joshi. **Facets of Indian Civilization. Recent Perspective.** Vol. # 1. Publisher Aryan Books International. New Delhi.
- Ball, V. 1881. A Manual of Geology of India. Part # 3. **Economic Geology.** GSI Calcutta.
- Barnes, J.W. and E.H. Bailey. 1972. *Geologists Discover Ancient Retort Evidence Points to Worlds Oldest Mine.* **World Mining** (US Edition). April pp. 49-55.
- Bhan, K.K. 1989. *Late Harappan Settlement of Western India with specific reference to Gujarat.* (Ed) J.M. Kenoyer. **Old Problem and New Perspective in the Archaeology of South Asia.** Wisconsin.
- Bhan, K.K., Nikita Mehta and K. Krishnan. 1999. A Preliminary Ethno-archaeological Study of Copper/ Brass Working in Rajasthan, Western India. (Forthcoming).
- Bhan, S. 1967. *New Light in the OCP.* Research Bulletin (Arts) # LVII (III). Punjab Uni. Chandigarh.
- Bhandarkar, D.R. 1920. **Archaeological Remains and Excavation at Nagari.** Memoirs of the Archaeological Survey of India, 4. Calcutta: Superintendent Govt. Printing.
- Bhattacharya, D.K. 1996. **An Outline of Indian Prehistory.** Palaka Prakashan. Delhi.
- Bhowmik, S.K. 1972. Early Copper and Bronze Technology of Gujarat. Thesis submitted to M.S. University of Baroda.
- Bola Nath 1967. Animal Remains from Adamgarth Rock shelter. Indian Museum Bulletin 2, 1: 28 – 37
- Bose, U. 1989. *Correlation Problems of the Proterozoic Stratigraphy of Rajasthan.* Indian Minerals Vol. 43 # 3&4 (July-Dec) pp.183-193.
-
- Bryson, R. and Swain, A. 1981. *Holocene Variations of Monsoon Rainfall in Rajasthan.* Quaternary Research 16: 13 5-145.
- Carlleyle, A.C.L. 1878. Report of a Tour in Eastern Rajputana in 1871 – 72 and 1972 – 73, Archaeological Survey of India Report VI: 107 – 108, 160 – 161.
- Casal, J. 1961. Fouilles de Mundigak. 2 Vols. Paris: Memoires de la Delegation Archaeologiquw Francais en Afghanistan, Tome – 17.
- Chakarbati, D. 1986. **The Pre – Industrial Mines of India.** Puratattava Vol. # 16 pp. 65-72.
- Chakarbati, D. and N. Lahiri. 1996. **Copper and Its Alloys in Ancient India.** Palaka Prakashan. Delhi.
- Chandak, G.J. and Sarkar, S.K. 1980. *Basemetal Mineralisation in the Precambrian Rocks of Rajasthan and Gujarat States, India - Environmental Concepts and Resources Appraisal.* Indian Minerals Vol. 34 # 3 (July-Sept) pp. 12-27. Geological Survey of India.
- Chang, K.C. 1983. *Settlement Patterns in Chinese Archaeology; A Case Study from the Bronze Age.* (Ed.) Evon Vogt & Richard Leventha. Prehistoric Settlement Patterns. Albuque que: University of New Mexico Press 361 – 374.
- Chang, K.C. 1958. *Study of the Neolithic Social Grouping; Example from the New World.* American Anthropologist, 60, 298 – 334.

- Chibber. H.L. 1925. *Old Copper Slag's from Ambamata and Kumbaria in Danta State.*_ Progs. Of the Indian Science Congress. 12th Session.
- Coghlan, S. and Constantini, L. 1951. *Notes on the Prehistoric Metallurgy of Copper and Bronze in the Old World.* Occasional Papers on Technology 4. Oxford. Pitt River Museum.
- Courty, M.A. 1989. *Integration of Sediment of Soil Formation in the Reconstruction of Protohistoric and Historic Landscape of the Ghaggar Plain, Northwest India.* **In South Asian Archaeology 1985.** (Eds.) K. Frifde and P. Sorensen. Curzon Press. London pp. 225-229.
- Craddock, P.T. 1995. **Early Metal Mining and Production.** Edinburgh Uni. Press.
- Craddock, P.T., L.K. Gurjar, and K.T.M. Hegde. 1983. *Zinc Production in Medieval India.*_**World Archaeology.** Vol. # 15 pp. 211-221.
- Crawford, A.R. 1970. Can. Jour. Earth Sci. 7: pp. 91-110. *The Precambrian Geochronology of Rajasthan and Bundelkhand, India.*
- Cunningham, A. 1873. Bairat, or Vairat, in Archaeological Report, 1964-65, pp. 242-49. Calcutta: Archaeological Survey of India.
- Dalal, K.F. 1980. A Short History of Archaeological Explorations in Bikaner and Bahawalpur along the 'Lost' Saraswati River, Indica 17(1): 1-40.
- Dales, G.F. and Kenoyer, J.M. 1986. **Excavation at Mohenjo Daro, Pakistan: the Pottery.** University Museum Press. Phildalphia.
- Das Gupta, S.K. 1975. *A revision of the Mesozoic Tertiary Stratigraphy of the Jaisalmer Basin, Rajasthan.*_Indian Jour. Earth Sci. 2(1): pp. 77-94.
- Das Gupta, S.P. 1968. *Structural History of the Khetri Copper Belt, Jhunjhunu and Sikar Districts, Rajasthan.*_Mem. Geol. Surv. Ind. Vol. #98.
- Deotare, B.C., M.D. Kajale, A.A. Kshirsagar and S.N. Rajaguru. 1998. *Geoarchaeological and Palaeoenvironmental Studies around Bap-Malar Playa, District Jodhpur, Rajasthan*, Current Science 75(3): 316-320.
- Deotare, B.C., M.D. Kajale, S.N. Rajaguru, S.Kusumgar, A.J.T. Jull and J.D. Donahue. 2004. *Palaeoenvironmental History of Bap-Malar and Kanod Playas of Western Rajasthan, Thar Desert*, Proceedings of the Indian Academy of Sciences (Earth and Planet. Sci), 113(3): 403-425.
- Dhaka, Ambika. 2002. Unpublished Copper Hoard from Varman, Distric Sirohi, Nucleus II (2) 121 – 134.
- Dhaka, Ambika. 2003. Unpublished Copper Hoard from Bhandar, District Pali, Puratattva 33: 127 – 130.
- Dhavalikar, M.K. 1984a. *Towards an Ecological Model for the Chalcolithic Cultures of Central India and Western India.* Journal of Anthropological Archaeology 3: 133-58.
- Dhavalikar, M.K. 1984b. *Sub-Indus Cultures of Central and Western India.* Mortimer Wheeler commemoration Volume. (Ed) B.B. Lal and S.P. Gupta. **Frontiers of the Indus Civilization.** New Delhi. Books and Books.
- Dhir, R.P., Kar, A., Wadhawan, S.K., Rajaguru, S.N., Misra, V.N., Singhvi, A.K. and Sharma, S.B. 1992. *Thar Desert in Rajasthan.* (Eds) Sanghvi, A.K. and Kar, A.) **Land, Man and Environment.**_Geol. Soc. Ind.: p. 191.
- Dikshit, K.N. 1984. *Late Harappan in Northern India.*_The Cultural Evolution of Civilization. (Ed) G.L. Possehl. **Ancient Cities of the Indus.** New Delhi. Vikas Pub. House.
- Fox, C. 1932. **Personality of Britain**, London.
- Francfort, H.P. 1984. *Evidence for Harappan Irrigation System in Haryana and Rajasthan.*_*The Eastern Anthropologist* 45.
- Gangadhar, K. 1996. Morpho-Tectonic and Hydrogeological Studies of the Upper Luni and Kantli River Basins, Western Rajasthan. Unpublished Thesis presented to M.S.University of Baroda. Geology Department.
- Gangadhar, K. and Tiwari, K.C. 1995. *The Jayal Gravels: Their Gnesis, Stratigraphic, and Tectonic Significance.* (Eds. Desai, N. and Ganapathi, S.). **Proceedings of National Seminar on Recent Researches in Geology of Western India**: pp. 143-151.
- Gaur, R.C. 1973. *The Ochre Coloured Pottery: A Reassessment Of The Evidence.* (Eds) J. E. Vanlohuizen-Deleeuw and J. M. M. Ubaghs. South Asian Archaeology 1973.
- Ghose, B. Singh, S., and Kar, A. 1977. *Geomorphology of the Rajasthan Desert.* **Desertification And Its Control.** Released on the occasion of The UN Conference On Desertification, Nairobi. 29th August- 9th September. Indian Council Of Agricultural Research. New Delhi.
- Ghosh, R.N. 1977. *Photogeological Studies on Ancient Water Regimes Rivers.* (Eds. Agrawal, D.P. and Pandey, B.M.) **Ecology and Archaeology of Western India**: pp. 157-168. Concept Pub House. Delhi.
- Gordon, C.V. 1934. **New Light on the most Ancient East; The Oriental Prelude to European Prehistory.** London; Routledge & Keygen Publication.
- Ghosh, A. 1989. **An Encyclopedia of India Archaeology.** Vol. # 1 & 2. Munshiram Manoharlal Pub. Delhi.
- Goudie, A., Allchin, B., and Hegde, K. T. M. 1973. *The Former Extensions Of The Great Indian Sand Desert.* The Geographical Journal Vol. 139 Part # 2. June 1973. Royal Geo. Soc. London.
- Gribble, C.D. 1988. **Rutley's Element of Mineralogy.** London.
- Grover, A.K. 1987. *Integrated Remote Sensing Studies of the Northern Part of Khetri Copper Belt and Adjoining Areas of Jhunjhunu and Sikar Districts of Rajasthan.* Unpublished GSI Report for F.S. 1986-87.
- Gupta, S. 1984. **Rajasthan District Gazetteers, Jhunjhunun**. Directorate of District Gazetteers, Government of Rajasthan, Jaipur.
- Handa, D. 1968. *Harappan, OCP, and Copper Hoards.*_(Ed) Giuseppe Tucci. East and West IsMEO. New Series, Vol. 18 #s 1-2. March – June 1968.

- Hanna, R. 1959. **Rang Mahal.** The Swedish Archaeological Expedition to India 1952-54. Acta Archaeological Lundensia. Series 4 #3. CWK. Gleerup Publisher Lund Sweden.
- Hegde, K.T.M. 1991. **An Introduction to Ancient Indian Metallurgy.** Geological Society of India. Bangalore.
- Hegde, K.T.M. and J.E. Ericson. 1985. *Ancient Indian Copper Smelting Furnaces.* (Eds.) P.T. Ctaddock and M.J. Hughes. **Furnaces and Smelting Technology in Antiquity.** British Museum Occasional Paper No. 48. London: British Museum Research Laboratory pp. 59 – 70.
- Hegde, K.T.M. 1965. Technical Studies in Chalcolithic Period Copper Metallurgy. Thesis submitted to M.S. University of Baroda.
- Heron, A.M. 1953. **The Geology of Central Rajputana.** Mem. Geol. Surv. India. 79: p.389.
- Hoffman. Randal, L. and Raghubans K. 2005. *Pb isotope analysis to determine the provenance of copper ore during Harappan period.*_Paper presented in South Asian Archaeological Conference, London.
- Hooja, R. 1988. **The Ahar Culture And Beyond.** Oxford: British Archaeological Review International Series # 412.
- Hooja, R. and Vijay Kumar. 1995. *Aspects of the Early Copper Age in Rajasthan.* (Ed.) B.Allchin. **South Asian Archaeology 1995.**
- Hooja, R. 1994. *Ganeshwar and the 'Frontier Theory': A Possible Framework for Understanding Ganeshwar-Harappan Interaction.* Shodhak, 23.Pt. C(69): 127-136.
- Imperial Gazetteer of India. 1909. Bombay Presidency Vol. II. Calcutta. Govt. of India.
- Indian Archaeological: A Review. 1953 - 54:37.
- Indian Archaeological: A Review. 1954 - 55: 58.
- Indian Archaeological: A Review. 1956 - 57: 5-8.
- Indian Archaeological: A Review. 1960 - 61: 13
- Indian Archaeological: A Review. 1962 - 63: 19-31.
- Indian Archaeological: A Review. 1963 - 64: 28-29.
- Indian Archaeological: A Review. 1972 - 73: 29.
- Indian Archaeological: A Review. 1993 - 94: 93-97.
- Indian Archaeology: A Review. 1959 - 60. Excavation at Gilund. Delhi: ASI
- Indian Archaeology: A Review. 1972 - 73. Excavation at Jodhpura (Rajasthan). Delhi: ASI
- Indian Archaeology: A Review. 1980 - 81. Excavation at Sunari. Delhi: ASI
- Indian Archaeology: A Review. 1960 - 61. Excavation at Adamgarh, District Hoshangabad. pp. 13-14. Delhi: ASI
- Jain, K.C. 1972. **Ancient Cities and Towns of Rajasthan (A Study of Culture And Civilization).** Motilal Banarsidas. Delhi.
- Jarrige, J. 1982. *Excavation at Mehrgarh: Their Significance for Understanding the background of the Harappan Civilization.* (Ed) Gregoory Possehl. **Harappan Civilization: A Contemporary Perspective.** Delhi: Oxford & IBH and the American Institute of Indian Studies: 79 – 84.
- Jarrige, J.F. 1984. *Chronology of the earliest periods of the Greater Indus as seen from Mehrgarh, Pakistan.* (Ed) B. Allchin. **South Asian Archaeology 1981.** Cambridge: Cambridge University Press. Pp. 21 – 28.
- Jensen, M and Bateman, A. 1981. **Economic Mineral Deposits.** Third Edition. John Wiley and Sons. New York Chichester Brisbane Toronto.
- Jodha, N.S. 1977. *Land-Tenture Problems And Policies In The Arid Region Of Rajasthan.* **Desertification and Its Control.** Released on the occasion of The UN Conference on Desertification, Nairobi. 29th August- 9th September. Indian Council of Agricultural Research. New Delhi.
- Joshi, J.P., Madhu Bala, and Jassu Ram 1984. *The Indus Civilization: A Reconsideration on the basis of Distribution Maps, in frontiers of the Indus Civilization:* Sir Mortimer Wheeler Commemoration Volume (B.B. Lal and S.P. Gupta Eds.) pp. 511-530. New Delhi: Books and Books.
- Kennedy, Kenneth A.R and Peggy Caldwell 1984. *"South Asian prehistoric Human Skeletal Remains and Burial Practices."* (Ed) John R. Lukacs. **The People of South Asia: The Biological Anthropology of India, Pakistan and Nepal.** 159-97. New York: Plenum Press.
- Kenoyer, J.M. 1991. *The Indus Valley Tradition of Pakistan and Western India.*_Journal of World Prehistory 5(4): 331-385.
- Kenoyer, J.M. and H.M. Miller. 1995. *Metal Technology of Indus Valley Tradition in Pakistan and Western India.*_In the Ancient Metallurgy of the Asian Old World. University Museum Monograph # 89. (Ed) V. Piggot. The University Museum Philadelphia.
- Kercross, D. 1950. *Ancient Mines and Miners of India.*_**Indian Minerals** Vol. #4 pp. 5-10.
- Krishnan, A. 1977. *A Climatic Analysis of the Arid Zone of North-Western India.* **Desertification and Its Control.** Released on the occasion of The UN Conference on Desertification, Nairobi. 29th August- 9th September. Indian Council of Agricultural Research. New Delhi.
- Krishnan, M.S. 1982. **Geology of India and Burma**. CBS Publishers & Distributors. India.
- Kumar, V. 1986. *Excavation at Ganeshwar District Sikar.* Indian Archaeological: A Review. 1983-84 pp.71-2.
- Kumar, V. 1977. Archaeological Discoveries. Pp. 33-36 in 81st Indian Science Congress: Souvenir, University of Rajasthan, Jaipur.
- Kumar, V. 1986. *Excavation at Ganeshwar District Sikar.* Indian Archaeological: A Review. 1983-84 pp.71-2.
- La Touche, T.D. 1902. **Geology of Western Rajputana**. Mem. Geol. Surv. India. 35: pp. 1-116.
- Lahiri, N. 1992. **The Archaeology of Indian Trade Routed upto C. 200 B.C. (Resources Use, Resources Access, and Lines of Communication).** Delhi. Oxford University Press.
- Lahiri, N. 1995. *Indian Metal and Metals Related Artifacts As Cultural Signifiers.*__An Ethno-

Archaeological Perspective. World Archaeology Vol. # 27(10) pp.116-132.

- Lal, B.B. 1958 "Birbhanpur: A Microlithic Site in the Damodar Valley, West Bengal." Ancient India 14:4-48.
- Lal, B.B. 1979. *Kalibangan and Indus Civilization.* (Eds.) D.P. Agrawal and Dilip Chakarbati. **In Essays in Indian Protohistory.** New Delhi.
- Lal, B.B. 1997. **The Earliest Civilization of South Asia.** Aryan Book International. New Delhi.
- Lal, B.B. 1979. Kalibangan and Indus Civilization. (Eds.) D.P.Agrawal and DilipChakarbati. **In Essays in Indian Protohistory.** New Delhi.
- Leshnik, L.S, 1968. *Prehistoric Explorations in North Gujarat and Parts of Rajasthan.* East and West (N.S) 118, 295-310 .
- Malhotra, S.P. 1977. *Socio-Demographic Factors and Nomadism in The Arid zone.* **Desertification and Its Control.** Released on the occasion of The UN Conference on Desertification, Nairobi. 29th August- 9th September. Indian Council of Agricultural Research. New Delhi.
- Meena, B. and Tripathi, A. 2000. *Excavation at Ojiyana.* Puratattva # 30. 1999 – 2000 pp. 67-73. Bulletin of the Indian Archaeological Society. New Delhi.
- Mehta, N., K.K. Bhan, and K. Krishnan. 1999. A Preliminary Ethno-Archaeological Study of Copper/ Brass Working in Rajasthan, Western India. M.A. Dissertation submitted to the Department of Archaeology. Baroda.
- Miller, H.M.L. 1994. *Metal Processing at Harappa and Mohenjodaro: Information from Nonmetal remains.* (Eds.) A Parpola and P.Koskikallio. **South Asian Archaeology 1993.** pp. 497 - 510 Tiedeakatemia, Hesinki, Finland.
- Misra, V. 1989. *Human Adaptation to The Changing Landscape of The Indian Arid Zone During The Quaternary Period.* (Ed) J. M. Kenoyer. **Old Problem and New Perspective in the Archaeology of South Asia.** Wisconsin Archaeological Report, Vol. 2.
- Misra, V. N. and Rajaguru, S. 1985. *Palaeo-environment and Prehistory Of The Thar Desert, Rajasthan, India.* (Eds) P. Sorensen and K. Frifelt. **South Asian Archaeology 1985.** Copenhagen: Scadinavia Institute of Asian Srudies.
- Misra, V.N. 1963. *Palaeolithic Industry of the Banas, Eastern Rajputana.* Journal of the Asiatic Society of Bombay, 34-35: 138-160.
- Misra, V.N. 1966. *Stone Age Research in Rajasthan A Review.* (Eds) D.Sen and A.K. Ghosh. **Studies in Prehistory.** Robert Bruce Foote Memorial Vol. Firma K.L. Mukhopadhyay. Calcutta.
- Misra, V.N. 1967. **Pre & Protohistory of the Berach Basin.** South Rajasthan. Pune: Deccan College Postgraduate and Research Institute.
- Misra, V.N. 1971. *Two Microlothic Sites in Rajasthan. A Preliminary Investigation.* The East Anthropologist XXIV/3 pp.237-288.
- Misra, V.N. 1973a *"Bagor: A Late Mesolithic Settlement in North-west India."* World Archaeology 5, 1:92-100.
- Misra, V.N. 1973b. *A New Prehistoric Ceramic From Rajasthan.* (Ed) Giuseppe Tucci. East and West. ISMEO. New Series Vol. 23 # 3-4. September – December 1973.
- Misra, V.N. 1984. *Climate, a Factor in the Rise and Fall of the Indus Civilization – Evidence from Rajasthan and Beyond.* (Ed) B.B. Lal and S.P. Gupta. **Frontiers of the Indus Civilization.** Books and Books, New Delhi.
- Misra, V.N. 1989. *Human Adaptions to the Changing Landscape of the Indian Arid Zone During the Quaternary Period,* (Ed.) J.M. Kenoyer in **Old Problems and New Prespectives in Archaeology of South Asia,** Wisconsin Archaeological Reports, Volume 2, pp.3-20. Madison: University of Wisconsin-Madison.
- Misra, V.N. 2006. *A Gazetteer of Archaeological Sites in Rajasthan (from Paleolithic to Early Historic).* Man & Environment Vol. XXXI, No.1. pp: 48-96. Pune.
- Misra, V.N. 1977a. *Bagor A Late Mesolithic Settlement in North-West India.* World Archaeology Vol. 5. #1.
- Misra, V.N. 1977b. *Evolution of the Patterns of Human Settlement in the Arid and Semi-Arid Regions.* Desertification And Its Control. Released on the occasion of The UN Conference On Desertification, Nairobi. 29th August- 9th September. Indian Council Of Agricultural Research. New Delhi.
- Misra, V.N. and Malti Nagar. 1963. *Two Stone Age Sites on the River Chambal, Rajasthan,* Bulletin of the Deccan College Research Institute, XXII: 156-169.
- Misra, V.N. et.al. 1994. *Excavation at Balathal. Their contribution to the Chalcolithic and Iron Age Cultures of Mewar, Rajasthan.* Man and Environment. Vol.XX #1, 1995.
- Misra, V.N., and M.S. Mate. 1965 **Indian Prehistory**. Poona: Deccan College Postgraduate and Research Institute.
- Misra, V.N. and Rajaguru, S. 1985. *Palaeoenvironment and Prehistory Of The Thar Desert, Rajasthan, India.* (P. Sorensen and K. Frifelt Eds). *South Asian Archaeology* 1985. Copenhagen: Scadinavia Institute of Asian Srudies.
- Misra, V.N., Rajaguru, S.N. and Raghavan, H. 1988. *Late Middle Pleistocene Environment and Acheulian Culture Around Didwana, Rajasthan. Palaeoclimate and Palaeoenvironmental Changes in Asia During the Last 4 million years.* Proc. Symp. Ind. Nat. Sci. Acad.: pp. 93-106. New Delhi.
- Misra, V.N., Rajaguru, S.N., Raghavan, H. and Gaillard, C. 1982. *Acheulian Occupation and Evolving Landscape Around Didwana in the Thar Desert, India.* Man and Environment 6: pp. 72-86.
- Mohanty, R., Mishra, A., Joglekar, P., Thomas, P., Kharakwal, J., and Panda, T. 2000. *Purani Marmi: A Late Ahar Culture Settlement in Chitaurgarh District,*

Rajasthan. Puratattva # 30. 1999-2000. Bulletin of the Indian Archaeological Society. New Delhi.

- Mohapatra, G.C., S.B. Bhatia and G.K. Sahu. 1963. *The Discovery of a Stone Age Site in the Indian Desert.* Research Bulletin (N.S.) of the Punjab University, 14(3-4): 205-223.
- Morgan, L. H. 1877. **Ancient Society.** London. Macmillan Publication
- Muktinath, 1969. *Phosphate Deposits in Rajasthan.* **Indian Minerals** 23 (7): pp. 29-42.
- Nigam, J. S. 1996. *Sothi Pottery at Kalibangan: A Reappraisal.* Puratattava # 26. 1995-96. Bulletin of the Indian Archaeological Society. New Delhi.
- Pant, G. and Maliekal, J. 1987. *Holocene Climatic Changes Over Northwest India: An Appraisal.* Climatic Change 10: 183-194.
- Possehl, G. L. 1980 **Indus Civilization in Saurashtra.** Delhi: B.R. Publishing Corp.
- Possehl, G.L. and Rissman, P.C. 1992. *The Chronology of Prehistoric India from Earliest Times to the Iron Age.* (Ed) Robert W. Ehrich. Chronology in Old World Archaeology. University of Chicago Press.
- Possehl, G.L., Vasant Shinde and Marata Ameri. 2004. *The Ahar-Banas Complex and the BMAC,* Man and Environment XXIX(2): 18-29.
- Puratattava # 5. 1972-72. Session on OCP.
- Puri, K.N. n.d. Excavations at Rairh During Samvat 1995-1996(1938-39 A.D.) Jaipur: State Archaeology Department.
- Raghav, K.S. 1986. *A Report on Quaternary Geology, Geomorphology, Geo-Environmental Appraisal and Land Use Around Northeastern Part of Aravalli Hill Gaps, Fringing The Thar Desert, Rajasthan.* Unpublished GSI Report for F.S. 1982-86.
- Raghav, K.S. and Grover, A.K. 1991. *Major Sub-Recent Changes in the Upper Reaches of Kantli River Course and Related Environment Implications. Rajasthan.* (Eds. Desai, N. Ganapathi,S. and Patel, P.K.) **Proceedings of National Seminar on Quaternary landscape of Indian sub-continent.** 1st - 3rd Dec. 1988. M.S. Uni. of Baroda: pp. 147-152.
- Raghubans, K. 2013. *Ancient Mines and Metal Processing Activities in Shekhawati Region of Rajasthan.* (Eds) Ajit Kumar, Rajesh SV, Abhayan GS and Vinod V. in Heritage Journal of Multidisciplinary Studies in Archaeology. Volume 1 special issue on Harappan Region Chalcolithic Culture of Greater Indus Region pp. 138 – 170. Department of Archaeology, University of Kerala, Thiruvananthapuram.
- Raghubans, K. 2013. *Explored Mesolithic Sites in Shekhawati Region of Rajasthan.* (Ed) Dr. D. Dayalan in Sivarsi: Perspective in Indian Archaeology, Art & Culture pp. 47 – 72. Agam Kala Prakashan, New Delhi.
- Raghubans, K. 2012. *Ganeshwar Culture an Archaeological Entity in Rajasthan.* Publication Awaited, paper presented in International Conference on Early Agro-Pastoral Cultures of South Asia organized by Archaeological Survey of India, Bangalore.
- Raghubans, K. 2004. *Geomorphic features and archaeological evidence in parts of Sikar and Jhunjhunun districts of Rajasthan.* (Eds) Tripathi, Agam Prasad, Awasthi in Triratna Heritage, Governance and Equity Vol. III pp. 1021 – 1042. Agam Kala Prakashan, New Delhi.
- Rakshit, P. and Sundaram, R.M. 1990. *Quaternaries of Rajasthan. Records*. Geol. Surv. Ind. 123 (7): p. 38.
- Reddi, N.H. 1982. Report on Systematic Hydrological Investigations in Parts of Jhunjhunu District, Rajasthan. Unpub. CGWB Report for 1978-79.
- Ritche, W.A. and R.E. Funk. 1973. **Aboriginal Settlement Pattern in the North East.** Albany, New York State Museum and Science Service, Mem. # 20.
- Roy, S.S., Malhotra, G. and Mohanty, M. 1998. **Geology of Rajasthan**. Geological Society of India. Bangalore.
- Sahani, Daya Ram. 1940. Report of Archaeological Remains and Excavations at Bairat. Jaipur: State Archaeology Department.
- Sanders, W. 1956. *The Central Mexican Symbiotic Region; A Study in Prehistoric Settlement Patterns.* (Ed.) G. Willey. Prehistoric Settlement Patterns in the New World. New York. Gren Foundation: 115 -127
- Sanders, W.T. 1965. The Culture Ecology of the Teotihuaces Valley. A Preliminary Report of the Results of the Teotihuaces Valley Project. Dept. of Sociology and Anthropology. The Pennsylvanian State University. Pennsylvania.
- Sankalia, H.D. 1956. *Nathdwara, a Palaeolithic Site in Rajputana,* Journal of the Palaeontological Society of India 1: 99-100.
- Sankalia, H.D. 1965 "Archaeology." Excavations at Langhnaj: 1944-63. Pt .1. Poona: Deccan College postgraduate and Research institute.
- Sankalia, H.D. 1970. **Some Aspects of Prehistoric Technology in India.** Indian National Science Academic. New Delhi.
- Sankalia, H.D. 1974 **The Prehistory and Protohistory of India and Pakistan.** 2d ed. Poona: Deccan College Postgraduate and Research Institute.
- Sankalia, H.D. 1988. **Archaeology of Rajasthan.** Sahitya Sansthan. Rajasthan. Vidhyapeth Udaipur.
- Sankalia, H.D., S.B. Deo and Z.D. Ansari. 1969. **Excavation at Ahar (Tambavati).** Deccan College Pune.
- Settar, S. and Korisettar, R. 2002. Indian Archaeology in Retrospect. Volume # 1 - 4. ICHR. Manohar Publisher. Delhi.
- Shaffer, J.G. 1991. *In Chronologies the Indus Valley. Baluchistan and Helmand Tradition: Neolithic through Bronzes Age.* (Ed.) R. Ehrich. Old World Archaeology (3rd Ed.) Vol. #1. pp. 441-464. University of Chicago Press. Chicago.
- Shankarnarayan, K.A. and Sen, A.K. 1977. *The Methodology Of Mapping Resources For Rational Utilization.* **Desertification And Its Control.** Released on the occasion of The UN Conference On Desertification, Nairobi. 29th August- 9th September. Indian Council Of Agricultural Research. New Delhi.

- Sharma, G.R. 1980. **Beginnings of Agriculture from Epi-Palaeolithic to Neolithic.** Allahabad. Abinash Prakashan.
- Sharma, G.R. and I.D. Clark 1983(Eds). Palaeoenvironments and Prehistory in the Middle Son Valley (Madhya Pradesh, North Central India). Allahabad: Abinash Prakashan.
- Sharma, G.R. Misra, V.D. and J.N. Pal. 1980. **Excavation at Mahagara.** Allahabad.
- Sharma, G.R., V.D. Misra, D. Mandal, B.B. Misra, and J.N. Pal. 1980. **Beginnings of Agriculture,** Allahabad: Abinash Prakashan.
- Singh, G. 1979. **Ancient Cites of the Indus** *The Indus Valley Culture* p.234. Vikas Publishing House Pvt.Ltd. New Delhi Bombay Bangalore Calcutta Kanpur.
- Singh, G. Joshi, R. Chopra, S. and Singh, A. 1974. *Late Quaternary History of Vegetation and Climate of the Rajasthan Desert, India.* Philosophical transaction of the Royal Society of London. Biological Sciences. Vol. 267 pp. 467 – 501. Pub. Royal Society, 6, Canton House Terrace, London.
- Singh, G., Joshi, R., and Singh, B. 1972. *Stratigraphy and Radiocarbon Evidence for the Age and Development of Three Salt lake Deposits in Rajasthan, India.* Quaternary Research 2: 496-505.
- Singh, G., Wasson, R., and Agrawal, D. 1990. *Vegetation and Seasonal Climatic Changes Since The Last Full Glacial In The Thar Desert, Northwestern India.* Review of Palaeobotany and Palynology 64:351-358.
- Singh, Harphool. 1993. *Copper Hoard-An Unpublished Find from Chithwari, Chomu (Jaipur),* Tulsi Pragya, XVIII(4): 141-143.
- Sinha, B.P. 1969-71. **Potteries in Ancient India.** Patan.
- Smith, V.A. 1905. **The Copper Age and Prehistoric Bronze Implements of India.** Indian Antiquity Vol. # 34 pp. 229-244.
- Statira, W. Korisettar, R. and Kale, V.S. 1995. **Quaternary Environments and Geoarchaeology of India.** Essays in Honour of Professor S.N. Rajaguru. Geological Society of India. Bangalore.
- Stein, A. 1989. An Archaeological Tour along the Ghaggar-Hakra River, In (S.P. Gupta Ed.), pp. 1-97. Meerut: Kusumanjali Prakashan.
- Stein, A. 1942. *A Survey of Ancient Sites Along the "Lost Saraswati River".* Geographical Journal, Vol. 99. The Royal Geographic Society, London.
- Steward, J.H. 1937. Ecological Aspects of Southern Society. Anthropos, 32: 87 – 104.
- Steward, J.H. 1955. **Theory of Culture Change, Urbana**. University of Illinois.
- Sukhwal, B.K. and P.K. Verdia. 1993. *Report on Base Metal Investigation Carried Out in Anjani, The, and Dariyawad Districts, Udaipur.*__Department of Geology and Mining, Udaipur.
- Sussman, C.R. Blumenshine, J.D.Clark, and B.B. Misra. 1983 *"Preliminary Report on Excavations and the Mesolithic Occupation Site at Baghor II Locality."* (Eds) G.R. Sharma and J.D. Clark, **Palaeoenvironments and Prehistory in the Middle Son Valley**, 161-96, Allahabad: Abinash Prakashan.
- Swain, A., Kutzbach, J., and Hastenrath, S. 1982. *Estimates of Holocene Precipitation for Rajasthan, India, Based on Pollen and Lake-Level Data.* Quaternary Research 19: 1-17. University of Chicago Press. Chicago.
- Todd, J. 1894. **Annals And Antiquities of Rajasthan: Or the Central and Western Rajput States of India.** Revision of the original Published in 1829 – 32. 2 Vols. Calcutta: S. K. Lahiri and Company.
- Trigger, B.G. 1967. Settlement Archaeology – its goals and promises, American Antiquity 32: 149 – 160.
- Verma, P.K. 1996. *Evolution and Age of the Great Boundary Fault of Rajasthan.* Memoir Geological Society of India # 36 pp. 197-212.
- Verma, R.P. and Srivastava, R.N. 1976. Geological and Geomorphological Evolution of the Part of the Kantli River Basin. Unpub. Geol. Surv. Ind. Report for F.S. 1974-75.
- Wasson, R., Smith, G., and Agrawal, D. 1984. *Late Quaternary Sediments, Minerals, And Inferred Geochemical History Of Didwana Lake, Thar Desert, India.* Palaeogeography, Palaeoclimatology, Palaeoecology. 46 (1984): 345-372. Elsevier Science Publisher B. V., Amsterdam.
- Willey, G.R. 1956. **Prehistoric Settlement Pattern in the New World.** New York, Viking Fund. Publication in Anthropology #23.
- Willey, G.R. 1956. *Problem Concerning Prehistoric Settlement Patterns in the Maya Lowlands.* (Ed.) Gordon, R. Willey. Prehistoric Settlement Patterns in the New World. Viking Publications 23. New York 107 – 114.
- Willey, G.R. 1953. **Prehistoric Settlement Pattern Studies in Viru Valley Peru**, Bureau of American Ethnology Bulletin 155.
- Willies, L. 1992. *Ancient and Post-Medieval Mining Sites in Khetri Copper Belt and Kolar Gold Field,* India. Bulletin of the Peak Distrcit Mines Historical Society, Vol II No 6. Winter.

APPENDIX 1: MINOR ANTIQUITIES

These minor antiquities highlight the salient features, which was not identified in the slot of prominent findings in the research area. These minor antiquities indicate extraneous material, material culture of the site, religious affinities and socio-religious environment of the site (Figure – 1 to 14).

Figure 1 - Human figure in black schist - GANESHWAR

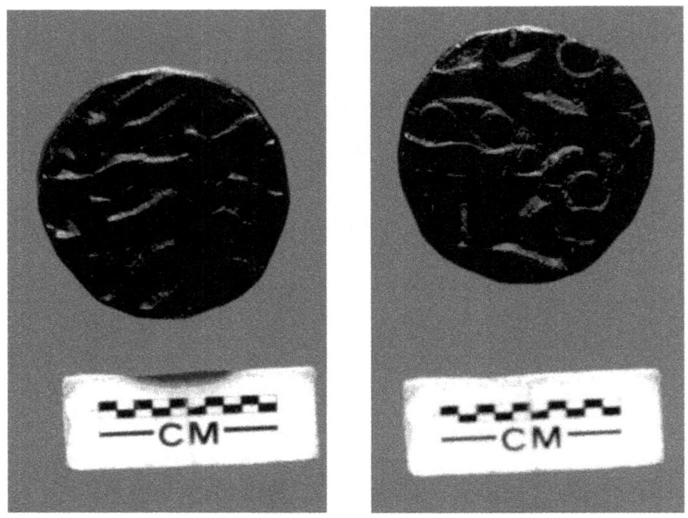

Figure 2 Tribal Coin in lead - GANESHWAR

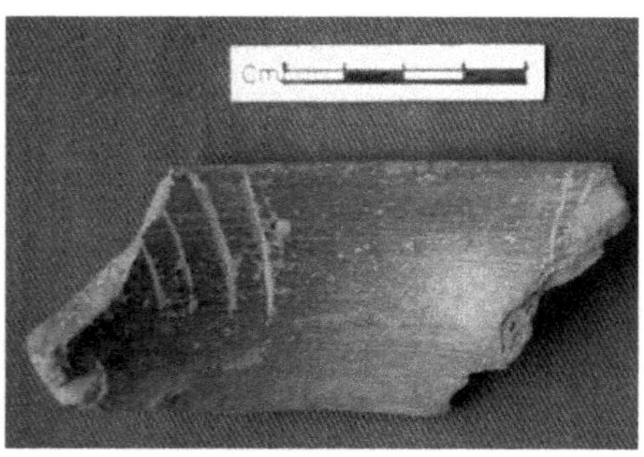

Figure 3 - Graffiti Sherds – GANESHWAR

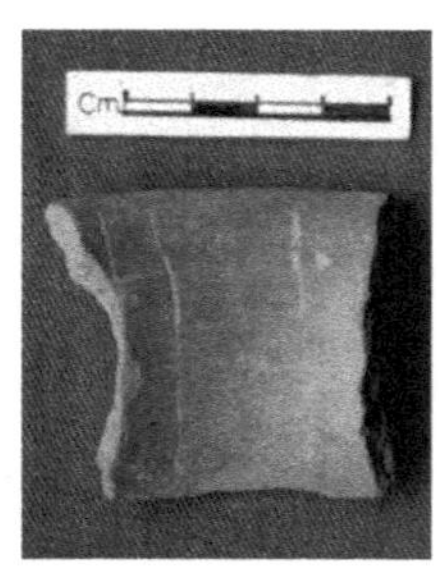

APPENDIX 1: MINOR ANTIQUITIES

Figure 4 - Terra cotta seal with three rows of Brahmi letters - KOT

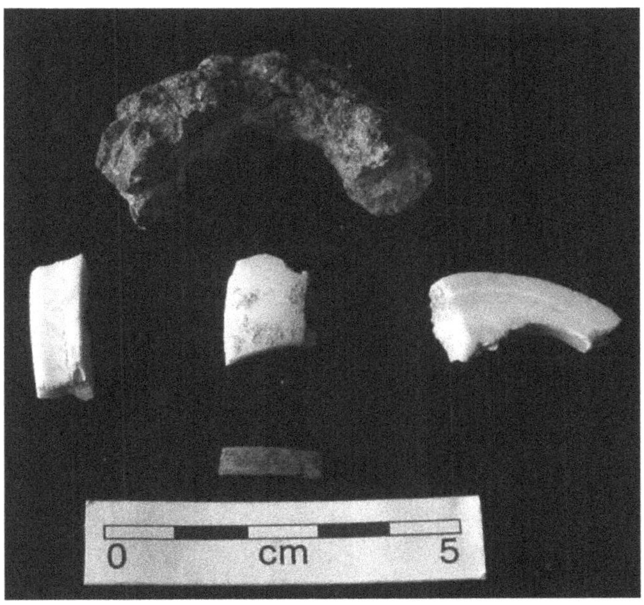

Figure 5 - Iron and Shell bangles – KOT

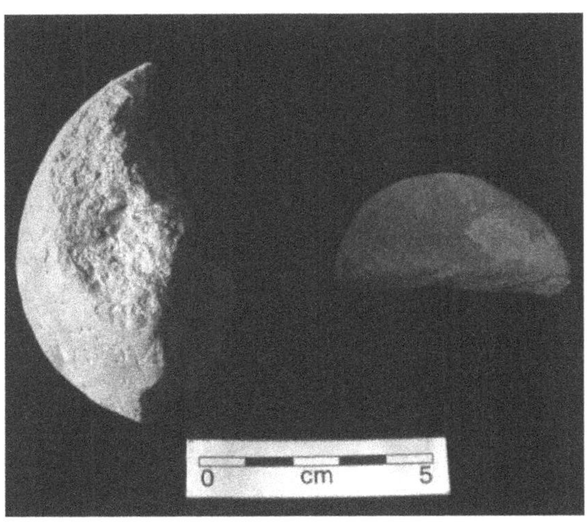

Figure 6 - Terra cotta anvil - KOT

APPENDIX 1: MINOR ANTIQUITIES

Figure 7 - Copper arrowhead and antimony rod – NARSHINGPUR

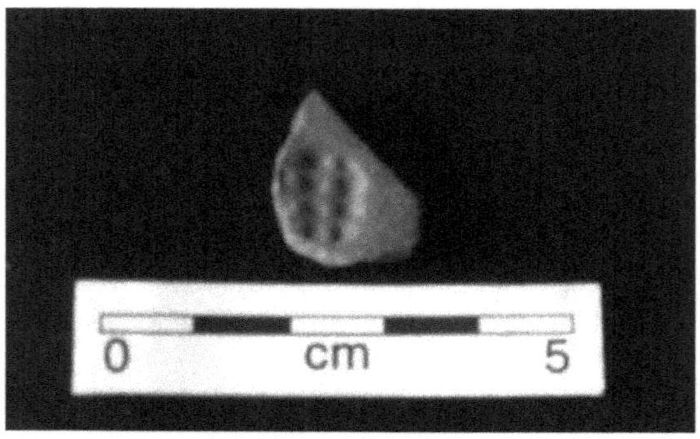

Figure 8 - Stamped Pottery – JOGI BADH

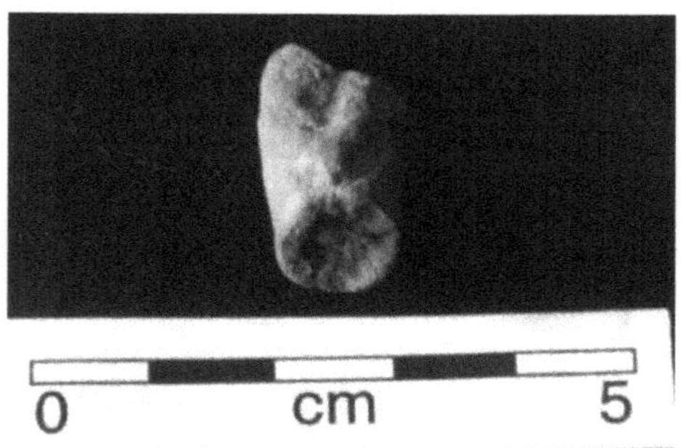

Figure 9 - Terra cotta figurine – Gidhali # 1

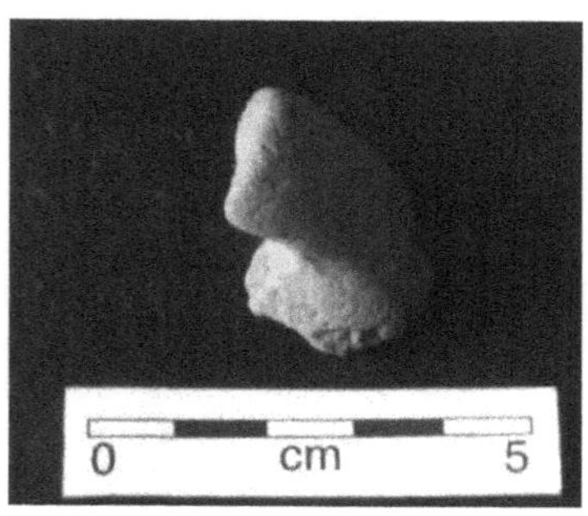

Figure 10 - Terra cotta figurine - Kakrana

APPENDIX 1: MINOR ANTIQUITIES

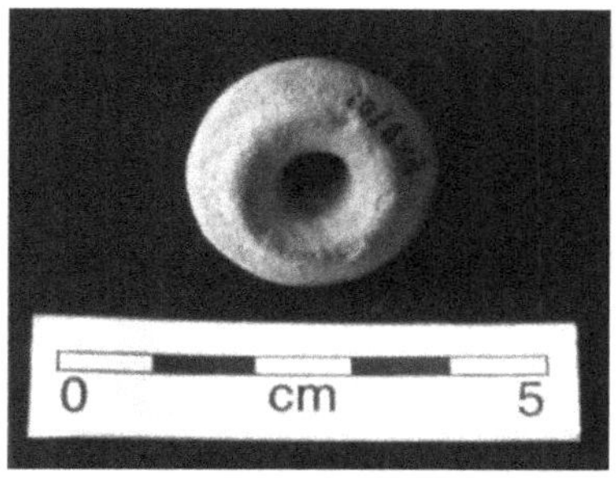

Figure 11 - Terra cotta bead – Payga Kundalia

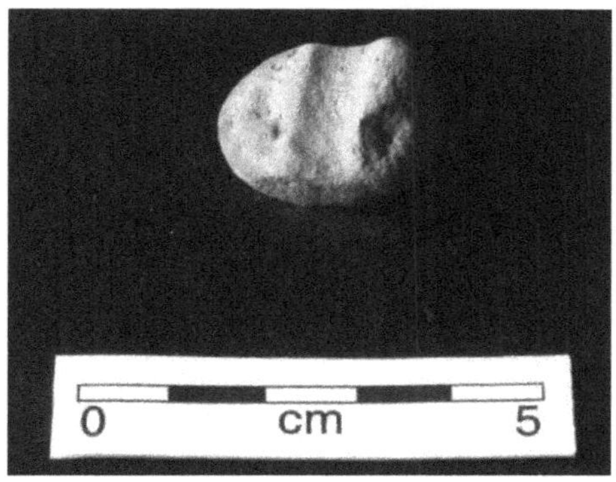

Figure 12 - Terra cotta bead – Jasi ka Bas

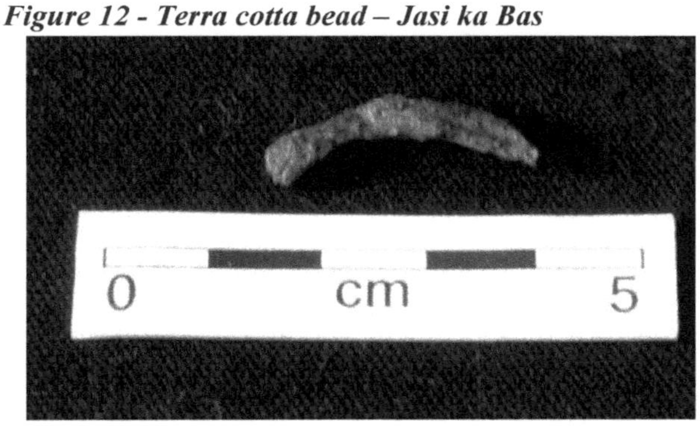

Figure 13 - Iron bangles – Gidhali # 2

Figure 14 - Shell bangles - Biharipur

APPENDIX 2: RESULT OF THE LEAD ISOTOPE ANALYSIS OF COPPER SAMPLES

A preliminary investigation of lead isotope analysis was carried out by Hoffman, Randal and Raghubans (2005) to determine the provenance of copper ore during Harappan period. The copper sample were derived from Harappa and were compared with seven regional copper source areas i.e. Ganeshwar and Singhana in Rajasthan, Chagai hills in Baluchistan, Shin Kai copper deposit in Waziristan, North Gujarat, Himachal Pradesh, Uttaranchal and Oman. The analysis resulted that Harappans perhaps acquired most of copper ore from the west of Harappa and some of its copper ore was also obtained from Rajasthan. This information has given a lead for further metallurgical research in the Shekhawati region.

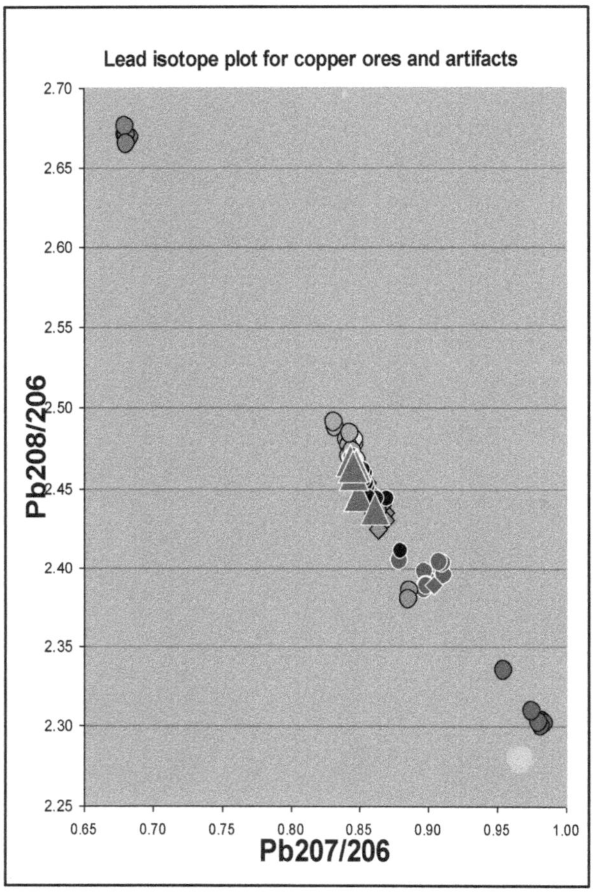

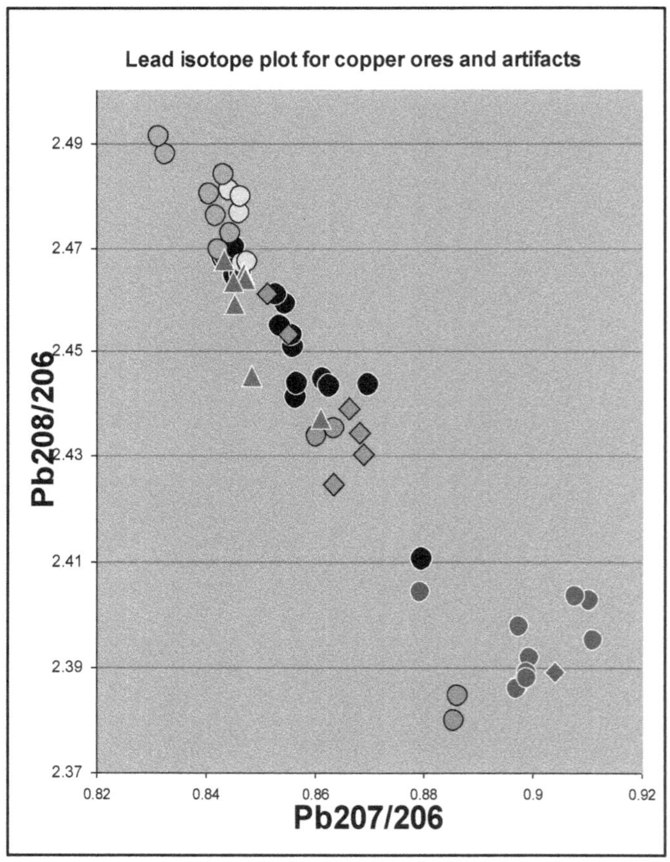

▲	Harappa	◇	Singhana	●	Ganeshwar
●	N. Gujarat	○	Chagai Hills	○	N. Waziristan
●	Uttaranchal	●	Himachal Pradesh	●	Oman

www.ingramcontent.com/pod-product-compliance
Lightning Source LLC
LaVergne TN
LVHW071208210926
848482LV00015B/754
* 9 7 8 1 4 0 7 3 1 3 1 3 9 *